Soil Biology

Volume 57

Series Editor
Ajit Varma, Amity Institute of Microbial Technology,
Amity University, Noida, Uttar Pradesh, India

The Springer series *Soil Biology* publishes topical volumes in the fields of microbiology, environmental sciences, plant sciences, biotechnology, biochemistry, microbial ecology, mycology and agricultural sciences.

Special emphasis is placed on methodological chapters or volumes. This includes coverage of new molecular techniques relevant to soil biology research or to monitoring and assessing soil quality as well as advanced biotechnological applications. Leading international authorities with a background in academia, industry or government will contribute to the series as authors or editors.

Key Topics: microbial-plant interactions; microbial communities; root symbiosis, mycorrhiza; rhizosphere environment; soil fauna, e.g. termites etc.; biochemical processes, soil proteins, enzymes, nucleic acids; degradation of biomaterials, decomposition, nutrient cycles; soil genesis, mineralization; bioremediation of contaminated sites; biotechnological applications of soil microorganisms.

More information about this series at http://www.springer.com/series/5138

Muhammad Zaffar Hashmi • Ajit Varma
Editors

Electronic Waste Pollution

Environmental Occurrence and Treatment Technologies

Springer

Editors
Muhammad Zaffar Hashmi
Department of Chemistry
COMSATS University Islamabad
Islamabad, Pakistan

Ajit Varma
Amity Institute of Microbial Technology
Amity University
Noida, Uttar Pradesh, India

ISSN 1613-3382 ISSN 2196-4831 (electronic)
Soil Biology
ISBN 978-3-030-26614-1 ISBN 978-3-030-26615-8 (eBook)
https://doi.org/10.1007/978-3-030-26615-8

This Springer imprint is published by the registered company Springer Nature Switzerland AG.
The registered company address is: Gewerbestrasse 11, 6330 Cham, Switzerland

Preface

Over the past 50 years, the production of small electronic devices including cell phones, printers, televisions, and personal computers has brought many societal benefits. Increased urbanization and globalization have resulted in rapid urban development leading to a rise in the use of electronic goods and appliances in our daily lives. These electronic goods and appliances enhance our living standards but have a definitive shelf life. The problem arises when it comes to the disposal of such electronics after the completion of their shelf lives. This is primarily because such electronics comprise precious and toxic metals which have severe human health and environmental effects if they are not disposed of carefully. The increased generation of such electronic wastes (e-waste) is a rising concern, particularly in the context of presence of both toxic and precious metals. This problem is of particular concern in developing countries with emerging economies, weak public health infrastructures, and rapidly growing populations.

The book consists of different sections such as history, types, trends, and monitoring of e-waste pollution; recycling of e-waste; exposure assessment and public health effects of e-waste pollution as well as e-waste effects on child health; metagenomics approaches to study the microbes in e-waste; techniques used and recycling of e-waste pollution; and toxicity and risk assessment approaches. The last seven chapters of the book focus on the treatment technologies; policy options; and challenges for e-waste.

The overall goal of this book, from a scientific perspective, is hence to promote the use of twenty-first-century tools to address a unique and rapidly growing global public health issue related to the disposal of high-technology electronic devices.

Islamabad, Pakistan — Muhammad Zaffar Hashmi

Acknowledgments

Special thanks to Prof. Dr. Chaofeng Shen, Department of Environmental Engineering, Zhejiang University China, for his kind suggestions to improve the quality of the book.

Special thanks to Higher Education Commission of Pakistan NRPU projects 7954 and 7964. Further thanks to Pakistan Science Foundation project PSF/Res/CP/C-CUI/Envr (151).

Contents

Contributors

Siddique Abdullah Bahria University, Islamabad, Pakistan

Wahid Abdul Department of Environmental Sciences, Bahauddin Zakerya University, Multan, Pakistan

A. Alhakamy Nabil Pharmaceutics Department, Faculty of Pharmacy, King Abdulaziz University, Jeddah, Saudi Arabia

Ahmad Ali Institute of Pure and Applied Biology, Bahauddin Zakerya University, Multan, Pakistan

Albar Hussain Mohammad Salem Ali Department of Community Health, Medical College, King Abdulaziz University, Jeddah, Saudi Arabia

Khan Mubarak Ali Department of Biotechnology, Faculty of Chemical and Life Sciences, Abdul Wali Khan University Mardan (AWKUM), Mardan, Pakistan

Shah Naseer Ali Department of Biosciences, COMSATS University Islamabad, Islamabad, Pakistan

Zakir Ali Department of Environmental Sciences, COMSATS University Islamabad, Islamabad, Pakistan

Farooq Amjad Department of Environmental Sciences, COMSATS University Islamabad, Islamabad, Pakistan

Padmavathy Anbarashan Guest Faculty, Department of Ecology and Environmental Sciences, Pondicherry University, Puducherry, India

Iqbal Aneela Institute of Environmental Sciences and Engineering, School of Civil and Environmental Engineering, National University of Sciences and Technology, Islamabad, Pakistan

Rasheed Anjum Centre for Climate Research and Development (CCRD), COMSATS University Islamabad, Islamabad, Pakistan

Srivastava Ankita Faculty of Bio-Sciences, Institute of Bio-Sciences and Technology, Shri Ramswaroop Memorial University, Lucknow, Uttar Pradesh, India

Besis Athanasios Environmental Pollution Control Laboratory, Department of Chemistry, Aristotle University of Thessaloniki, Thessaloniki, Greece

Ashfaq Ayesha Institute of Environmental Sciences and Engineering (IESE), Islamabad, Pakistan

B. Balabanova Faculty of Agriculture, University "Goce Delčev", Štip, Republic of Macedonia

Riaz Hamza Bin Institute of Environmental Sciences and Engineering (IESE), Islamabad, Pakistan

Rehman Bushra Department of Environmental Sciences, Faculty of Biological Sciences, Quaid-i-Azam University, Islamabad, Pakistan

Nabi Deedar Institute of Environmental Sciences and Engineering (IESE), Islamabad, Pakistan

Nazar Ehtisham Department of Environmental Sciences, University of Gujrat, Gujrat, Pakistan

Dsouza Elvis EDPC Polymer Industries, Vasai, Maharashtra, India

Zulfiqar Farhad Department of Economics, COMSATS University, Islamabad, Pakistan

Ali Huma Department of Biotechnology, Bacha Khan University, Charsadda, KP, Pakistan

Ismail Iqbal Mohammad Ibrahim Center of Excellence in Environmental Studies, King Abdulaziz University, Jeddah, Saudi Arabia
Department of Chemistry, Faculty of Science, King Abdulaziz University, Jeddah, Saudi Arabia

Ali Ijaz Department of Biosciences, COMSATS University Islamabad, Islamabad, Pakistan

Kaleem Imdad Department of Biosciences, COMSATS University Islamabad, Islamabad, Pakistan

Rashid Muhammad Imtiaz Center of Excellence in Environmental Studies, King Abdulaziz University, Jeddah, Saudi Arabia

Liaqat Irfan Centre for Climate Research and Development (CCRD), COMSATS University Islamabad, Islamabad, Pakistan

Nawaz Ismat Department of Environmental Sciences, Biotechnology Program, COMSATS University Islamabad, Abbottabad, Pakistan

Khan Jadoon Department of Microbiology, Quaid I Azam University Islamabad, Islamabad, Pakistan

Shahzad Khurram Central Cotton Research Institute (CCRI), Multan, Pakistan

Manzoor Maria Institute of Environmental Sciences and Engineering, School of Civil and Environmental Engineering, National University of Sciences and Technology, Islamabad, Pakistan

Abbas Mazhar Department of Management Sciences, COMSATS University Islamabad, Islamabad, Pakistan

Iqbal Mazhar Department of Environmental Sciences, Faculty of Biological Sciences, Quaid-i-Azam University, Islamabad, Pakistan

Raza Mohsin Centre for Climate Research and Development, COMSATS University, Islamabad, Pakistan

Abid Muhammad Centre for Climate Research and Development, COMSATS University, Islamabad, Pakistan

Arshad Muhammad Institute of Environmental Sciences and Engineering, School of Civil and Environmental Engineering, National University of Sciences and Technology, Islamabad, Pakistan

Mubeen Muhammad Department of Environmental Sciences, COMSATS University Islamabad, Islamabad, Pakistan

Jamal Muhsin Department of Microbiology, Abdul Wali Khan University Mardan (AWKUM), Mardan, Pakistan

Ali Nadeem Center of Excellence in Environmental Studies, King Abdulaziz University, Jeddah, Saudi Arabia

Malik Riffat Naseem Ecotoxicology and Environmental biology Laboratory, Department of Environmental Sciences, Faculty of biological Sciences, Quaid I Azam University Islamabad, Islamabad, Pakistan

Ullah Nazif Department of Biotechnology, Faculty of Chemical and Life Sciences, Abdul Wali Khan University Mardan (AWKUM), Mardan, Pakistan

Chandra Niharika Faculty of Biotechnology, Institute of Bio-Sciences and Technology, Shri Ramswaroop Memorial University, Lucknow, Uttar Pradesh, India

Gaonkar Omkar Nuevo Chakra, Mumbai, Maharashtra, India
SRM Research Institute, SRM Institute of Science and Technology, Kattankulathur, Kancheepuram District, Tamil Nadu, India

Chakraborty Paromita Department of Civil Engineering, SRM Institute of Science and Technology, Kattankulathur, Kancheepuram District, Tamil Nadu, India
Nuevo Chakra, Mumbai, Maharashtra, India

SRM Research Institute, SRM Institute of Science and Technology, Kattankulathur, Kancheepuram District, Tamil Nadu, India

Murtaza Rabbia Centre for Climate Research and Development (CCRD), COMSATS University Islamabad, Islamabad, Pakistan

Ganguly Rajiv Department of Civil Engineering, Jaypee University of Information Technology, Waknaghat, District Solan, Himachal Pradesh, India

Rajakumaran Revathy Environmental and Water Resource Engineering Division, Department of Civil Engineering, IIT Madras, Chennai, Tamil Nadu, India

Akram Rida Department of Environmental Sciences, COMSATS University Islamabad, Islamabad, Pakistan

Sultana Syeda Riffat Department of Agronomy, University of Agriculture Faisalabad, Faisalabad, Pakistan

Rex Ronnie Department of Civil Engineering, SRM Institute of Science and Technology, Kattankulathur, Kanchipuram District, Tamil Nadu, India
SRM Research Institute, SRM Institute of Science and Technology, Kattankulathur, Kancheepuram District, Tamil Nadu, India

Eqani Syed Ali Musstjab Akber Shah Health and Environment Division, Department of Biosciences, COMSATS Institute of Information Technology, Islamabad, Pakistan

Fahad Shah Department of Agriculture, University of Swabi, KPK, Pakistan

Khurram Shahzad Center of Excellence in Environmental Studies, King Abdulaziz University, Jeddah, Saudi Arabia

Ahmad Shakeel Department of Agronomy, Bahauddin Zakariya University, Multan, Pakistan

Noreen Sibgha Institute of Pure and Applied Biology, Bahauddin Zakerya University, Multan, Pakistan

Kathi Srujana Guest Faculty, Department of Ecology and Environmental Sciences, Pondicherry University, Puducherry, India

Jaiswal Sukriti Faculty of Biotechnology, Institute of Bio-Sciences and Technology, Shri Ramswaroop Memorial University, Lucknow, Uttar Pradesh, India

Siddique Sunbal Department of Meteorology, COMSATS University Islamabad, Islamabad, Pakistan

Kumar Sunil Faculty of Bio-Sciences, Institute of Bio-Sciences and Technology, Shri Ramswaroop Memorial University, Lucknow, Uttar Pradesh, India

Khan Tariq Department of Biotechnology, University of Malakand, Chakdara Dir Lower, Pakistan

Ahmed Toqeer Centre for Climate Research and Development (CCRD), COMSATS University Islamabad, Islamabad, Pakistan

Jan Azmat Ullah Department of Biosciences, COMSATS University Islamabad, Islamabad, Pakistan

Khan Muhammad Usman Department of Environmental Sciences, Faculty of biological Sciences, Kohat University of Science & Technology, Kohat, Pakistan

Nasim Wajid Department of Environmental Sciences, COMSATS University Islamabad, Islamabad, Pakistan
Department of Agronomy, University College of Agriculture and Environmental Sciences, Bahawalpur, Pakistan

Anjum Raja Muhammad Waqas Department of Biosciences, COMSATS University Islamabad, Pakistan

Rasheed Yasir Department of Biosciences, COMSATS University Islamabad, Islamabad, Pakistan

Hashmi Muhammad Zaffar Department of Chemistry, COMSATS University Islamabad, Islamabad, Pakistan

Abbreviations

ATSDR	Agency for Toxic Substances and Disease Registry
BFRs	Brominated flame retardants
DP	Dechlorane plus
E-waste	Electronic waste
EPR	Extended producer responsibility
ME	Minor elements
NBFRs	New brominated flame retardants
OH-PBDEs	Hydroxylated polybrominated diphenyl ethers
OPEs	Organophosphate ester
PAHs	Polycyclic aromatic hydrocarbons
PBDEs	Polybrominated diphenyl ethers
PCBs	Polychlorinated biphenyls
PCDDs/Fs	Polychlorinated dibenzo-*p*-dioxins and dibenzo-furans
PFOA	Perfluorooctanoic acid
POP	Persistent organic pollutants
RCRA	Resource Conservation and Recovery Act
REE	Rare earth elements
RT-PCR	Reverse transcription polymerase chain reaction
UNEP	United Nations Environmental Programme
WEEE	Waste electrical and electronic equipment
WHO	World Health Organization

Chapter 1
History and Major Types of Pollutants in Electronic Waste Recycling

Sunbal Siddique and Abdullah Siddique

Abstract Electronic waste is a global problem that requires a global solution and attention. It could be one of the possible key ways to reduce electronic waste generation. Although recycling reduces greenhouse gases emissions, also water and soil pollution associated with making new products from raw materials, however, safe and sound recycling is a big environmental issue. One more challenging aspect of reducing illegal export of electronic products to underdeveloped and less developed countries in order to ensure safe recycling is the imposing of strict laws and policies against it. Since this import-export business occurs without taking into account any international treaties, laws, etc. against hazardous materials transport, it worsens environmental as well as health issues. Educating people about how to recycle, reuse, and dispose electronics at all levels, combination of new technological solutions and increased public awareness regarding environmental effects of electronic waste will teach them how to behave more responsibly toward recycling.

Keywords History · Types · E-waste · Pollutants

1.1 Introduction

1.1.1 History of Electronic Waste Recycling

Mid of 1970s embarks the action and need for disposing electronic waste. In 1976 Resource Conservation and Recovery Act (RCRA) made it illegal to dispose off electronic waste in America. Similarly, the Basel Convention of 1989 made it illegal to dispose off electric waste in less developed countries. Since it became illegal to

S. Siddique (✉)
Department of Meteorology, COMSATS University Islamabad, Islamabad, Pakistan
e-mail: sunbal.siddique@comsats.edu.pk

A. Siddique
Bahria University, Islamabad, Pakistan

M. Z. Hashmi, A. Varma (eds.), *Electronic Waste Pollution*, Soil Biology 57,
https://doi.org/10.1007/978-3-030-26615-8_1

dispose the electronic waste in many areas of the world, there emerged a dire need to develop new solutions that are legalized and safe. This led to the practice of legal and safe recycling of worn-out electronics.

One of the waste disposal incidents that drew the attention and need for legislation formation regarding disposal of e-waste in a safe way internationally was in 1980s when a Liberian ship called the "Khian Sea" was hired to ship away 14,000 tons of incinerated electronic waste ash from Philadelphia city. The ash was supposed to land in New Jersey; however, the authorities there refused to receive it. Reluctantly, the ship had to head southward for the Caribbean Sea, dumping 4000 tons of ash there. The rest was ultimately disposed at different locations as the ship headed toward Southeast Asia. Internationally, the public responded harshly over this incident/act and led to the formation of Basel Convention. Ultimately, safe disposal of electronic waste laws were adopted. This also led to the growth of recycling industry from a mere local hobby to a large and lucrative industry. This industry is responsible for collecting reusable parts from discarded electronic equipment and then recycles them for the benefit of local businesses. This industry not only has created jobs but also is a contributor of saving many local environments from serious harm since dumping of electronic waste straight away into a landfill or nearby water body is truly a dreadful act.

1.1.2 Classification of Electronic Waste

The European Union has classified electronic waste into ten underlined categories: huge domestic electric appliances, small domestic electric appliances, IT equipment, media/entertainment/recording equipment, lightning appliances, electric construction tools of wood and metal, toys/leisure equipment along with sports items, medical equipment, surveillance equipment, and electric dispensers for hot and cold drinks, solid products, and money, along with all other machines used for automatic delivery of different products.

Almost 8% of the municipal waste contains electronic waste. It is one of the fastest growing waste categories by now (Widmer et al. 2005). Constituents of electronic waste are presenting an alarming state due to their toxicity and quantity. The use of electronic equipment has been increased drastically since last two decades. Nevertheless, swift changes and versatility in information and communication technologies along with decreasing trend in prices have caused reduced lifespan of electronic products. Majority of used computers have already reached to the end of their quality/purposeful life or soon they will. Their number has also exceeded 1 billion (Bastiaan et al. 2010).

According to the US EPA (Bastiaan et al. 2010), on an average, each home in America uses 34 electronic products, generating more than 5×10^6 tons of electronic waste on annual basis. In the European Union, each citizen generates about 15 kg of electronic waste on an average annually, producing an estimated total of 7×10^6 tons (Bastiaan et al. 2010). In developing countries, such as China and India, per capita electronic waste production is still only about 1 kg $year^{-1}$ (Widmer et al. 2005), but this is increasing rapidly. Approximately 41.8 million metric tons of waste electrical and electronic equipment (WEEE) was generated in year 2014 (Baldé et al. 2015), whereas results of another independent study found that almost 4 Ktons of WEEE is being discarded every hour on global level (Ketai et al. 2008). It was estimated that the global production of *electronics will reach up to 50 million metric tons by 2018* (Baldé et al. 2015) and will increase up to 52.2 million tons by 2021.

Newly industrialized and developing countries have record increase in their electronic waste since they import electronic waste from developed countries. Precisely, 50–80% of this waste is produced in developed world countries but shipped to third world markets either for recycling or reuse purpose (Widmer et al. 2005). Of the 20–50 million tons of electronic waste that is produced every year, almost 80% of it goes to Asian and African countries to be recycled or disposed (BAN et al. 2002; Odeyingbo et al. 2017). Moreover, USA is the only developed country that pulled out from signing the UN Basel Convention, whose motive is to ban the export of e-waste to developing countries.

Numerous countries have implemented extended producer responsibility (EPR) regulations to find the solution of "the electronic waste problem" since last decade. EPR necessitate "producers" of electric products to collect back their equipment and recycle them once they have served their purpose. The challenge related to electronic waste recycling is not limited to technical issues only. Other challenges include politics, legislation, economics, and awareness in society and cultural behavior. Another obstacle to recycling is lack of consumer's interest and awareness regarding recycling of electronics scraps benefits and impacts on the environment. This reckless behavior of consumers preference & behavior plays an important role in collection of electronic products, for instance, only 10% of mobile phones are collected for recycling in Germany (Nokia 2011). Tons of electronic scrap are recycled in unauthorized workshops where cheapest procedures are practiced. These include sorting the waste, firing, incinerating it, and washing it with acids. These practices straightforwardly cause harm to the workers' and local peoples' health.

Despite regulations, laws, and policies implementation for collection and recycling, merely1/3 of electronic scrap is collected separately and treated appropriately in the European Union (EU). A fraction of 2/3 waste in still believed to be a fate of landfill or other cheap/unauthorized workshops/treatment sites inside the European Union or outside the European Union often through illegal exports (European Commission). Collection and recycling of paper material, glass, plastic, and metal products has already become a common norm in numerous countries. However, only 20% of the electronic waste produced is properly recycled. Developing countries, chiefly, China, Vietnam, Brazil, Ghana, Mexico, Nigeria, and Philippines, recycle approximately 80% of the electrical waste generated globally through informal methods (Perkins et al. 2014). China is the leading electronic manufacturing country in the world; however it has become the largest electronic waste dumping ground (Chia et al. 2011). Unauthorized and unsafe recycling activities have released tons of toxic pollutants into the environmental matrix of developing countries (Robinson 2009). Literature is full of examples from Taizhou, Delhi, Guiyu, Lagos, and Accra. These countries are destined as hubs for electronic waste dumping and recycling since the past two decades from around the world (Chen et al. 2011).

Electronics become a waste when its construction and condition is no more competent of providing the projected performance with respect to the purpose allocated to them by the producers. Recycling is a process which converts the discarded material back into something useful in order to minimize contamination of the environment (Pongrácz et al. 2008). In essence, materials recovered from recycling serve as a secondary material source for using in the production of new products. It trims down the expenditure and utilization of primary raw materials, significantly saving energy and ultimately helping to decrease air pollution and GHG emissions (due to incineration) and water pollution (due to landfilling). Basically, recycling is performed to remove the most valuable materials from the waste and reuse them. At first, an electronic product/waste is disassembled followed by sorting of their parts according to material categories.

Effective and successful recycling of electronic waste is a crucial step in maintaining environmental health and economic recovery of useful materials for reuse. Since electronic waste is being produced at an enormous rate, recycling and recovering of precious/useful materials for reuse has become the most important economical and environmental issues globally.

Safe and standardized recycling of electronic scrap is a necessity. Developing and developed countries should have same risk threshold levels for hazardous, secondary e-waste. However, it could vary for children and adults due to their physical differences and pronounced vulnerabilities of children. Unfortunately, health consequences due to both direct exposures (during recycling) and indirect exposures (through environmental contamination) especially in developing countries are either ignored or poorly studied. Similarly, policy frameworks for protecting vulnerable populations do exist but rarely followed and applied.

Electronic waste also contains some precious materials of an economic value if recycled. Unluckily, most of the electronic waste is recycled in the unregulated manner, hence resulting inconsiderable exposure risk to toxics for the recyclers. Mostly these recyclers are women and children. Roughly, only 25% waste is recycled in formal recycling centers where occupational health hazards are considered and sufficient health prevention and control measures are ensured for the workers.

StEP (Solving the E-waste Problem Initiative) reported 45.6 million metric tons of electronic waste generation in 2012 (Duffert, WHO). Similarly, the United Nations Environmental Program (UNEP) found that 100 Empire State buildings could be filled with just the amount of electronic waste generated in year 2012. It averages to more than 6.8 kg for every living person. Approximately 7 billion humans inhabit the earth, and it is estimated that there are at least 6 billion mobile phones. 11.1 million tons of electronic waste was generated by China alone in year 2012. Similarly, the electronic waste generated by the United States in year 2012 was 10 million tons (StEP 2013). One can easily infer that on an average each American generated 29.5 kg of e-waste. In contrast, per person in China produces even less than 5 kg. An investigation on electronic waste generation by the United Nations in the year 2012 reported that by 2017 global e-waste will increase from 49.7 million to 65.4 million tons per annum; an increase of 33% (UNU 2013). In India, discarded e-waste compromising of cell phones only is expected to increase 18-fold by 2020 (Schluep et al. 2009). Market competition, customer demand, innovations, and inventions are some of the reasons for an exponential increasing in e-waste production (Gagliardi).

1.1.2.1 Metals

Nigeria is one of the worst hit countries by e-waste recycling practice since it is a source of metal pollution in their country. Unfortunately, 25–75% of used PCs imported by Nigeria in 2005 were nonfunctional or unrepairable. Universally, ferrous metal is present in largest quantity in electronic waste, comparatively to copper and other valued metals (Lundgren et al. 2012). E-waste also contains neodymium, indium, and gallium (rare metals) in minute quantities, for instance, the quantity/concentration of the metals in a cell phone is about 0.15 wt.%, making their recovery a challenge. Seventy percent of Hg (mercury) and Cd (cadmium) in US landfills is due to electronic waste (Kang and Schoenung 2005). Although these metals are no more being used in most of the modern electronics, the older products which have reached the end of their productive life may contaminate the environment through landfills dumping.

Similarly, heavy metals released during recycling/disassembling of electronics penetrate the soils or can reach any water source, contaminating it. When this water is used for irrigating purpose, it easily enters any plant body. Plants take up metals from the soil and accumulate them in their tissues (Zhao et al. 2009). When HM-toxicated food is consumed by humans, it automatically transfers heavy metal toxicity from the environment to the human body.

Lead is a heavy metal, and its behavior/performance during recycling process is of utmost concern and importance, and it is one of the most toxicant metals in electronic scrap (Chen et al. 2011). It is a developmental neurotoxicant, and blood Pb levels of ≥ 5 μg/dL in early childhood are responsible for having negative effects (CDC 2012).

Cd (cadmium) also has adverse neurodevelopmental effects. The biological half-life of Cd is ~26 years, and mostly, it accumulates in the kidneys (Friberg 1984), causing significant damage to the kidneys (Honda et al. 2010). It is found in light-sensitive resistors, corrosion-resistant alloys (for marine and aviation environments), and nickel-cadmium rechargeable batteries. Although sale of Cd-containing batteries has been banned in the European Union, it is negotiable for medical purpose. If such batteries are not recycled according to hazardous waste recycling guidelines, policies, or protocols, these can leach into the soil, causing harm to the microbes and disturbing the soil ecosystem. In case of inhalation, Cd is malignant for lungs as well. Kidney is also one of Cadmium's target organs. Cadmium is also associated with deficits in cognition, learning, behavior, and neuromotor skills in children (Chen et al. 2011).

Similarly, manganese (Mn) is also a neurotoxin for both adults and children (ATSDR 2012). Some evidence suggests a link between Mn exposure and attention deficit hyperactivity disorder (ADHD), autism, and other behavioral abnormalities (Sanders et al. 2015). Copper results in liver damage; and chronic exposure to cadmium increases the risk of lung cancer and kidney damage (Chan et al. 2007). Children are particularly susceptible to HM (heavy metal) exposure due to high gastrointestinal uptake and the permeable blood-brain barrier (Ogunseitan 2013).

Americium usually found in smoke alarms is highly carcinogenic. Similarly, mercury health effects include sensory impairment, dermatitis, and memory loss and muscles weakness. Exposure while being pregnant (inside uterus) causes fetal deficits in motor function, attention, and verbal domains (Chen et al. 2011). Uptake by animals causes death and reduced fertility along with slow growth and development.

Hexavalent chromium is used in metal coating to protect from corrosion. Inhaling hexavalent chromium can lead to cancer (Chen et al. 2011).

Sulfur is found in lead-acid batteries and causes damage to liver, kidney, and heart. It is also responsible for irritation in eyes and throat. Sulfur dioxide upon mixing with rainwater forms sulfuric acid responsible for acid rain.

1.1.2.2 Plastic Material

Electronic waste also includes components such as polymers (plastics), glass, gold, copper, and silver (Widmer et al. 2005). Fifteen percent of the non-flame retardant plastics and 5% of flame retardant polymers are found in WEEE. Collectively they account for 20% plastic material in electronic waste (Ma et al. 2016). Styrene polymeric components hold maximum share in WEEE plastic stream (Chatterjee 2015).

1.1.2.3 Organic Pollutants

Commonly termed as persistent organic pollutants (POPs), have the ability to travel throughout the globe. Usually, polluted air travels toward the poles, but it is sometimes carried on by the trade winds from Asia to North America (Yienger 2000). PBDEs, PCBs, and PCDD/PCDFs are highly toxic man-made pollutants (Martin et al. 2003). Chemicals, for instance, polychlorinated biphenyls (PCBs) in capacitors/transformers, polybrominated diphenyl ethers (PBDEs) in circuit board, and polychlorinated dibenzo-*p*-dioxins and dibenzo-furans (PCDDs/PCDFs) as the combustion products of PVC in electric wire, can either be emitted or formed during electronic waste recycling procedures (Wang et al. 2005). Airborne PBDEs were found to be approximately 10 times higher when compared with the controls (Sjodin et al. 2001). Similarly, significant amounts of PBDEs were detected in the serum of employees working at electronic waste-recycling factories (Sjödin et al. 2001). In addition, PCDD/PCDFs were also reported in sediments, ash, and hair samples of electronic waste-recycling sites and workers (Söderström and Marklund 2002). In short, when recycling and disposal methods are unregulated, chances of significant damage to the local environment and human health are high.

In Southeast China, one of the largest recycling e-waste sites is in Taizhou of the Zhejiang Province. This area encompasses approximately thirty villages actively involved in e-waste recycling. Since the late 1990s, plastics and rare heavy metals had been obtained by shattering and disassembling e-wastes including televisions, mobile phones, computers, etc. Discarded e-waste powder is stacked around these villages. Up to 25,479 ng g^{-1} dry weight of PBDEs in soil, 3526 ng g^{-1} dry weight in sediments, and biotic samples have been reported from this area (Liang et al. 2008).

Polybrominated diphenyl ethers (PBDEs) contain 209 congeners. They are environmentally persistent and lipid-soluble; hence they easily bioaccumulate in environmental matrixes and biomagnify through the food chain. Commercially, penta-, octa-, and deca-brominated diphenyl ethers are most commonly utilized. Presently, three commercial forms of polybrominated diphenyl ethers are banned or restricted in the EU, America, and Canada. On the other hand, commercial mixtures of deca-BDE are even today widely used in China and numerous other third world countries of Asia (Ni et al. 2013). Contact with polybrominated diphenyl ethers, no matter, in minute quantities, has significant effects on the development of brain in animal neonates, raising distress on possible health risks for humans, predominantly on children (Fromme et al. 2016).

Approximately 20–50 million tons of electronic waste is produced globally annually (UNEP 2005). Yearly, more than 2 m tons of electronic scrap is processed for recycling of metals in Taizhou, a main electronic waste recycling site in People's Republic of China. Taizhou practices substandard recycling techniques by burning lots of wires in an open environment in order to extract the coatings on metals. Circuit boards are melted over coal grills to remove important chips (Yu et al. 2006); the useless plastics left after fragmentation are discarded recklessly into the surroundings. These unchecked practices and poor electronic waste recycling surveillances have caused PBDEs to spread easily in to the environment. Therefore, significant amounts of PBDEs were discovered in the environmental matrices of electronic waste recycling sites in Taizhou, including sediments and mud snails (Yang et al. 2009), soil and plants (Zhao et al. 2009), the atmosphere (Liua et al. 2008), and even poultry (Liang et al. 2008).

Hydroxylated polybrominated diphenyl ethers (OH-PBDEs) are structural analogs of PBDEs. They cause a number of side effects in living organisms, such as disturbance of thyroid gland homeostasis, oxidative phosphorylation disturbance, changed estradiol synthesis, and neurotoxic effects (Marchesini et al. 2008). They have been reported in wildlife tissues (Verreault et al. 2005) and human blood (Stapleton et al. 2008). PBDEs enter human's body through dust inhalation and dietary ingestion (Ni et al. 2013) where diet being one of the primary exposure sources in adults (Fernandez-Gonzalez et al. 2015).

1.1.2.4 Halogenated Compounds

Since ages, chemically stable halogenated organic compounds have made them attractive to be used by the electronics manufacturers. On the other hand, BFRs, CFRs, PVC, and other halogenated organics environmental issues.

Brominated flame retardants (BFRs) are used as flame retardants in plastics of most electronics and include PBBs, PBDE, DecaBDE, OctaBDE, and PentaBDE. BFRs are responsible for impaired development of the nervous system. It disrupts functioning of the thyroid gland and liver (Birnbaum and Staskal 2004).

Polyvinyl chloride (PVC) is an insulation material in electrical cables. Chlorine in PVCs tends to bioaccumulate (Thornton 2000).

Perfluorooctanoic acid (PFOA) is used as an antistatic additive in industrial applications and found in electronics and also found in nonstick cookware (PTFE). PFOAs are synthetic materials formed during environmental degradation. PFOA increases risk of spontaneous abortion (miscarriage) in pregnant women, stillbirth, preterm birth, and low birth weight (Wu et al. 2012).

Beryllium oxide is used as a filler in some thermal interface materials such as thermal grease used on heat sinks for CPUs and power transistors (Becker et al. 2005), magnetrons, X-ray-transparent ceramic windows, heat transfer fins in vacuum tubes, and gas lasers. Beryllium sensitization, chronic beryllium disease, and acute beryllium disease along with lung cancer are some of the health issues caused by beryllium oxide (Health effects 2012).

TCE—trichloroethylene—is highly carcinogenic and has been used by semiconductor industries in industrial solvents. Similarly, TCA—trichloroethane—solvent is used by semiconductor industries to clean chips.

Halogenated organics have an ability of behaving like precursors for generating dioxin, a recognized carcinogenic agent in humans that is toxic in even minute quantities. There is no safe level for human exposure to dioxins (US EPA 1994).

Typically, dioxins are generated when halogenated organics of electronics such as the BFRs, CFRs, and PVC are incinerated during informal recycling in the developing world (Costner 1998). These airborne dioxins cause trans-boundary pollution by entering both terrestrial and aquatic food chains. For instance, some deceitful recyclers sent the plastic casings from e-waste to China for incineration, and the dioxins released during this process are ultimately reversed to the United States.

Halogenated compounds can easily skip into the environment at any phase of their life cycle starting from chemical manufacturing, incorporation into products, usage, disposal, and/or even during recycling (Santillo and Johnston 2003). Halogens are capable of producing toxic byproducts as well, ironically, throughout their lifetime, especially when they are originally synthesized by the chemical manufacturers. For example, PVC is synthesized by using vinyl chloride, a known carcinogen (ATSDR).

Halogens used in electronics, for instance, "additive" flame retardants that are loosely chemically bound into the plastics, over the passage of time can slowly migrate out of the product. Such halogens have been discovered in the dust of home environments (Blake et al. 2004), where children (Stapleton et al. 2008) and pets are the most vulnerable recipients of these compounds.

During 2007, atmospheric concentration of Guiyu in China was found to be heavily contaminated with chlorinated dioxins and furans. This area is famous for recycling electronic products (Li et al. 2007). In Guiyu, daily intake by breast feeding infants of dioxins, according to The WHO's (World Health Organization) estimation was found to be exceeding guidelines by 11–25 times (Chan et al. 2007).

When bromine and chlorine are present in significant quantities while combusting under the informal practices for instance as practiced in Guiyu, China, or in any other developing country, mixed halogenated dioxins and furans are generated (Söderström and Marklund 2002). Japan's atmosphere (Hayakawa et al. 2002), rain, soil, and river sediments (Ohta et al. 2002) and marine sediments in Hong Kong and Korea tested positive for the presence of these compounds (Terauchi et al. 2009). Burning of electronics could generate mixed halogenated dioxins and furans whose toxicity level can be more than 2,3,7,8 TCDD, usually termed by WHO as the most toxic chlorinated dioxin. Similarly, mixed halogenated dioxins and furans responsible for high degree of binding to the aryl hydrocarbon (AH) and having same positions halogenated atoms can be formed in thousands that are related to 2,3,7,8 TCDD's toxicity (Korach 1998).

References

ATSDR (2012) Toxicological profile for manganese (draft for public comment). U.S. Department of Health and Human Services, Public Service, Atlanta, GA

Baldé CP, Wang F, Kuehr R, Huisman J (2015) The global e-waste monitor. United Nations University, IAS – SCYCLE, Bonn

Becker G, Lee C, Lin Z (2005) Thermal conductivity in advanced chips: emerging generation of thermal greases offers advantages. Adv Packag, 2–4

Bastiaan C, Zoeteman J, Krikke HR, Venselaar J (2010) Handling WEEE waste flows: on the effectiveness of producer responsibility in a globalizing world. Int J Adv Manuf Technol 47:415–436

Birnbaum LS, Staskal DF (2004) Brominated flame retardants: cause for concern? Environ Health Perspect 112(1):9–17. https://doi.org/10.1289/ehp.6559

Blake A, McPherson A, Thorpe B (2004) Brominted flame retardants in dust on computers: the case for safer chemicals and better computer design. Clean Production Action and the Electronic Take Back Coalition, New York

CDC Response to Advisory Committee on Childhood Lead Poisoning Prevention Recommendations (2012) Low level lead exposure harms children: a renewed call for primary prevention. U. S. Centers for Disease Control and Prevention, Atlanta, GA. (updated 7 Jun 2012)

Chan JKY, Xing GH, Xu Y, Liang Y, Chen LX, Wu SC, Wong CKC, Leung CKM, Wong MH (2007) Body loadings and health risk assessment of polychlorinated Dibenzo-p-dioxins and Dibenzofurans at an intensive electronic waste recycling site in China. Environ Sci Technol 41 (22):7668–7674

Chatterjee S (2015) Sustainable recycling technology for electronic waste: DeitY initiatives

Chen A, Dietrich KN, Huo X, Ho SM (2011) Developmental neurotoxicants in e-waste: an emerging health concern. Environ Health Perspect 119(4):431–438

Chia X, Martin P, Mark YLW, Reuter MA (2011) Informal electronic waste recycling: a sector review with special focus on China. Waste Manag 31:731–742

Costner P (1998) Correlation of chlorine input and PCDD/PCDF at a full-scale hazardous waste incinerator. Organohalogen Compounds 36:147–152

Fernandez-Gonzalez R, Yebra-Pimentel I, Martinez-Carballo E, Simal-Gandara J (2015) A critical review about human exposure to polychlorinated dibenzo-p-dioxins (PCDDs), polychlorinated Dibenzofurans (PCDFs) and polychlorinated biphenyls (PCBs) through foods. Crit Rev Food Sci Nutr 55:1590–1617

Friberg T (1984) Cadmium and the kidney. Environ Health Perspect 54:1–11

Fromme H, Hilger B, Kopp E, Miserok M, Völkel W (2016) Polybrominated diphenyl ethers (PBDEs), hexabromocyclododecane (HBCD) and novel brominated flame retardants in house dust in Germany. Environ Int 64:61–68

Hayakawa K, Takatsuki H, Watanabe I, Sakai S (2002) Polybrominateddiphenyl ethers (PBDEs), polybrominated dioxins/furans (PBDDs/DFs) and monobromo-polychlorinated dioxins/furans (MoBPXDDs/DFs) in atmosphere and bulk deposition in Kyoto, Japan. Organohalogen Compounds 59:299–302

Health Effects (2012) United States Department of Labor. Archived from the original on 12 October 2016. Retrieved 30 October 2016.

Honda R, Swaddiwudhipong W, Nishijo M, Mahasakpan P, Teeyakasem W, Ruangyuttikarn W, Satarug S, Padungtod C, Nakagawa H (2010) Cadmium induced renal dysfunction among residents of rice farming area downstream from a zinc-mineralized belt in Thailand. Toxicol Lett 198(1):26–32

Kang H-Y, Schoenung JM (2005) Electronic waste recycling: a review of US infrastructure and technology options. Resour Conserv Recycl 45:368–400

Ketai H, Li L, Wenying D (2008) Research on recovery logistics network of waste electronic and electrical equipment in China. In: Industrial electronics and applications ICIEA. 3rd IEEE conference, pp 1797–1802

Korach K (1998) Reproductive and developmental toxicology. CRC, Boca Raton, FL, p 1998

Li H, Yu L, Sheng G, Fu J, Peng P'a (2007) Severe PCDD/F and PBDD/F pollution in air around an electronic waste dismantling area in China. Environ Sci Technol 41(16):5641–5646

Liang SX, Zhao Q, Qin ZF, Zhao XR, Yang ZZ, Xu XB (2008) Levels and distribution of polybrominated diphenyl ethers in various tissues of foraging hens from an electronic waste recycling area in South China. Environ Toxicol Chem 27:1279–1283

Liua H, Qunfang Z, Yawei W, Qinghua Z, Zongwei C, Guibin J (2008) E-waste recycling induced polybrominated diphenyl ethers, polychlorinated biphenyls, polychlorinated dibenzo-p-dioxins and dibenzo-furans pollution in the ambient environment. Author links open overlay panel. Environ Int 34(1):67–72

Ma C, Yu J, Wang B, Song Z, Xiang J, Hu S, Sun L (2016) Chemical recycling of brominated flame retarded plastics from e-waste for clean fuels production: a review. Renew Sust Energ Rev 61:433–450

Marchesini GR, Meimaridou A, Haasnoot W, Meulenberg E, Albertus F, Mizuguchi M et al (2008) Biosensor discovery of thyroxine transport disrupting chemicals. Toxicol Appl Pharmacol 232:150–160

Martin M, Richardson BJ, Lam PKS (2003) Harmonisation of polychlorinated biphenyl (PCB) analyses for ecotoxicological interpretations of Southeast Asian environmental media: what's the problem? Mar Pollut Bull 46:159–170

Ni K, Yonglong L, Tieyu W, Yajuan S, Kurunthachalam K, Li X, Qiushuang L, Shijie L (2013) Polybrominated diphenyl ethers (PBDEs) in China: policies and recommendations for sound management of plastics from electronic wastes. J Environ Manag 115:114–123

Nokia Corporation (2011) Consumer study

Odeyingbo O, Nnorom IC, Deubzer O (2017) Person in the Port Project: assessing import of used electrical and electronic equipment into Nigeria. UNU-ViE SCYCLE and BCCC, Africa, Bonn

Ogunseitan OA (2013) The Basel convention and e-waste: translation of scientific uncertainty to protective policy. Lancet Glob Health 1:313–314

Ohta S, Nakao T, Nishimura H, Okumura T, Aozasa O, Miyata H (2002) Contamination levels of PBDEs, TBBPA, PCDDs/DFs, PBDDs/DFs and PXDDs/DFs in the environment of Japan. Organohalogen Compounds 57:57–60

Pongrácz E, Ylä-Mella J, Tanskanen P, Keiski RL (2008) Proceedings of the 23rd international conference on solid waste technology and management, Philadelphia, PA; March 30–April 2, 2008. Journal of Solid Waste Technology and Management, Department of Civil Engineering, Widener University, Chester, PA, pp 359–365

Perkins DN, Brune Drisse M-N, Nxele T, Sly PD (2014) E-waste: a global hazard. Ann Glob Health 80(4):286–295

Robinson BH (2009) E-waste: an assessment of global production and environmental impacts. Sci Total Environ 408:183–191

Sanders AP, Birgit CH, Robert OW (2015) Perinatal and childhood exposure to cadmium, manganese, and metal mixtures and effects on cognition and behavior: a review of recent literature. Curr Environ Health Rep 2(3):284–294

Santillo D, Johnston P (2003) Playing with fire: the global threat presented by brominated flame retardants justifies urgent substitution. Environ Int 29:725–734

Schluep M, Hagelüken C, Kuehr R et al. (2009) Sustainable innovation & technology transfer industrial sector studies: recycling–from e-waste to resources. United Nations Environment Programme (UNEP) and StEP–solving the e-waste problem. http://www.unep.org/pdf/Recycling_From_e-waste_to_resources.pdf. Accessed 14 Oct 2014

Sjodin A, Carlsson H, Thuresson K, Sjolin S, Bergman A, Ostman C (2001) Flame retardants in indoor air at an electronics recycling plant and at other work environments. Environ Sci Technol 35:448–454

Söderström G, Marklund S (2002) PBCDD and PBCDF from incineration of waste-containing brominated flame retardants. Environl Sci Technol 36(9):959–1964

Stapleton HM, Allen JG et al (2008) Alternate and new brominated flame retardants detected in U.S. house dust. Environ Sci Technol 42(18):6910–6916

StEP Initiative (2013) StEP annual report. http://step-initiative.org/tl_files/step/StEP_AR/StEP_AR.html. Accessed 14 Oct 2014.

Terauchi H, Takahashi S, Lam P K S, Min B-Y, Tanabe S (2009) Polybrominated, polychlorinated and monobromo-polychlorinated dibenzo-p-dioxins/dibenzofurans and dioxin-like polychlorinated biphenyls in marine surface sediments from Hong Kong and Korea. Environ Poll, vol 157. pp724–730

The Basel Action Network (BAN), The Silicon Valley Toxics Coalition (SVTC), Toxics Link India, SCOPE (Pakistan), Greenpeace China (2002) Exporting harm: the high-tech trashing of Asia. Seattle,WA, and San Jose, CA; February 25th, 2002

Thornton J (2000) Pandora's poison. MIT, Cambridge, MA

U.S. Environmental Protection Agency (1994) Health assessment document for 2,3,7,8-tetrachlorodibenzo-p-dioxin (tcdd) and related compounds. Vol. III of III. [EPA/600/BP-92/001c] (Cincinnati, Ohio: U.S. Environmental Protection Agency, August)

UNEP (2005) E-waste, the hidden side of IT equipment's manufacturing and use. Early warnings on Emerging Environmental Threats No. 5. United Nations Environment Programme

United Nations University (UNU), Solve the E-waste Problem (StEP), Massachusetts Institute of Technology (MIT), National Center for Electronics Recycling (NCER) (2013) World e-waste map reveals national volumes, international flows. https://www.vie.unu.edu/file/get/11505.pdf. Accessed 14 Oct 2014

Verreault J, Gabrielsen GV, Chu SG, Muir DCG, Andersen M, Hamaed A et al (2005) Flame retardants and methoxylated and hydroxylated polybrominated diphenyl ethers in two Norwegian Arctic top predators: glaucous gulls and polar bears. Environ Sci Technol 39:6021–6028

Wang DL, Cai ZW, Jiang GB, Leung A, Wong MH, Wong WK (2005) Determination of polybrominated diphenyl ethers in soil and sediment from an electronic waste recycling facility. Chemosphere 60:810–816

Widmer R, Oswald-Krapf H, Sinha-Khetriwal D, Schenellmann M, Böni H (2005) Global perspectives on e-waste. Environ Impact Assess Rev 25:436

Wu K, Xu X, Peng L, Liu J, Guo Y, Huo X (2012) Association between maternal exposure to perfluorooctanoic acid (PFOA) from electronic waste recycling and neonatal health outcomes. Environ Int 41(1):–8

Yang ZZ, Zhao XR, Qin ZF, Fu S, Li XH, Qin XF, Xu XB, Jin ZX (2009) Polybrominated diphenyl ethers in mudsnails (Cipangopaludina cahayensis) and sediments from an electronic waste recycling region in South China. Bull Environ Contam Toxicol 82:206–210

Yienger JJ (2000) The episodic nature of air pollution transport from Asia to North America. Geophys Res J 105(22):26931–26946

Yu XZ, Gao Y, Wu SC, Zhang HB, Cheung KC, Wong MH (2006) Distribution of polycyclic aromatic hydrocarbons in soils at Guiyu area of China, affected by recycling of electronic waste using primitive technologies. Chemosphere 65:1500–1509

Zhao YX, Qin XF, Li Y, Liu PY, Tian M, Yan SS, Qin ZF, Xu XB, Yang YJ (2009) Diffusion of polybrominateddiphenyl ether (PBDE) from an e-waste recycling area to the surrounding regions in Southeast China. Chemosphere 76(11):1470–1476

Chapter 2
Biomonitoring of Electronic Waste Polluted Environment

Naseer Ali Shah, Azmat Ullah Jan, and Ijaz Ali

Abstract The environment is polluted by different wastes, while informal recycling of e-wastes is a major source of environmental pollution. These wastes include organic and inorganic compounds along with heavy metals and trace elements. These pollutants have a negative effect on living organisms, i.e., human, animals, and plants. Living organisms are used as biomonitoring agents to measure the level of e-wastes or pollutant in their bodies. Biomonitoring tells us about the sources, amount, geographical distribution, and harmful effects of contaminant. Biomonitoring involves the indicator species of plants, animals, and humans. This chapter is focused on human and animal species. Various bioindicators of the human such as blood/serum, hair, urine, breast milk, and nail have been used to monitor the exposure to e-wastes. Blood is the best bioindicator because it circulates throughout the body and has contact with all tissues. Different species of vertebrates such as chickens, ducks, fish, frogs, and mice have been studied for the measurement of e-wastes in their bodies. E-wastes were abundantly found in the muscles and liver of mice, chicken, duck, and frogs as well as in chicken's egg. The high concentrations of these wastes in animals such as chicken are negatively correlated with human health.

Keywords Biomonitoring · E-waste · Pollution · Environment

2.1 Introduction

Biomonitoring is the use of living organisms to measure the level of chemicals or wastes in their bodies. Biomonitoring is the direct measurement of a body exposure to contaminants; it tells us about the amount of the contaminants that get into the body from the environment. Biomonitoring is used to identify and eliminate

N. A. Shah (✉) · A. U. Jan · I. Ali
Department of Biosciences, COMSATS University Islamabad, Islamabad, Pakistan
e-mail: drnaseer@comsats.edu.pk

M. Z. Hashmi, A. Varma (eds.), *Electronic Waste Pollution*, Soil Biology 57,
https://doi.org/10.1007/978-3-030-26615-8_2

different sources of contaminants, to prove the effectiveness of restriction or ban, to evaluate contaminants that cause abnormalities, to identify the geographical distribution of contaminants, and to find the relationship between personal habits and chemical exposure (Esteban and Castaño 2009).

The e-waste polluted environment often contains heavy metals and organic compounds, mostly polybrominated and polychlorinated biphenyls (PBBs and PCBs), polycyclic aromatic hydrocarbons (PAHs), polybrominated diphenyl ethers (PBDEs), hexabromocyclododecanes (HBCDs), dechlorane plus (DP), polybrominated and polychlorinated dibenzodioxins (PBDDs and PCDDs), and polybrominated and polychlorinated dibenzofurans (PBDFs and PCDFs), and some other compounds are also found (Deshpande 2013).

Biomonitoring involves indicator species of plants, animals, and humans that accumulate pollutants in their body tissues and fluids such as blood and urine. In this chapter, we are mainly focused on human and animals (vertebrates) that accumulate e-wastes in their bodies.

2.2 Human Biomonitoring

Humans are exposed to e-wastes either occupationally by working in the informal e-waste recycling sites or nonoccupationally, which are living in the e-waste polluted area. Different strategies were used to analyze the human exposure to e-wastes (Deshpande 2013).

In humans, various bioindicators were used such as blood/serum, hair, urine, breast milk, and nail to monitor the exposure to many pollutants. Blood is the best bioindicator because it circulates throughout the body and contacts with all tissues of the body (Esteban and Castaño 2009) (Table 2.1).

2.2.1 Human Blood/Serum

Pollution caused by heavy metals, minor elements (ME), and rare earth elements (REE) can become in those sites where heavy metals are improperly processed. This is because of the high use of these elements in the electronic devices. Exposure to some of the toxic elements causes hematological problems. A study was conducted on many countries of the African continent to determine the association of these elements with anemia. The blood samples have been collected from sub-Saharan immigrants with anemia and analyzed for 48 elements including ME and REE by ICP-MS. The numbers of exposed individuals with anemia and control without anemia were 63 and 78, respectively. The higher concentrations of Ce, As, La, Ba, Ag, Bi, Pb, Eu, Er, Pr, Ga, Nd, Nb, Ta, Sn, Sm, Tl, Th, U, and V were found in the exposed individuals with anemia than control. The median of anemic and control participants was 1.21 and 0.32 ng/mL, respectively, and P value is 0.016. The summed blood level of REE and ME was significantly higher in the individuals of

Table 2.1 Biomonitoring of e-waste in human

Bioindicator	E-waste	References
Human blood/ serum	High level of As, Ag, Bi, Ba, Ce, Ga, Er, Eu, La, Nd, Nb, Pb, Pr, Sn, Sm, Th, Ta, Tl, V, and U	Henríquez-Hernández et al. (2017)
	High concentration of Pb	Huo et al. (2007)
	High concentration of DBDPE and PBDE and a dominating BFRs (BDE209) in serum	Liang et al. (2016)
	Significantly high level of polybrominated diphenyl ethers (PBDEs)	Sheng and Fu (2008)
Breast milk and placenta	High level of PCDD/Fs	Xing et al. (2007)
Human hair	High level of metals in the following order Ni < Cd < Cu < Pb < Zn	Zheng and Yuan (2011)
	High level of dechlorane plus (DP)	He and Yuan (2010)
	High level of PBDE and DBDPE and BDE209 was dominating BFRs	Liang et al. (2016)
Human scalp hair	The high level of Pb, Cd, and Cu	Wang et al. (2009)
Human urine	A significantly higher level of PAH metabolites	Feldt et al. (2014)
	Significantly high concentration of Pb, Fe, and Sb	Asante et al. (2012)

those countries where e-wastes were informally recycled (Henríquez-Hernández et al. 2017).

In Guiyu, Huo et al. (2007) found that the blood Pb level was higher in exposed preschool children than control. The mean value of Pb in exposed was 15.3 μg/dL and 9.94 μg/mL in control children. These results indicate that blood is a good bioindicator for Pb in the e-waste polluted sites (Huo et al. 2007).

In 2008, Sheng and Fu analyzed the level of polybrominated diphenyl ethers (PBDEs) in human serum. They collect 49 samples from exposed peoples in a village at the e-waste polluted site and other control samples from a village situated 50 km away from the area of e-wastes. The significantly higher ($P < 0.05$) level of PBDEs was found in the serum of exposed individuals than control. The levels of PBDEs in exposed and control individuals are median PBDEs, 382 ng/g lipid weight, in the range of 77–8452 ng/g lipid weight, and 158 ng/g, in the range of 18–436 ng/g lipid weight, respectively. This study has shown that the level of PBDEs was higher in the serum of peoples exposed to e-wastes in the workplace (Sheng and Fu 2008).

2.2.2 *Human Breast Milk*

Breast milk is a very common biomonitoring matrix in human that provides information about both mother and her child exposure. Numerous studies were conducted on the measurement of different chemicals in the polluted environment. Since 1976, WHO has reported the presence of persistent organic pollutants (POPs) in the breast milk through the Global Environment Monitoring System/Food Contamination Monitoring and Assessment Program (GEMS/Food Program). In many studies, persistent organic pollutants (POPs) had been found at high concentration in the breast milk. In 2007, the World Health Organization (WHO) also carried out surveys to measure PCDFs, PCDDs, and dioxin-like PCBs in breast milk (Esteban and Castaño 2009).

A study on breast milk in e-waste recycling site of China reported that the daily intake of polychlorinated dibenzofurans (PCDD/Fs) and polychlorinated dibenzo-p-dioxin was two times higher in exposed breastfed infants than control (Xing et al. 2007). Mercury was also reported in breast milk by several studies at polluted sites. Further studies are needed to investigate other different types of chemicals, like PBDEs and phthalates which may exist in breast milk (Esteban and Castaño 2009).

2.2.3 *Human Hair*

Human hairs are used for environmental biomonitoring; these are stable bioindicators as they are easy to collect, store, and transport. An expert panel was organized by Agency for Toxic Substances and Disease Registry (ATSDR, USA) to standardize the procedure for hair analysis (Esteban and Castaño 2009).

Zheng and Yuan (2011) collected the hair samples from residents of Longtang town which were exposed occupationally and nonoccupationally to e-waste and control samples from rural town analyzed by flame atomic absorption spectrometer (FLAAS; Hitachi Z-2300, Japan) for the concentration of heavy metals. The concentrations of heavy metals in both exposed and control individuals were in the following order Zn > Pb, Cu > Cd > Ni. A multivariate analysis found Cd, Cu, and Pb at a significantly higher concentration in occupationally exposed workers than control group (Zheng and Yuan 2011). Human hair is also an excellent indicator of mercury. A study reported that 80% methyl mercury of the total mercury was found in the hair of exposed individuals (Esteban and Castaño 2009).

Wang et al. (2009) found the trace elements in order of Pb > Cu > Mn > Ba > Cr > Ni > Cd > As > V in scalp hair samples collected from exposed individuals of e-waste recycling area at Taizhou (TZ) and a control site. The GM's values for Cd, Cu, Mn, and Pb were significantly higher in exposed than nonexposed samples, which are 0.518 mg/g, 39.8 mg/g, 5.32 mg/g, and 49.5 mg/g, respectively, for exposed samples. The concentration of Cu and Pb was 143 and 251 mg/g in the exposed group and 5.27–14.0 mg/g and 1.09–15.9 in control group, respectively. The concentrations of Cd, Cu, and Pb were found in hair and analyzed through multivariate analysis which was significantly correlated with e-waste-related activities (Wang et al. 2009).

Liang et al. (2016) collected samples from the population of Wenling of Zhejiang province in China which were exposed occupationally and nonoccupationally to e-waste and a control group. These samples were studied for the concentration of PBDEs and DBDPE. The results revealed that the concentrations of PBDEs and DBDPE were significantly higher in the hair of occupationally exposed worker than nonoccupational and control. The mean concentrations of total PBDEs (∑PBDEs) and DBDPE in the hair of e-waste recycling workers were 292.9 ng/g dw and 82.5 ng/g dw, respectively. In occupational individuals, the level of PBDEs was one order and five times of the magnitude, and DBDPE was three and eight times higher than nonoccupational and control group, respectively. The mean concentration of ∑8PBDEs (BDE28, 47, 100, 99, 153, 154, 183, 209) in hair of recycling workers was also higher than previously reported studies (Liang et al. 2016).

He and Yuan (2010) analyzed the level of dechlorane plus (DP) and dichlorination product, 1,6,7,8,9,14,15,16,17,17,18-octadeca-7,15-diene (*anti*-Cl11-DP) in hair samples collected from population of control and e-waste polluted sites in South China. The concentration of anti-Cl11-DP and DP in exposed group was detected in the range of nd (not detected) to 0.23 ng/g and 0.02–58.32 ng/g, respectively. The ∑DP concentration was three times higher in e-waste exposed population than control (He and Yuan 2010).

Further studies are needed to explore other chemicals which are found in hair, to find out the correlation between the high level of contaminants in hair with blood/urine samples and their effects on health, and to further allocate different exposure pathways (Wang et al. 2009).

2.2.4 Human Urine

Urine is a good matrix for biomonitoring of e-wastes in a polluted area. In 2014, Feld et al. studied the level of PAH metabolites in urine samples collected from e-waste recycling population of Africa and controls from the suburb of Accra. The concentration of PAH metabolites has been measured on creatinine base to regulate the hydration status of the individuals. The concentration of all PAH metabolites in urine of normal (nonexposed) vs exposed individuals, of 1-OH-phenanthrene was 0.55 vs 0.85 μg/g creatinine ($P < 0.001$), of 4-OH-phenanthrene 0.11 vs 0.22 μg/g creatinine ($P < 0.001$), of 3-OH-phenanthrene 0.63 vs 0.99 μg/g creatinine ($P < 0.001$), of 2-/9-OH-phenanthrene 0.37 vs 0.55 μg/g creatinine ($P = 0.005$), and of 1-OH-pyrene 0.54 vs 1.33 μg/g creatinine ($P < 0.001$). These results show a significantly high concentration of all measured PAH metabolites in participants from e-waste recycling site than controls (Feldt et al. 2014).

Similarly, trace elements were analyzed in the urine of e-waste recycling workers by Asante et al., in 2012 at Accra, Ghana. A significantly high concentration of Sb, Pb, and Fe was found in the e-waste recycling workers than nonexposed individuals (Asante et al. 2012). Further studies are recommended to assess the impacts of e-wastes on the health of residents of e-waste recycling area (Feldt et al. 2014).

2.2.5 *Human Nail*

Nails are used as a biomarker and diagnostic tool in the medical sciences (Parizanganeh et al. 2014). Both fingernails and toenails can be used as biomarkers; some authors suggest that toenails are better than fingernails because they are less exposed to external contamination (Barbosa et al. 2005).

In a study, fingernail samples have been collected from e-waste affected and non-affected site of Zanjan Province in northwest Iran and analyzed by inductively coupled plasma-optical emission spectroscopy (ICP-OES). The levels of toxic metals were significantly higher in exposed samples than nonexposed and available values in literature. The level of metals in exposed samples was Cd = 1.18, Pb = 15.15, As = 15.47, Ni = 18.22, and Zn = 68.46 mg/kg (Parizanganeh et al. 2014). The level of elements in nails was compared with another matrix. The concentration of mercury in nail was compared with blood and Cd and Pb levels in nail with hair in exposed individuals. A significant correlation was found between these matrices (Esteban and Castaño 2009). Further studies are needed to find the correlation between different biomarkers.

2.3 Animal Biomonitoring

Different tissues and organs of the animals were reported as bioindicators of e-wastes in a polluted environment (Table 2.2).

Different vertebrates (chickens, ducks, fish, frogs, and mice) have been studied at e-waste recycling area in Taizhou, China, for the concentration of polybrominated diphenyl ethers (PBDEs) in liver, brain, and muscle tissues. The tissue levels of PBDEs in frogs, fish, mice, chickens, and ducks were ranged 2.75–108, 0.02–32.0, 0.45–206, 0.06–18.8, and 1.83–112 ng/g wet weight, respectively. PBDEs were distributed preferentially in the muscles and liver of mice, chicken, duck, and frogs. The level of PBDEs was higher in the brain of fish followed by frogs and lowest in the mammals and birds because the blood-brain barrier (BBB) of fish provides little protection against PBDEs compared with higher vertebrates (Zhao et al. 2016). In

Table 2.2 Animals are bioindicators of e-waste polluted environment

Bioindicator	E-waste	References
Fish brain	Level of PBDEs was high	Zhao et al. (2016)
Liver and muscles of chickens, frogs, ducks, and mice	High level of PBDEs	
Various tissues (brain, liver, testis, and egg) of adult frogs (Rana limnocharis)	High PBDE levels	Liu et al. (2011)
Chicken muscle and eggs	High PBDE levels	Xiaofei et al. (2011)

other study tissue samples of adult frogs (Rana limnocharis) collected from e-waste recycling area of China were investigated for PBDE concentration. The total PBDE (∑PBDE) was found in the range of 17.10–141.11 ng/g wet weight (Liu et al. 2011).

Xiaofei et al. (2011) investigated the concentration of PBDEs in eggs and tissues of chicken at e-waste recycling area in South China. The mean concentration of PBDEs in eggs was 563.5 ng/g w and in tissues ranged from 15.2 to 3138.1 ng/g lipid weight. The human dietary intake through eggs and tissues of chicken were also investigated that the mean value was 13.5 ng/ (day kg, b.w). In this study, the concentration of PBDEs in chicken tissues and human intake was 2–3 orders and one order higher than previously reported literature, respectively (Xiaofei et al. 2011). Further studies are recommended to confirm the PBDE transport across BBB in vertebrates and to investigate other e-wastes in animals and their effects on human.

2.4 Conclusion

These studies reported the e-waste in the tissues of humans and animals exposed into e-waste. Human blood/serum, breast milk, hair, urine, and nail have been used as bioindicators for e-wastes such as heavy metals, DBDPE, PBDEs, PCDD/Fs, DBDPE, BFRs, DP, and PAHs. These pollutants were found at high concentration in humans. E-wastes, e.g., PBDEs, were also detected at high level in different tissues of animals, i.e., chickens, ducks, fish, frogs, and mice. The human dietary intake of e-wastes from affected chicken meat and eggs had been reported. Further studies are recommended to investigate the effects of e-waste on human health.

References

Asante KA, Agusa T, Biney CA, Agyekum WA, Bello M, Otsuka M et al (2012) Multi-trace element levels and arsenic speciation in urine of e-waste recycling workers from Agbogbloshie, Accra in Ghana. Sci Total Environ 424:63–73

Barbosa F, Tanus-Santos JE, Gerlach RF, Parsons PJ (2005) A critical review of biomarkers used for monitoring human exposure to lead: advantages, limitations, and future needs. Environ Health Persp 113(12):1669–1674

Deshpande PC (2013) Assessment of pollutant emissions and occupational exposure in the informal e-waste recycling and disposal facilities. Report Submitted (December)

Esteban M, Castaño A (2009) Non-invasive matrices in human biomonitoring: a review. Environ Int 35(2):438–449

Feldt T, Fobil JN, Wittsiepe J, Wilhelm M, Till H, Zoufaly A et al (2014) High levels of PAH-metabolites in urine of e-waste recycling workers from Agbogbloshie, Ghana. Sci Total Environ 466–467:369–376

He L, Yuan J (2010) Dechlorane Plus in human hair from an e-waste recycling area in South China: comparison with dust. Environ Sci Technol *44*(24):9298–9303

Henríquez-Hernández LA, Boada LD, Carranza C, Pérez-Arellano JL, González-Antuña A, Camacho M et al (2017) Blood levels of toxic metals and rare earth elements commonly

found in e-waste may exert subtle effects on hemoglobin concentration in sub-Saharan immigrants. Environ Int 109(July):20–28
Huo X, Peng L, Xu X, Zheng L, Qiu B, Qi Z et al (2007) Elevated blood lead levels of children in Guiyu, an electronic waste recycling town in China. Environ Health Perspect 115(7):1113–1117
Liang S, Xu F, Tang W, Zhang Z, Zhang W, Liu L (2016) Brominated flame retardants in the hair and serum samples from an e-waste recycling area in southeastern China: the possibility of using hair for biomonitoring. Environ Sci Pollut Res Int 130:14889–14897
Liu PY, Du GD, Zhao YX, Mu YS, Zhang AQ, Qin ZF et al (2011) Bioaccumulation, maternal transfer and elimination of polybrominated diphenyl ethers in wild frogs. Chemosphere 84 (7):972–978
Parizanganeh A, Zamani A, Bijnavand V, Taghilou B (2014) Human nail usage as a bio-indicator in contamination monitoring of heavy metals in Dizajabaad, Zanjan province-Iran. J Environ Health Sci Eng 12(1):1–9
Sheng G, Fu J (2008) Elevated serum polybrominated diphenyl ethers and thyroid-stimulating hormone associated with lymphocytic micronuclei in Chinese workers from an E-waste dismantling site. Environ Sci Technol 42(6):2195–2200
Wang T, Fu J, Wang Y, Liao C, Tao Y, Jiang G (2009) Use of scalp hair as indicator of human exposure to heavy metals in an electronic waste recycling area. Environ Pollut 157 (8–9):2445–2451
Xiaofei Q, Qin Z-F, Li Y, Zhao Y, Xia X, Yan S et al (2011) Polybrominated diphenyl ethers in chicken tissues and eggs from an electronic waste recycling area in southeast China. J Environ Sci (china) 23:133–138
Xing GH, Xu Y, Liang Y, Chen LX, Wu C, Wong CKC et al (2007) Body loadings and health risk assessment of polychlorinated dibenzo- p-dioxins and dibenzofurans at an intensive electronic waste recycling site in China. Environ Sci Technol 41(22):7668–7674
Zhao Y, Li Y, Qin X, Lou Q, Qin Z (2016) Accumulation of polybrominated diphenyl ethers in the brain compared with the levels in other tissues among different vertebrates from an e-waste recycling site. Environ Pollut 218:1334–1341
Zheng J, Yuan XLJ (2011) Heavy metals in hair of residents in an e-waste recycling area, south China: contents and assessment of bodily state. Arch Environ Contam Toxicol 61(4):696–703

Chapter 3
New Brominated Flame Retardants in the Environment of Developing Countries

Nadeem Ali, Khurram Shahzad, Ehtisham Nazar, Nabil A. Alhakamy, Iqbal Mohammad Ibrahim Ismail, Muhammad Imtiaz Rashid, and Syed Ali Musstjab Akber Shah Eqani

Abstract The environmental occurrence of new brominated flame retardants (NBFRs) in the emerging economies has been reviewed in the current chapter. The strict regulations about the manufacturing and usage of polybrominated diphenyl ethers (PBDEs) as flame retardants lead to the development and increased use of NBFRs to satisfy the fire and safety guidelines. These NBFRs are assumed to be harmless for the usage, but current studies have shown their occurrence in different environmental media and have raised questions on their use. In this chapter, the presence of NBFRs in different environmental matrices such as air, dust, food, water, humans, etc. has been reviewed. Finally, the investigation gaps were figured out with commendations for forthcoming research work.

Keywords New brominated flame retardants · Developing countries · Human exposure

---

N. Ali (✉) · K. Shahzad · M. I. Rashid
Center of Excellence in Environmental Studies, King Abdulaziz University, Jeddah, Saudi Arabia
e-mail: nabahadar@kau.edu.sa

E. Nazar
Department of Environmental Sciences, University of Gujrat, Gujrat, Pakistan

N. A. Alhakamy
Faculty of Pharmacy, Pharmaceutics Department, King Abdulaziz University, Jeddah, Saudi Arabia

I. M. I. Ismail
Center of Excellence in Environmental Studies, King Abdulaziz University, Jeddah, Saudi Arabia

Faculty of Science, Department of Chemistry, King Abdulaziz University, Jeddah, Saudi Arabia

S. A. M. A. S. Eqani
Health and Environment Division, Department of Biosciences, COMSATS University, Islamabad, Pakistan

M. Z. Hashmi, A. Varma (eds.), *Electronic Waste Pollution*, Soil Biology 57,
https://doi.org/10.1007/978-3-030-26615-8_3

Table 3.1 List of important NBFRs discussed in this chapter

Name	Abbr.	Formula	Application
1,2-Bis(2,4,6-tribromophenoxy) ethane	BTBPE	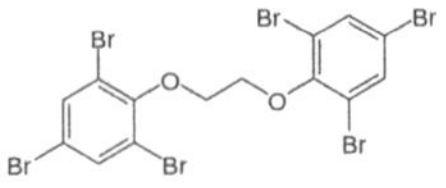	Substitute for octa-BDE and primarily used in BTBPE production which is ABS, HIPS, thermosoftening plastics, thermosetting polymers or resins, polycarbonate, and coatings
Decabromodiphenyl ethane	DBDPE	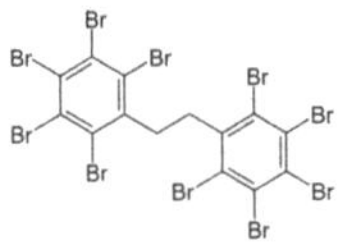	Substitute for deca-BDE and used as FR in textile products (cotton and polyester), ABS, HIPS, polypropylene (PP), etc.
Bis(2-ethylhexyl)-3,4,5,6-tetrabromophthalate	TBPH	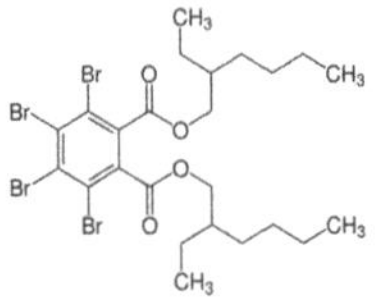	Substitute for penta-BBE and used as FR additive in polyvinyl chloride (PVC), neoprene (polychloroprene pr PC rubber), insulations for wire and cables, plastic film and sheeting, the backing of carpet, fabric coatings, wall coverings, and adhesives
2-Ethylhexyl-2,3,4,5-tetrabromobenzoate	TBB	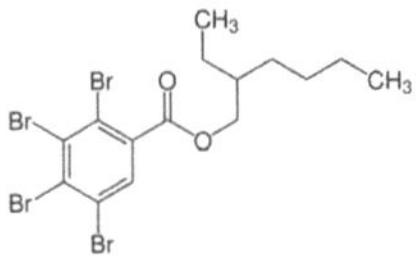	The mixture of TBB with TBPH is used as FR additive product known as Firemaster 550. The main application of this mixture is in the production of polyurethane foam (PUF)
Hexabromobenzene	HBB	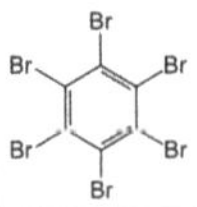	Used as Fr additive for paper, furniture or other wooden surfaces, textile materials, electro-electronic, and plastic goods

3.1 Introduction

In recent years, the use of new brominated flame retardants (NBFRs) has increased significantly following the strict regulations on the usage of polybrominated biphenyls ethers (PBDEs) (Covaci et al. 2011; Wei et al. 2015). Although many of the NBFRs are in use for many years, their consumption as alternative flame retardants increased markedly in recent years (Covaci et al. 2011; Ali et al. 2017). Several scientific research works have reported the industrial utilization of NBFRs as additive FRs in the polymer, textile, furniture, electro-electronics, automobile, petrochemicals, and construction (Harju et al. 2009). These NBFRs are additives in consumer products; therefore, their release into the environment from the treated products will increase during manufacture, use, and reprocessing of the same materials in the future (Ali et al. 2017; Zheng et al. 2015).

The name, formula, and application of selected NBFRs discussed in this chapter are given in Table 3.1. The precise data about the manufacturing volume for different

NBFRs is hard to find as suggested by the Norwegian Climate and Pollution Agency (Harju et al. 2009). Recent figures showed that production volume of these NBFRs is increasing sharply, e.g., production volume of tetrabromobisphenol A-bis 2,3-dibromopropyl ether (TBBPA–DBPE) and decabromodiphenyl ethane (DBDPE) is increasing up to 80% in China (Xiao 2006). However, information on the production of most of the NBFRs are not available in developing countries. NBFRs are used as additives in the polymeric materials; consequently, they release from these materials into the surroundings during manufacture, consumption, and waste recycling of the treated products (Covaci et al. 2011). The NBFRs enter the primeval atmospheric and aquatic environment via long-range atmospheric transportation and precipitation (Li et al. 2015; Ogawa et al. 2010; Wu et al. 2011). Several recently published articles have stated the existence of NBFRs in numerous environmental specimens including air, dust, water, sediments, and human sample, etc. (Covaci et al. 2011). Majority of the specified research work has been conducted in the advanced countries, while inadequate figures are accessible of their existence in developing countries. This is partly because of knowledge deficiency as well as readiness and practical skills that the systematic data breaches occur on the existence of these compounds in the environment of underdeveloped countries. The lack of data makes it tough for the law enforcement authorities to take ecological legislative actions in their corresponding jurisdiction. This chapter presents accessible data on the occurrence of different NBFRs in the ecological systems in different emerging countries. Finally, the existing data gaps were discussed and recommended the upcoming studies in the future.

3.2 Environmental Occurrence

3.2.1 Indoor Dust

Indoor dust is described as "indoor-pollution archive," and trapped chemicals in it degrade slowly due to the lack of moisture and direct sunlight (Butte and Heinzow 2002). Most of the modern building structures are built in limited space and in most cases lack direct sunlight and cross ventilation. The recent literature reveals that both dust ingestions and inhalation are the main exposure pathways for numerous environmental contaminants, particularly the ones which are unleashed from the in-house consumer goods and specific cohort, e.g., kids (Ali et al. 2013a; Mercier et al. 2011).

Important NBFRs reported in dust from residential and commercial settings of different countries are given in Table 3.2. In most of the studies, DBDPE was predominant NBFR in indoor dust; however in Indian commercial, residential, and e-waste samples, the main NBFR was BTBPE (Devanathan et al. 2011). The reported results for India were based on analysis of dust samples which were collected from two of the fastest-growing, technologically advanced cities, i.e., Chennai and Bangalore. As a result, these cities generate enormous amounts of

Table 3.2 NBFRs reported in indoor dust collected from various developing countries

Country	Microenvironment	NBFRs			
		TBB	BTBPE	TBPH	DBDPE
China					
Longtang (Zheng et al. 2015)	E-waste	7.5 (nd–192)	28 (5.8–374)	88 (10–268)	1160 (nd–6540)
	E-waste	29 (nd–75)	148 (7.2–12,700)	160 (8.2–652)	10,900 (nd–96,800)
	E-waste	311 (4.9–862)	101 (2.8–432)	7120 (nd–17,600)	26,300 (11,300–181,000)
Dali	E-waste	60 (nd–178)	40 (3.8–227)	193 (nd–928)	6400 (1920–23,400)
Guiyu	E-waste	60 (nd–178)	3870 (20–27,600)	49 (nd–779)	45,400 (197–201,000)
Beijing (Cao et al. 2014)	Office 1	22.5 (4–29)	8420 (3190–11,700)	321 (43–714)	2320 (750–3140)
	Office 2	14.3 (4–66.4)	46.5 (6–523)	315 (209–464)	4620 (1990–16,200)
	Office 3	15 (7–26.4)	13.6 (5–83)	508 (258–928)	3010 (2020–5800)
Beijing (Qi et al. 2014)	Household	0.85 (nd–6300)	2.5 (0.2–220)	29 (nd–1600)	280 (nd–16,000)
Guangzhou (Wang et al. 2010)	E-waste	–	20 (nd–998)	–	63 (13.5–1144)
	Household	–	6.5 (nd–211)	–	2733 (100–47,000)
Iraq (Al-Omran and Harrad 2015)	Household	5.3 (nd–28)	14.1 (nd–51)	64.2 (nd–294)	125 (33–269)
Egypt (Hassan and Shoeib 2015)	Household	0.81 (0.11–369)	0.24 (nd–2.63)	0.12 (nd–1.77)	–
	Workplaces	7.14 (0.3–154)	1.3 (0.5–2.95)	0.09 (nd–0.24)	–
	Car	5.8 (0.4–91)	2.4 (1.3–13.4)	0.6 (nd–8.3)	–
Pakistan					
Faisalabad (Ali et al. 2013a)	Urban house	0.4 (nd–4.8)	15 (1–192)	5.8 (1.6–167)	90 (2.5–465)
	Car	0.5 (nd–175)	10.5 (3–100)	6.5 (nd–105)	65 (6–5420)

(continued)

e-waste which require an ever-increasing number of e-waste shredding/reprocessing facilities in the locality. These results indicate that electronic goods might be the main source of BTBPE input to the Indian dust, where it is used as FR in the casing of computers and other electronic devices (Devanathan et al. 2011). Similarly, in the

Table 3.2 (continued)

Country	Microenvironment	NBFRs TBB	BTBPE	TBPH	DBDPE
Gujrat (Ali et al. 2012)	Mosque	0.03 (nd–4)	2.16 (nd–52)	5.23 (nd–31)	14.6 (nd–296)
	Rural house	0.03 (nd–4.5)	3.15 (nd–397)	3.5 (nd–141)	14 (5–850)
Faisalabad and Islamabad (Ali et al. 2014)	Electronic shops	1 (nd–15)	17 (0.5–17,150)	20 (0.6–950)	140 (2–52,150)
	Clothing shops	0.7 (nd–1.2)	6.5 (0.5–20)	9 (nd–35)	31 (nd–155)
	University office	1.3 (nd–16)	9.5 (2–125)	19 (3–225)	60 (10–120)
Saudi Arabia (Ali et al. 2016)	Household	16 (nd–250)	5 (nd–18)	25 (nd–330)	800 (110–1650)
	Car	12 (3–8700)	3 (nd–70)	32 (nd–2120)	275 (45–6020)
	AC filter	25 (10–5500)	4 (nd–33)	10 (2–55)	650 (250–1200)
Kuwait (Ali et al. 2013a)	Household	6.6 (0.6–550)	6.8 (0.9–535)	54 (7.2–1835)	220 (40–2175)
	Car	13 (2.0–3450)	4 (0.7–21)	85 (12–3700)	202 (84–8200)
Vietnam (Tue et al. 2013)	Urban house	–	7.1 (1.2–100)	–	40 (17–150)
	Sub-urban house	–	3.1 (1.6–6.7)	–	46 (15–110)
	E-waste	–	56 (5.4–620)	–	230 (39–470)
Thailand (Ali et al. 2011b)	E-waste	13 (nd–59)	4360 (3–22,600)	180 (79–1300)	4650 (380–44,000)
India (Devanathan et al. 2011)	Commercial	–	220	–	67
	E-waste	–	65,000	–	120
	Residential	–	48	–	15

Indian dust, the levels of DBDPE were lower BTBPE which is contrary to other studies from developing countries (Table 3.2). Nonetheless, the study considered a small number of samples, so the obtained results cannot represent the overall status of NBFRs in Indian indoor environment. Similarly, significant levels of DBDPE and BTBPE were stated in indoor dust collected from diverse e-waste reprocessing units in Thailand, whereas concentrations of TBB, TBPH, and TBBPA-DBPE have multifold lesser values. This indicates lower usage of TBBPA-DBPE as a substitute to deca-BDE (Ali et al. 2011a, b).

Nowadays, one of the main BFRs being used in China is DBDPE, where recent studies showed a significant increase in levels of DBDPE in the indoor environment of China (Cai 2008). The pioneer study carried out by collecting dust samples from e-waste handling area revealed the existence of NBFRs in the Chinese environment. The reported values of BTBPE and DBDPE were 14.6–232 ng/g and <LOQ and 139 ng/g, respectively. Recent studies performed in China using dust specimens from homes (Qi et al. 2014; Wang et al. 2010), offices (Cao et al. 2014), and e-waste (Wang et al. 2010; Zheng et al. 2015) handling facilities showed increasing levels of NBFRs (Table 3.2). High levels of BTBPE and DBDPE (up to 400 times) in e-waste dust were found by Zheng et al. (2015) than Wang et al. (2010) from exactly the same places. It presented that reusing of NBFR-treated e-waste is increasing, while its use in the electronic products is also on the rise. Also, lower concentrations of pentabromotoluene (PBT), pentabromoethylbenzene (PBEB), and 1,2,3,4,5-pentabromobenzene (PBB) were also stated in Chinese indoor dust collected from home and e-waste handing facility (Qi et al. 2014; Wang et al. 2010). The concentration of various NBFRs reported in Chinese occupational settings (Cao et al. 2014) was greater than the ones reported from Pakistani (Ali et al. 2014) and Egyptian (Hassan and Shoeib 2015) settings (Table 3.2). The predominant NBFR in dust samples from Pakistani and Chinese offices was DBDPE, while in Egyptian workplaces, TBB was the major one. The variation in the profile of NBFRs in dust samples from different working sites refers to the diversity of fire safety guidelines and the existence of diverse emission sources or preferences to use various FRs as supplements to the consumer goods in different countries. Ali et al. (2013a, 2016) reported the presence of NBFRs in car dust samples collected from the Kingdom of Saudi Arabia (KSA), Kuwait, and Pakistan, where DBDPE was the main the chemical followed by TBB, TBPH, BTBPE, and HBB (Table 3.2). In these studies, concentrations of BTBPE were lower than DBDPE, which has been extensively used in car electronics. Similarly, levels of TBB and TBPH were also high; these chemicals are utilized as additives in the foam used to make car seats, which might be their emission points (Ali et al. 2013a, 2016). AC filter dust is an important indicator of indoor air quality; Ali et al. (2016) reported the existence of NBFRs in the dust samples collected from AC filters in different households of Jeddah, Saudi Arabia. The variation in the profiling of NBFRs (Table 3.2) in indoor dust collected from several countries can describe the prevailing fire safety regulations as well as implementation of up-to-date fire safety protocols along with phasing out the PBDEs (Ali et al. 2013a). China is one of the vital manufacturers of NBFRs, but other developing countries do not produce them; instead, they import these chemicals from the major producers to fulfill their consumption demand. Similarly, the occurrence of NBFRs in these developing countries might also show their emission from the import of consumer products loaded with NBFRs. Zheng et al. (2015) concluded that increasing levels of NBFRs in Chinese indoor environments will further increase in future due to their increased consumption in the consumer goods and with increasing recycling activities; thus, chronological data is vital to realize the dynamics of NBFRs in these environments.

3.2.2 Air and Tree Bark

Air is a vital compartment of the ecosystem which serves as a sink for different carbon-based contaminants emitting in the surrounding environment. The fingerprints of levels and distribution of NBFRs in various areas have been revealed by conducting several studies on ambient air (Marklund et al. 2005; Shi et al. 2009). The distribution pattern of NBFRs from different surroundings varies depending on the emission patterns of NBFRs from various sources such as industries, waste recycling units, building materials used for construction, etc. (Marklund et al. 2005; Shi et al. 2009). Shi et al. (2009) analyzed gas and suspended particulate matter samples collected from Pearl River Delta (PRD) from an area located near an e-waste handling facility. The levels for BTBPE (pg/m^3) (ranged 3.8 and 67), TBBPA-DBPE (ranged 131–1240), and DBDPE (ranged 402–3578 pg/m^3) were reported. Li et al. (2015) collected passive air samples from 11 different Chinese cities and reported the occurrence of NBFRs between 2005–2007 and 2008–2010. They found air samples collected in the later years have higher levels of NBFRs. In later year air samples, DBDPE was the main compound with an average concentration of 14 ng/sample and detection rate of recurrence of 74%, while in the prior samples, DBDPE was present in about 25% of the specimens only. Other NBFRs, namely, PBBZ (0.33 ng/sample), PBT (0.47 ng/sample), PPBEB (0.13 ng/sample), and BTBPE (0.35 ng/sample), were the other more frequently detected NBFRs in these air samples (Li et al. 2015). Few other articles from China also stated low levels of PBBZ, PBEB, BTBPE, and PBT in air samples collected from the northern region of China (Lin et al. 2013; Zhao et al. 2013). DBDPE (average concentrations of 10 ng/sample) were detected in air samples collected from Northern China (Zhao et al. 2013). These results were similar to the air sample results exhibited by the study conducted in various cities of China from 2008 to 2010 (Li et al. 2015). In two separate studies, Tian et al. (2011a, b) reported different NBFRs in the troposphere of the southern China region. The concentrations (pg/m^3) of BTBPE 398, DBDPE 2240, and PBT 125 were observed in the specimens gathered from e-waste handling site. This study reveals that the surrounding environment plays a significant part in the environmental pollution caused by these chemicals. Tree bark is a valuable reflexive bioindicator for the airborne pollutants because of its chemical composition, containing high lipid constituents, the physical structure having a permeable surface, and no metabolic activities (Samecka-Cymerman et al. 2006). Qiu and Hites (2008) found BTBPE (ranged from 3.1 to 38 ng/g) and DBDPE (up to 1000 ng/g lipid) in tree bark from PRD region China. In a further study, Yuan et al. (2016) described detectable levels (ng/g lw) of PBBz (0.09–2.2), PBT (0.06–0.78), and HBB (0.07–2.0) in all tree bark samples gathered from Kaiyuan City, China, which might present widespread existence of these chemicals in the investigated area.

3.2.3 Water

Recently a number of studies are available in the literature which reported the occurrence of NBFRs in water specimens gathered from wastewater treatment units, river, marine, rain, and groundwater. Most of the data is available from the developed countries and few studies from the emerging countries (Olukunle and Okonkwo 2015; Wei et al. 2015). Khan et al (2016) stated that the occurrence of NBFRs along with other pollutants in the shallow groundwater samples from Pakistan is caused by the infiltration of industrial as well as domestic wastewater from drainage system along with the use of low-quality plastic pipes for the water supply network and wastewater drainage system. In this study, ∑NBFRs (nd–1.39 pg/L) were reported in the water specimens collected from industrialized manufacturing areas. DBDPE (36.3%) was the most abundant NBFRs, while other contaminants include TBPH, TBB, TBE, and HBB having occurrence values in the decreasing order, such as (31.8%), (11.4%), (11.3%) and (9.2%), respectively. In Pakistan, the FRs levels in the water were linked to the industrial processing areas. On the other hand, these chemicals might have been released in the water from plastic pipes and storage tanks which are normally used for the storage of portable in Pakistan (Khan et al. 2016). Olukunle and Okonkwo (2015) found BTBPE (4.4–15 pg/L) and TBB (8.7–142 pg/L) in landfill leachates ($n = 18$) of Gauteng Province, which is the key industrial region of South Africa. While, DBDPE was below detection limits and TBPH was found in only one sample. A few recent studies have reported about the open disposing of e-waste in the PRD region, China. Due to which, excessive levels of environmental pollutants including DBDPE and TBBPA-DBPE have been reported in its environs (Shi et al. 2009). DBDPE (ranged between 9.1 and 990 ng/L) was the major chemical, while BTBPE (nd–26.3), HBB (nd–1.45), and PBT (nd–0.45) were the other important NBFRs reported in the water samples collected from sewage network of industrial and urban region ending in the Dongjiang River, China. Ruan et al. (2009) analyzed the water specimens collected from the Liuyang River which flows in southern China close to the polymer manufacturing facility. This study reported the presence of tris(2,3,-dibromopropyl) isocyanurate (TBC) (ranging between 2.3 and 163 ng/L). This NBFR is used as FR supplement in various types of polymeric material. The identification of TBC in water samples raises the awareness of its occurrence in the ambient environment.

3.2.4 Soil, Sludge, and Sediment

The soil, sediments, and sludge are key components of the environment and offer effective awareness about the fingerprints of various pollutants in the ecosystem. In eThekwini Metropolitan Municipality, South Africa, coastal sediments analysis revealed that DBDPE (average, 3208 ng/g) and TBB (average, 545 ng/g) were detected in >90% of the specimens (La Guardia et al. 2013). Shi et al. (2009)

studied different NBFRs in the soil, sediments, and sewage sludge specimens gathered from southern China. Sewage sludge and sediments were more contaminated than farmland. The maximum concentration (ng/g dw) of BTBPE, DBDPE, and TBBPA-DBPE was measured in sewage sludge which was 1.7, 2000, and 8950; their values in sediments were 22, 364 and 2300, while in farmland soil samples, their values were measured 0.11, 36, and 60, correspondingly (Shi et al. 2009). Zhang et al. (2009) studied DBDPE in sediment samples collected from the surface (19 ng/g dw) as well as from core (430 ng/g dw) in the industrial region of southern China. It was recorded that DBDPE was present in the recent sediment layers, which might reflect the increasing demand of DBDPE to fulfill the local requirement of the region. Wu et al. (2012) analyzed sediment cores gathered from 12 lakes of China and stated DBDPE (1.02–3.64 ng/g); the study also reveals an increasing chronological trend of DBDPE from the bottommost to the top surface which was associated with urbanization and industrial development. Chen et al. (2013) described DBDPE (ranged <dl to 1730 ng/g) occurrence in 97% of the analyzed sediment samples. Interestingly, the majority of the sediments showed elevated levels (ng/g) of DBDPE compared to BDE-209, while BTBPE (0.20–8.47 ng/g) were lower than PBDEs. Zheng et al. (2015) analyzed soil samples from China and found BDE-209 (nd–5900 pg/g) as the 2nd most abundant flame retardant after DBDPE as the major flame retardant (25–18,000 pg/g) and observed a positive correlation between these chemicals and population density. The distinguished emission patterns proposed the usage of DBDPE as replacement of the deca-BDE in the specified area. High average concentrations (ng/g dw) were reported for BTBPE (107) and DBDPE (96) in the dust specimens gathered from an area close to e-waste handling unit, which eventually affects the nearby farmland soil (Shi et al. 2009). Zheng et al. (2011) recorded extremely high average levels, which are multiple times more than those described by Shi et al. (2009) of BTBPE (297), DBDPE (3030), and HBB (68.5) in the soil specimens collected nearby e-waste handling facility. This shows a shift on the way to the use of NBFRs.

3.2.5 Fish and Birds

The occurrence of NBFRs in fish and avian samples has been testified by some studies. Ogawa et al. (2010) carried out a study using muscle samples from various Asian countries. It was reported that DBDPE was present in the specimens of green mussel (*Perna viridis*) and blue mussel (*Mytilus edulis*) gathered from 17 out of 67 places with concentrations measured ranging from <0.3 to 22 ng/g lw. Similarly, a study performed on three commercially grown fish species grown nearby an e-waste reprocessing facility in the southern region of China showed BTBPE 0.15 ng/g lw (Shi et al. 2009). Numerous NBFRs were studied in freshwater food web samples gathered from an e-waste reprocessing place, in southern China, including mud carp (*Cirrhinus molitorella*) (n = 12), northern snakehead (*Ophicephalus argus*) (n = 6), Chinese mystery snail (*Cipangopaludina chinensis*) (*n* = 43), crucian carp (*Carassius auratus*) (*n* = 18), Chinese water snake (*Enhydris chinensis*) (*n* = 2), and prawn (*Macrobrachium nipponense*) (*n* = 7) (Wu et al. 2010). The important NBFRs found in the freshwater food web included DBDPE (<3.8–338), BTBPE (1.71–518), PBEB (3.98–25.6), HBB (197–3099), and PBT (1.20–3.60) (Wu et al. 2010). Li et al. (2015) reported the presence of DBDPE in almost 79% of food samples (including meat, fish, and vegetables), at concentrations (pg/g WW) ranging between ND and 350, collected from food marts of Binhai Economic Development Zone, China. The occurrence of DBDPE in vegetables was reported for the first time, and this indicated widespread contamination of this chemical in the studied region. Mo et al. (2012) collected common kingfishers (*A. at this. n* = 22) from e-waste reprocessing places of southern China and measured high values of BTBPE (between 2.7 and 38) and DBDPE (between 4.5 and 52) in these samples. The biomagnification factors (BMFs) for DBDPE and BTBPE in this study were up to 0.77 and 3.60, respectively, which show possibly high ecological risks to these composites (Mo et al. 2012). Similarly, Labunska et al. (2015) collected edible fish and shrimp samples from the districts of Luqiao and Wenling situated in Taizhou City, the major e-waste reprocessing region in Eastern China, and reported DBDPE in shrimps (45.3 ng/g lw) and Tobin fish (62.2 ng/g lw). In another study, Labunska et al. (2015) analyzed meat, liver, and eggs of chicken and ducks to study TBB, BTBPE, PBEB, and TBPH. The reported levels (ng/g lw) of various NBFRs in the muscle samples were TBB = 24.7; 24.2, TBPH = 9.0; 7.2, BTBPE = 1.5; 4.6, and PBEB = <0.17; 0.9 for chicken and duck, respectively. These levels were lower than measured for equivalent avian livers which are TBB = 35.0; 38.4, TBPH = 10.6; 13.7, BTBPE = 15; 11.7, and PBEB = 2.3; 3.4 for chicken and duck, correspondingly. TBB (average, 4.3 ng/g lw) and TBPH (average, 1.1 ng/g lw) were present in all ovum samples collected from e-waste reprocessing sites (Labunska et al. 2015). Shi et al. (2009) studied BTBTPE and DBDPE in the liver and muscles of watercock from southern China. In a study on waterbirds, the samples gathered from an immense e-waste reprocessing region in southern China, Luo et al. (2009) measured maximum concentration 800 ng/g lw of DBDPE in the muscle specimens. The observed values of DBDPE by Luo et al. (2009) were 1.7 times higher than those reported in eggs of aquatic birds picked from the Yellow River Delta of National Nature Reserve in North China (Gao et al. 2009). Long-tailed shrike (LTS), oriental magpie-robin (PMR), and light-vented bulbul (LVB) samples gathered from seven different places in southern China detected DBDPE (nd–130 ng/g lw) (Sun et al. 2009). Interestingly birds living in the urban areas exhibited higher DBDPE levels than those from the e-waste site. Sun et al. (2009) hypothesized that in urban areas emissions of NBFRs mainly come from the local industrial activities. These findings showed that NBFRs have already passed into the food web. It is of huge apprehension because the manufacturing and consumption of these chemicals are still on the rise.

3.2.6 Terrestrial Biota and Human

The use of NBFRs was considered safe for both human and animals; however, few present-day studies have reported their detection in both animals and humans. Hu et al. (2008) studied liver, brain, kidney, gonad, and adipose tissues of confined pandas and reported DBDPE in 87% of giant panda while 71% of red pandas. The levels of DBDPE in giant panda gonad tissue (863 ng/g lw) were at a conservative estimate 20 times more than in other tissues of giant and red pandas. Likewise, the concentration of DBDPE in livers of giant panda have higher levels than those estimated in the red panda. Serum and hair samples from domesticated cats and dogs in Pakistan were found positive for bromophenols (BPs), DBDPE, BTBPE, and TBPH (Ali et al. 2013b). BP was the major studied chemical in cat and dog serum ranging from 20 to 335 ng/g lw and <5 to 150 ng/g lw, respectively. Similarly, DBDPE was the main NBFR reported in the hair samples of cat (ranged <2–18 ng/g dw) and dog (ranged <2–7) (Ali et al. 2013b). Zhu et al. (2009) collected serum specimens from office cleaning guys, students from university, and policemen in the city of Tianjin (North of China) during 2006 and reported HBB (ranged between 0.11 and 1.50 ng/g lw) in about 20% of the samples. Ali et al. (2013a) detected BTBPE (ranged <0.2–8 ng/g lw) and BPs (ranged <5–45 ng/g lw) in the human serum ($n = 85$) sample from Pakistan. More than 90% serum sampled contained BPs, while BTBPE was detected in <20% samples only. Zheng et al. (2011) conducted a study on human hair samples collected from different localities, i.e., e-waste reprocessing unit employees ($n = 30$), people living near the e-waste reprocessing unit ($n = 82$), inhabitants of the Guangzhou city ($n = 29$), and inhabitants of countryside of Tuantan town ($n = 32$), South China, and reported significant amounts of DBDPE, BTBPE, and HBB in collected samples. Hair samples from the workers employed in e-waste reprocessing units which are in direct contact with the e-waste showed higher levels than nonoccupationally exposed residents. The levels of selected NBFRs ranged between 5.9–365 (average 24.2) for DBDPE, 0.15–29.2 (1.21) for BTBPE, and nd–12 (2.12) ng/g dw for HBB in occupationally exposed workers. This study demonstrated that workers directly dealing with e-waste in the reprocessing units as well as residents of the nearby zone are more likely to be exposed to higher levels of NBFRs. Also, DBDPE and BTBPE amounts in hair and dust samples present significant correlation which advocated that dust ingestion is one of the main sources of human exposure (Zheng et al. 2011). In another study from China on human hair samples, Yuan et al. (2016) informed PBBs (0.025–0.095), HBB (nd–0.12), and PBT (0.031–0.17) in the pooled hair samples. The detection rate of NBFRs was 100% for PBT and PBB while 64% for HBB in these hair samples. It suggests the pervasive existence of these contaminants in the study region. Yuan et al. (2016) suggested that the main source of PBT was external exposure, while both extrinsic and intrinsic exposures contribute to the occurrence of HBB and PBBs in human hair. Shi et al. (2016) determined 6 NBFRs in 29 pooled human milk samples that ranged between 2.48 and 23.9 ng/g. DBDPE with average levels of 8.06 ng/g lw was the major compound in these samples.

Shi et al. (2016) found no obvious spatial distribution patterns for these human milk samples.

3.3 Concluding Remarks and Future Perspectives

In this chapter, we arranged available data on the existence of NBFRs in the atmosphere of developing countries. The discussed chemicals were considered safe alternatives for the regulated FRs, but their documented presence in the environment presents few questions about their environmental safety. Data on the manufacture and dynamics of the specified chemicals are still inadequately known in most of the developing countries. As a result, it is very difficult to make a risk assessment for these chemicals. This requires a more detailed investigation to apprehend the dynamics of these substances in the atmosphere of emerging countries. For this large scale, well-planned studies are needed on different environmental samples especially biomonitoring studies, which will help in designing mitigation strategies to define exposure pathways. Toxicological studies on most of the chemicals are missing or very few available in the literature. The toxicological data is also needed to establish the updated RfDs for these chemicals, and studies are needed to know the synergetic effects of NBFRs. There are scarce data regarding the epidemiological studies on these chemicals and different health endpoints (including clinical parameters). For this large scale, well-planned studies are needed on both abiotic and biota environmental samples. Biomonitoring studies using serum, blood, urine, and breast milk are lacking but very important in designing mitigation strategies to delineate exposure pathways. There is a lack of knowledge on the degradation, biotransformation, long-range transportation, phase distribution, and environmental fate of NBFRs. All these information are vital and required to prioritize the merit of future studies on these chemicals. Another important point, the majority of the discussed data in this chapter was available from China, and data from most of the developing countries are missing. Developing countries are in the race of rapid industrialization, and little concerns are shown on the environmental issues; this is partly due to the lack of funding, research infrastructure, and knowledge among public and policymakers. The profiling of the specified chemicals may differ significantly in different developing countries, because of their geographical position, safety regulations, and socioeconomic situations. The knowledge on these chemicals is increasing in recently with an increasing number of scientific studies annually, but still, there are many data gaps existing, and this warrants more efforts in the future.

References

Ali N, Harrad S, Goosey E, Neels H, Covaci A (2011a) "Novel" brominated flame retardants in Belgian and UK indoor dust: implications for human exposure. Chemosphere 83:1360–1365. https://doi.org/10.1016/j.chemosphere.2011.02.078

Ali N, Harrad S, Muenhor D, Neels H, Covaci A (2011b) Analytical characteristics and determination of major novel brominated flame retardants (NBFRs) in indoor dust. Anal Bioanal Chem 400:3073–3083. https://doi.org/10.1007/s00216-011-4966-7

Ali N, Van den Eede N, Dirtu AC, Neels H, Covaci A (2012) Assessment of human exposure to indoor organic contaminants in Pakistan. Indoor Air 22:200–211. https://doi.org/10.1111/j.1600-0668.2011.00757.x

Ali N, Ali L, Mehdi T, Dirtu A, Al-Shammari F, Neels H (2013a) Levels and profiles of organochlorines and flame retardants in car and house dust from Kuwait and Pakistan: implication for human exposure via dust ingestion. Environ Int 55:62–70. https://doi.org/10.1016/j.envint.2013.02.001

Ali N, Malik RN, Mehdi T, SAMAS E, Javeed A, Neels H, Covaci A (2013b) Organohalogenated contaminants (OHCs) in the serum and hair of pet cats and dogs: bio-sentinels of indoor pollution. Sci Total Environ 449:29–36. https://doi.org/10.1016/j.scitotenv.2013.01.037

Ali N, Mehdi T, Malik RN, Eqani SA, Kamal A, Dirtu AC, Neels H, Covaci A (2014) Levels and profile of several classes of organic contaminants in matched indoor dust and serum samples from occupational settings of Pakistan. Environ Pollut 193:269–276. https://doi.org/10.1016/j.envpol.2014.07.009

Ali N, Eqani SAMAS, Ismail IMI, Malarvannan G, Kadi MW, Albar HMS, Rehan M, Covaci A (2016) Brominated and organophosphate flame retardants in indoor dust of Jeddah, Kingdom of Saudi Arabia: implications for human exposure. Sci Total Environ 569:269–277. https://doi.org/10.1016/j.scitotenv.2016.06.093

Ali N, Shahzad K, Rashid MI, Shen H, Ismail IMI, Eqani SAMAS (2017) Environ Sci Pollut Res 24:18721. https://doi.org/10.1007/s11356-017-9336-3

Al-Omran LS, Harrad S (2015) Polybrominated diphenyl ethers and "novel" brominated flame retardants in floor and elevated surface house dust from Iraq: implications for human exposure assessment. Emerg Contam 2:7–13. https://doi.org/10.1016/j.emcon.2015.10.001

Butte W, Heinzow B (2002) Pollutants in house dust as indicators of indoor contamination. Rev Environ Contam Toxicol 175:1–46

Cai YY (2008) Current major brominated flame retardants. In: Handbook of modern technology of flame retardance. pp 35–45 (in Chinese)

Cao Z, Xu F, Covaci A, Wu M, Wang H, Yu G et al (2014) Distribution patterns of brominated, chlorinated, and phosphorus flame retardants with particle size in indoor and outdoor dust and implications for human exposure. Environ Sci Technol 48:8839–8846. https://doi.org/10.1021/es501224b

Chen SJ, Feng AH, He MJ, Chen MY, Luo XJ, Mai BX (2013) Current levels and composition profiles of PBDEs and alternative flame retardants in surface sediments from the Pearl River Delta, southern China: comparison with historical data. Sci Total Environ 444:205–211. https://doi.org/10.1016/j.scitotenv.2012.11.104

Covaci A, Harrad S, Abdallah MAE, Ali N, Law RJ, Herzke D et al (2011) Novel brominated flame retardants: a review of their analysis, environmental fate and behaviour. Environ Int 37:532–556. https://doi.org/10.1016/j.envint.2010.11.007

Devanathan G, Subramanian A, Isobe T, Kajiwara N, Suzuki G, Asante KA, Takahashi S, Tanabe S (2011) Organohalogen contaminants in dust samples from different indoor environments in India: implications on human exposure. Organohalog Compd 73:1101–1104

Gao F, Luo XJ, Yang ZF, Wang XM, Mai BX (2009) Brominated flame retardants, polychlorinated biphenyls, and organochlorine pesticides in bird eggs from the Yellow River Delta, North China. Environ Sci Technol 43:6956–6962. https://doi.org/10.1021/es901177j

Harju M, Heimstad ES, Herzke D, Sandanger T, Posner S, Wania F (2009) Emerging "new" brominated flame retardants in flame retarded products and the environment. Report 2462. Norwegian Pollution Control Authority, Oslo, 113 pp

Hassan Y, Shoeib T (2015) Levels of polybrominated diphenyl ethers and novel flame retardants in microenvironment dust from Egypt: an assessment of human exposure. Sci Total Environ 505:47–55. https://doi.org/10.1016/j.scitotenv.2014.09.080

Hu GC, Luo XJ, Dai JY, Zhang XL, Wu H, Zhang CL et al (2008) Brominated flame retardants, polychlorinated biphenyls, and organochlorine pesticides in captive giant panda (Ailuropoda melanoleuca) and red panda (Ailurus fulgens) from China. Environ Sci Technol 42:4704–4709. https://doi.org/10.1016/j.scitotenv.2014.09.080

Khan MU, Li J, Zhang G, Malik RN (2016) First insight into the levels and distribution of flame retardants in potable water in Pakistan: an underestimated problem with an associated health risk diagnosis. Sci Total Environ 565:346–359. https://doi.org/10.1016/j.scitotenv.2016.04.173

La Guardia MJ, Hale RC, Newman B (2013) Brominated flame retardants in sub Saharan Africa: burdens in inland and coastal sediments in the eThekwini metropolitan municipality, South Africa. Environ Sci Technol 47:9643–9650. https://doi.org/10.1021/es4020212

Labunska I, Abdallah MAE, Eulaers I, Covaci A, Tao F, Wang M, Santillo D, Johnston P, Harrad S (2015) Human dietary intake of organohalogen contaminants at e-waste recycling sites in Eastern China. Environ Int 74:209–220. https://doi.org/10.1016/j.envint.2014.10.020

Li P, Wu H, Li Q, Jin J, Wang Y (2015) Brominated flame retardants in food and environmental samples from a production area in China: concentrations and human exposure assessment. Environ Monit Assess 187:1–11. https://doi.org/10.1007/s10661-015-4947-y

Lin Y, Qiu X, Zhao Y, Ma J, Yang Q, Zhu T (2013) Polybromobenzene pollutants in the atmosphere of North China: levels, distribution, and sources. Environ Sci Technol 47:12761–12767. https://doi.org/10.1021/es403854d

Luo XJ, Zhang XI, Liu J, Wu JP, Luo Y, Chen SJ et al (2009) Persistent halogenated compounds in water birds from an e-waste recycling region in South China. Environ Sci Technol 43:306–311. https://doi.org/10.1021/es8018644

Marklund A, Andersson B, Haglund P (2005) Traffic as a source of organophosphorus flame retardants and plasticizers in snow. Environ Sci Technol 39:3555–3562. https://doi.org/10.1021/es0482177

Mercier F, Glorennec P, Thomas O, Le Bot B (2011) Organic contamination of settled house dust, a review for exposure assessment purposes. Environ Sci Technol 45:6716–6727. https://doi.org/10.1021/es200925h

Mo L, Wu JP, Luo XJ, Zou FS, Mai BX (2012) Bioaccumulation of polybrominated diphenyl ethers, decabromodiphenyl ethane, and 1,2-bis(2,4,6-tribromophenoxy) ethane flame retardants in kingfishers (Alcedo atthis) from an electronic waste-recycling site in South China. Environ Toxicol Chem 31:2153–2158. https://doi.org/10.1002/etc.1929

Ogawa S, Isobe T, Ramu K, Subramanian A, Takahashi S, Tanabe S (2010) Monitoring of contamination by non-PBDE brominated flame retardants in Asian coastal waters using mussels as an bioindicator. Interdiscip Stud Environ Chem 5:205–211

Olukunle OI, Okonkwo OJ (2015) Concentration of novel brominated flame retardants and HBCD in leachates and sediments from selected municipal solid waste landfill sites in Gauteng Province, South Africa. Waste Manag 43:300–306. https://doi.org/10.1016/j.wasman.2015.07.009

Qi H, Li W, Liu LY, Li YF (2014) Levels, distribution and human exposure of new non-BDE brominated flame retardants in the indoor dust of China. Environ Pollut 195C:1–8. https://doi.org/10.1016/j.envpol.2014.08.008

Qiu X, Hites RA (2008) Dechlorane plus and other flame retardants in tree bark from the north eastern United States. Environ Sci Technol 42:31–36. https://doi.org/10.1016/j.envpol.2014.08.008

Ruan T, Wang Y, Wang G, Wang P, Fu J, Yin Y et al (2009) Identification and evaluation of a N heterocyclic brominated flame retardant tris(2, 3-dibromopropyl) isocyanurate in environmental

matrices near a manufacturing plant in Southern China. Environ Sci Technol 43:3080–3086. https://doi.org/10.1021/es803397x
Samecka-Cymerman A, Kosior G, Kempers AJ (2006) Comparison of the moss Pleurozium schreberi with needles and bark of Pinus sylvestris as biomonitors of pollution by industry in Stalowa Wola (southeast Poland). Ecotoxicol Environ Saf 65:108–117. https://doi.org/10.1016/j.ecoenv.2005.05.009
Shi T, Chen SJ, Luo XJ, Zhang XL, Tang CM, Luo Y et al (2009) Occurrence of brominated flame retardants other thanpolybrominated diphenyl ethers in environmental and biota samples from southern China. Chemosphere 74:910–916. https://doi.org/10.1016/j.chemosphere.2008.10.047
Shi Z, Zhang L, Li J, Zhao Y, Sun Z, Zhou X, Wu Y (2016) Novel brominated flame retardants in food composites and human milk from the Chinese Total Diet Study in 2011: concentrations and a dietary exposure assessment. Environ Int 96:82–90. https://doi.org/10.1016/j.envint.2016.09.005
Sun Y, Wu J, Zhou L, Lin Y, Fu S (2009) Changes of soil nutrient contents after prescribed burning of forestland in Heshan City, Guangdong Province. Chin J Appl Ecol 20(3):513–517
Tian M, Chen SJ, Wang J, Shi T, Luo XJ, Mai BX (2011a) Atmospheric deposition of halogenated flame retardants at urban, e-waste, and rural locations in Southern China. Environ Sci Technol 45:4696–4701
Tian M, Chen SJ, Wang J, Zheng XB, Luo XJ, Mai BX (2011b) Brominated flame retardants in the atmosphere of e-waste and rural sites in southern China: seasonal variation, temperature dependence, and gas-particle partitioning. Environ Sci Technol 45:8819–8825
Tue NM, Takahashi S, Suzuki G, Isobe T, Viet PH, Kobara Y et al (2013) Contamination of indoor dust and air by polychlorinated biphenyls and brominated flame retardants and relevance of non-dietary exposure in Vietnamese informal e-waste recycling sites. Environ Int 51:160–167
Wang J, Ma YJ, Chen SJ, Tian M, Luo XJ, Mai BX (2010) Brominated flame retardants in house dust from e-waste recycling and urban areas in South China: implications on human exposure. Environ Int 36:535–541
Wei GL, Li DQ, Zhuo MN, Liao YS, Xie ZY, Guo TL, Li JJ, Zhang SY, Liang ZQ (2015) Organophosphorus flame retardants and plasticizers: sources, occurrence, toxicity and human exposure. Environ Pollut 196:29–46
Wu J, Guan Y, Zhang Y, Luo X, Zhi H, Chen S, Mai B (2010) Trophodynamics of hexabromocyclododecanes and several other non–PBDE brominated flame retardants in a freshwater food web. Environ Sci Technol 44:5490–5495
Wu JP, Guan YT, Zhang Y, Luo XJ, Zhi H, Chen SJ, Mai BX (2011) Several current-use, non-PBDE brominated flame retardants are highly bioaccumulative: evidence from field determined bioaccumulation factors. Environ Int 37:210–215
Wu F, Guo J, Chang H, Liao H, Zhao X, Mai B, Xing B (2012) Polybrominated diphenyl ethers and decabromodiphenylethane in sediments from twelve lakes in China. Environ Pollut 162:262–268
Xiao J (2006) A perspective on the development of brominated flame retardants in China. http://www.polymer.cn/html/industrynews/200612/15/_2007529102655763.htm
Yuan H, Jin J, Bai Y, Li Q, Wang Y, Hu J (2016) Concentrations and distributions of polybrominated diphenyl ethers and novel brominated flame retardants in tree bark and human hair from Yunnan Province, China. Chemosphere 154:319–325
Zhang XL, Luo XJ, Chen SJ, Wu JP, Mai BX (2009) Spatial distribution and vertical profile of polybrominated diphenyl ethers, tetrabromobisphenol a, and decabromodiphenylethane in river sediment from an industrialized region of South China. Environ Pollut 157:1917–1923
Zhao Y, Ma J, Qiu X, Lin Y, Yang Q, Zhu T (2013) Gridded field observations of polybrominated diphenyl ethers and deca-bromodiphenyl ethane in the atmosphere of North China. Environ Sci Technol 47:8123–8129
Zheng J, Luo XJ, Yuan JG, Wang J, Wang YT, Chen SJ, Mai BX, Yang ZY (2011) Levels and sources of brominated flame retardants in human hair from urban, e-waste, and rural areas in South China. Environ Pollut 159:3706–3713

Zheng Q, Nizzetto L, Li J, Mulder MD, Sanka O, Lammel G, Bing H, Liu X, Jiang Y, Luo C, Zhang G (2015) Spatial distribution of old and emerging flame retardants in Chinese forest soils: sources, trends and processes. Environ Sci Technol 49:2904–2911

Zhu L, Ma B, Hites RA (2009) Brominated flame retardants in serum from the general population in northern China. Environ Sci Technol 43:6963–6968

Chapter 4
Status of Organophosphate Esters in the Environment of Developing Countries and Their Impact on Human Health

Nadeem Ali, Muhammad Imtiaz Rashid, Nabil A. Alhakamy, Hussain Mohammad Salem Ali Albar, Khurram Shahzad, and Syed Ali Musstjab Akber Shah Eqani

Abstract The presence of organophosphate ester (OPEs) in the environment of developing countries has been reviewed in the current chapter. The strict regulations about the manufacturing and usage of flame retardants (FRs) like polybrominated diphenyl ethers (PBDEs) lead to the increased production and use of OPEs as FRs and plasticizers to satisfy the fire and safety guidelines. These OPEs are assumed to be environmentally safe but recent research focused on their environmental presence have raised questions on their use and impact on human health. In this chapter, data is compiled on the occurrence of OPEs in different environmental matrices including air, dust, food, water, humans, etc., from developing countries. In the end, the investigation gaps were figured out with commendations for future research work.

Keywords Organophosphate esters · Developing countries · Indoor dust · Air · Water · Soil · Humans

N. Ali (✉) · M. I. Rashid · K. Shahzad
Center of Excellence in Environmental Studies, King Abdulaziz University, Jeddah, Saudi Arabia
e-mail: nabahadar@kau.edu.sa

N. A. Alhakamy
Pharmaceutics Department, Faculty of Pharmacy, King Abdulaziz University, Jeddah, Saudi Arabia

H. M. S. A. Albar
Department of Community Health, Medical College, King Abdulaziz University, Jeddah, Saudi Arabia

S. A. M. A. S. Eqani
Health and Environment Division, Department of Biosciences, COMSATS University, Islamabad, Pakistan

M. Z. Hashmi, A. Varma (eds.), *Electronic Waste Pollution*, Soil Biology 57,
https://doi.org/10.1007/978-3-030-26615-8_4

4.1 Introduction

Organophosphate esters (OPEs) are class of chemicals which are utilized as plasticizers and flame retardants (FRs); important OPEs are listed in Fig. 4.1, with their name, abbreviations, and structural formulas. Additionally, OPEs have been applied as stabilizers with anti-foaming properties and used as additives to cutting oils, transmission oil, hydraulic fluids, cellulose, floor polishes, lacquers, polyvinylchloride, and lubricants (Wei et al. 2015). OPEs are classified as halogenated, which are mainly used as FRs, and non-halogenated OPEs, predominantly used as plasticizers (Andresen and Bester 2006; Wei et al. 2015). The OPEs production was approximately 100,000 tons worldwide in 1992; however, in 2007, approximately 70,000 tons were produced in China alone and its annual production is increasing at a rate of 15% (WHO 1997; Ou 2011; Wang et al. 2010). Since OPEs are additives in many polymeric materials, studies have reported their occurrence in the environment; this might reflect their leaching from the treated products during their production process, usage, and recycling of e-waste into the surroundings (Wei et al. 2015). Moreover, long-range atmospheric transport of OPEs from urban proximities is responsible for their occurrence in the pristine areas (Bollmann et al. 2012; Marklund et al. 2005; Moller et al. 2012). The consumption of OPEs is increasing sharply around the world after regulating the use of PBDEs (EFRA 2007). The increasing consumption of OPEs is apparent by their occurrence in the environment, as reported by a number of scientific studies from the different part of the world (Ali et al. 2017; Wei et al. 2015). Increasing production and consumption of these chemicals have resulted in their pervasive occurrence in the environment which may pose a serious risk to the environmental health (Wei et al. 2015). Hence, humans get exposed to these chemicals via inhalation, dermal contact, ingestion of dust, or dietary intake and pose threats to human health. Many of the OPEs, for example, tris(2-chloroisopropyl) phosphate (TCIPP), tris(2,3-dichloropropyl) phosphate (TDCPP), and tris(2-butoxyethyl) phosphate (TBOEP), are listed as suspected carcinogenic with studied tumor growth in different body organs such as thyroid, brain, liver, testes, and kidney (WHO 1998, 2000; Van der Veen and De Boer 2012). Other health problems such as reproductive toxicity, various allergies, endocrine disruption, atopic dermatitis, and neurotoxicity are associated with exposure to different OPEs (WHO 1991; Stapleton et al. 2009). This has resulted in listing some of OPEs of high concern under EU Registration, Evaluation, Authorisation and Restriction of Chemicals (REACH) regulations. In this chapter, we arranged data published from different developing countries on OPEs in their environment and pointed out the existing research gaps for future scientific studies.

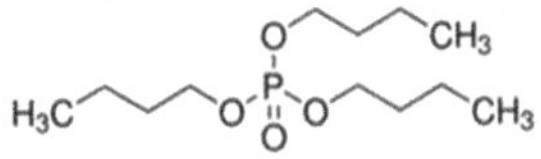

Tri–n–butyl phosphate
(TnBP)

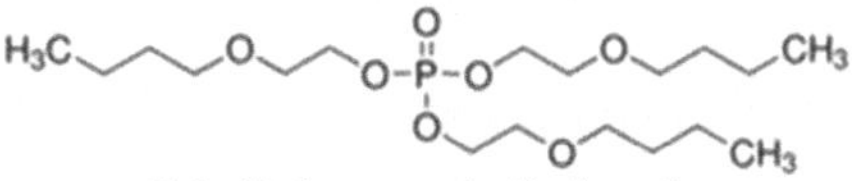

Tris (2–butoxyethyl) phosphate
(TBOEP)

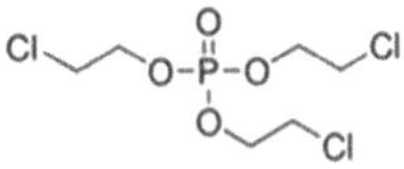

Tris (2–chloroethyl) phosphate
(TCEP)

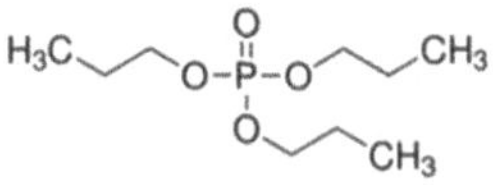

Tri-propyl phosphate
(TPP)

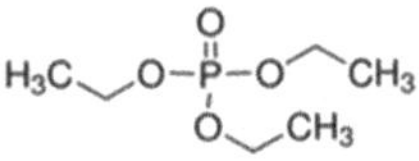

Tri-ethyl phosphate
(TEP)

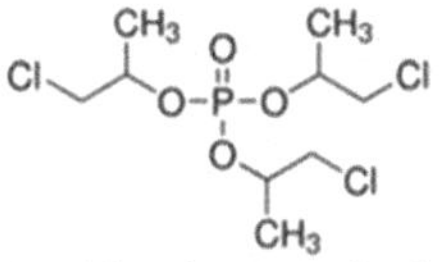

Tris (2–chloroisopropyl) phosphate
(TCPP)

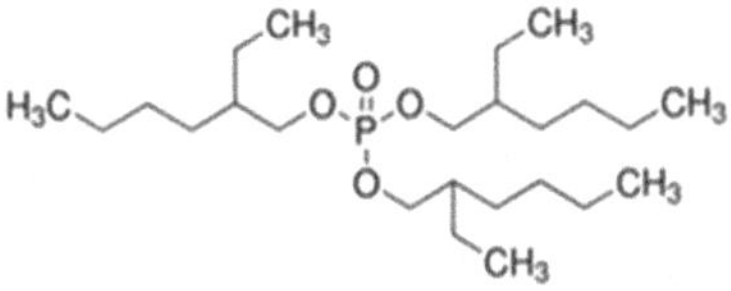

Tris (2–ethylhexyl) phosphate
(TEHP)

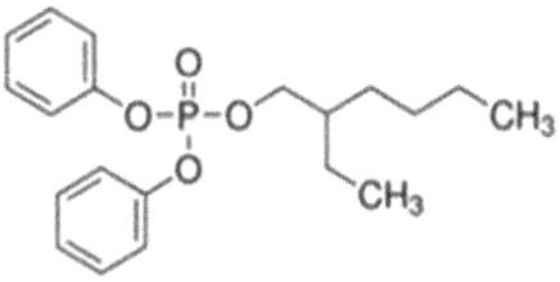

2–Ethylhexyl diphenyl phosphate
(EHDPP)

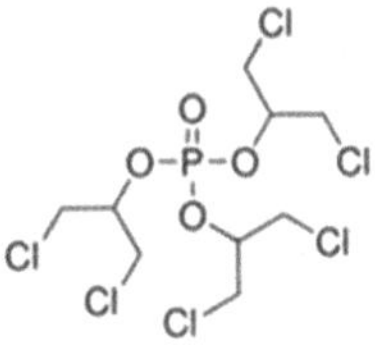

Tris (1,3–dichloro–2–propyl) phosphate
(TDCPP)

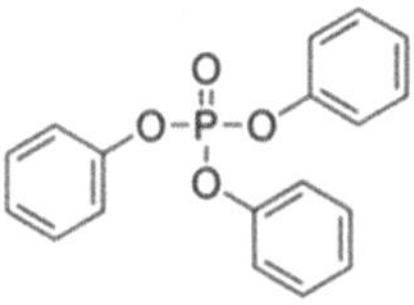

Triphenyl phosphate
(TPhP)

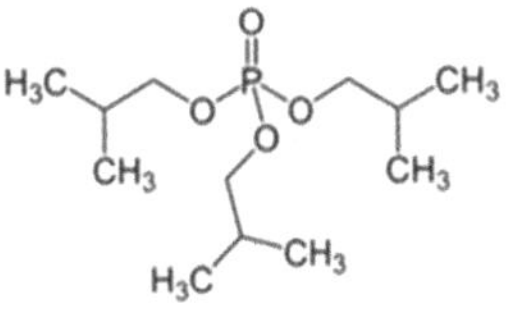

Tri–iso–butyl phosphate
(TiBP)

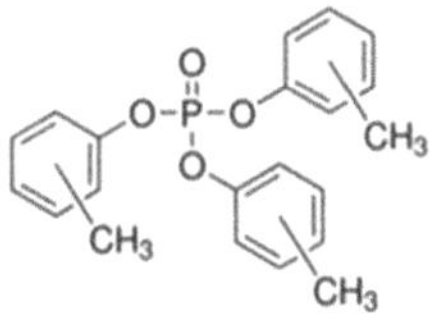

Tricresyl phosphate
(TCP)

Fig. 4.1 Name, abbreviation, and structural formulas for important OPEs discussed in this chapter

4.2 Environmental Occurrence

4.2.1 Indoor Dust

Indoor dust is gathering a great interest from the researchers to study indoor health; in fact, they have suggested household dust as an archive of indoor pollution because of the number of contaminants buildup in the dust over the period of time (Butte and Heinzow 2002). Indoor house dust is a mixture of many things, for example, fungal spores, wear and tear particles from the indoor inventories, pet dander, particulate matters resulting from indoor aerosols and cross ventilation, volatile organic compounds (VOCs), tracked in the soil, and traces of metals (Butte and Heinzow 2002). Due to the lack of moisture and direct sunlight, many of the chemicals trapped in the indoor dust degrade slower, and in some cases not at all (Paustenbach et al. 1997). A number of studies have suggested that for many chemicals especially those with persistent, toxic, and bioaccumulation properties, inhalation and involuntary intake of dust is one of the major exposure pathways especially for toddlers and people with medical conditions, for example, respiratory issues (Dirtu et al. 2012; Mercier et al. 2011).

Organophosphates esters (OPEs) are utilized in the building materials as flame retardants (FRs) and plasticizers, electronics, textile industry, and furnishings. Discharge of the aforementioned organic pollutants from these industrial products leads to their widespread presence in indoor dust of different microenvironments including mosques, private houses, stores, vehicles, e-waste sites, and public offices. (Abdallah and Covaci 2014; Ali et al. 2013, 2014, 2016; Kim et al. 2013). These recent research studies carried out in various developing countries have shown high levels of OPEs compared to the newly identified brominated FRs and PBDEs in the indoor dust samples (Abdallah and Covaci 2014; Ali et al. 2013, 2016). Such observations indicated the increased usage of OPEs as FRs and plasticizers than brominated chemicals in various industries that are linked to different household items. The concentration and profile of OPEs in the indoor dust are very much dependent on the building materials, type of housing, and quantity/age of the furniture present indoor as well as electrical home appliances (Wei et al. 2015; Ali et al. 2017). This may explain the variation in profile and concentrations of the OPEs present in the indoor dust of the different countries (Fig. 4.2; Table 4.1). Many of the studies on indoor dust indicated that OPEs concentrations and profiles varied greatly; however, in all these studies, the major share was coming from chlorinated-OPEs (Cl-OPEs) (Table 4.1). Studies from China, Saudi Arabia, and Kuwait reported high concentrations of OPEs in the indoor dust than those from Egypt, the Philippines, and Pakistan (Abdallah and Covaci 2014; Ali et al. 2013, 2016, 2018; He et al. 2015; Kim et al. 2013). China produced and consumed a higher share of OPEs whereas Saudi Arabia and Kuwait are importing most of its consumed OPEs products from abroad (Ali et al. 2013, 2016, 2018; He et al. 2015). The higher concentration of OPEs present in Kuwait and Saudi Arabian dust can be explained by the fact that imported products might be treated with OPEs, hence elevating the levels in their

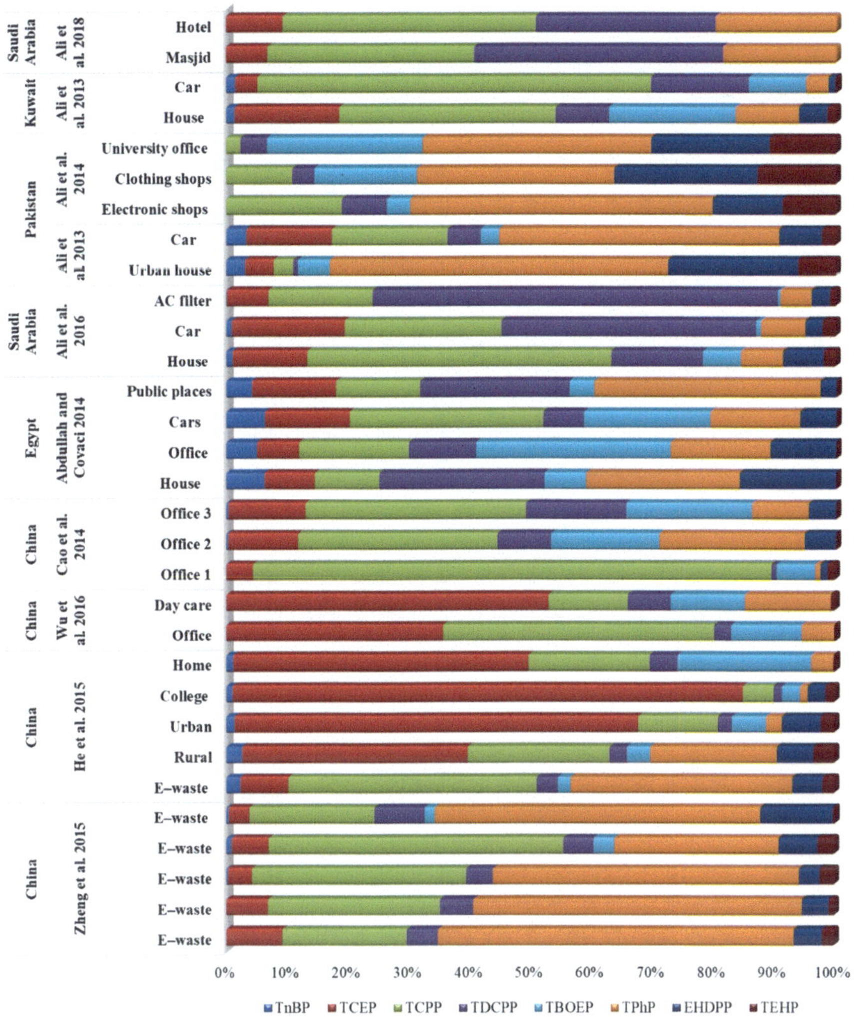

Fig. 4.2 Profiling of OPEs in indoor dust of different microenvironments of different countries

indoor environment (Ali et al. 2013, 2016, 2018). On the other hand, weak legislation about the use of FRs in consumer products can explain the low concentrations of OPEs (Table 4.1) in the dust of other developing countries.

The most frequently, and with high concentration OPEs, reported in the dust samples collected from various electronic waste dumping sites of Dali, Guiyu, and Longtang, China, were TPhP and TCPP (Zheng et al. 2015). Moreover, these accounted for >70% of the total OPEs found in these dust samples (Zheng et al. 2015). In this study, authors found that TPhP had the highest concentration ranged between 31 and 332000 ng/g, and this was followed by TCPP, TDCPP, TCEP, and

Table 4.1 Concentrations (median, mini–maxi (ng/g of dust)) of OPEs in indoor dust of developing countries

Country	Microenvironment	TnBP	TCEP	TCPP	TDCPP	TBOEP	TPhP	EHDPP	TEHP
China, Longtang (Zheng et al. 2015)	E-waste	nd–92	177 (nd–671)	386 (48–4940)	97 (nd–636)	–	1110 (86–15,800)	88 (28–731)	41 (nd–936)
	E-waste	nd–29	322 (nd–528)	1300 (527–2840)	252 (nd–1500)	–	2500 (122–16,500)	201 (34–1380)	49 (nd–131)
	E-waste	16 (nd–53)	269 (131–977)	2310 (878–4800)	288 (128–1610)	–	3320 (119–6030)	221 (111–630)	170 (57–1500)
Dali	E-waste	61 (19–2100)	394 (76–1740)	3110 (640–8790)	317 (nd–1730)	220 (nd–1060)	1740 (31–6660)	406 (28–4550)	189 (68–566)
Guiyu	E-waste	78 (24–789)	633 (149–6920)	3760 (854–10,000)	1500 (228–14,100)	301 (nd–1860)	9810 (371–332,000)	2180 (222–6900)	75 (13–2550)
Guangzhou City and Qingyuan (He et al. 2015)	E-waste	280 (20–500)	930 (180–1560)	4770 (110–22,300)	410 (110–7020)	240 (40–810)	4270 (90–27,400)	580 (150–2390)	250 (50–1090)
	Rural	140 (40–1460)	1930 (50–9360)	1220 (240–10,700)	150 (nd–2770)	200 (30–1760)	1090 (260–5340)	310 (60–1280)	190 (80–1850)
	Urban	80 (30–150)	3780 (1550–9700)	750 (160–2930)	130 (nd–9630)	320 (nd–3050)	150 (10–800)	360 (30–3470)	140 (30–1370)
	College	100 (60–140)	7940 (2780–20,800)	480 (60–2300)	130 (60–3710)	280 (40–770)	120 (20–940)	270 (nd–570)	160 (nd–320)
Beijing (Wu et al. 2016)	Home	122 (nd–1018)	4992 (2231–30,847)	2048 (220–13,804)	479 (nd–1531)	2259 (493–41,917)	376 (122–1829)	–	37 (nd–350)
	Office	nd (nd–809)	12879 (185–71,167)	16133 (308–158,014)	1007 (102–5568)	4189 (1034–14,117)	1928 (31–38,646)	–	89 (nd–1156)
	Day care	nd (nd–660)	1986 (391–17,805)	486 (71–30,157)	267 (nd–3741)	454 (85–4561)	531 (41–3514)	–	30 (nd–13,113)

Beijing (Cao et al. 2014)	Office 1	33 (nd–83)	5610 (1180–11)	107,000 (36,200–191,000)	1110 (268–3980)	7880 (1830–13,000)	1100 (350–1890)	1440 (463–1840)	1740(533–2410)
	Office 2	47 (24–73)	1980 (1330–3120)	5540 (4240–7680)	1520 (1010–2710)	3030 (1720–4820)	4050 (1420–15,100)	873 (728–1170)	nd
	Office 3	54 (27–123)	2160 (1400–3900)	6090 (4190–9480)	2770 (1500–7650)	3480 (2130–6150)	1580 (862–3130)	746 (583–944)	nd
Egypt Assiut (Abdallah and Covaci 2014)	House	17 (nd–26)	22 (nd–132)	28 (nd–123)	72 (nd–557)	18 (nd–305)	67 (8–289)	42(nd–102)	–
	Office	23 (nd–46)	31 (nd–125)	80 (nd–700)	49 (nd–490)	143 (nd–1244)	73 (11–337)	48 (nd–74)	–
	Cars	59 (nd–258)	127 (nd–572)	291 (nd–1425)	61 (nd–283)	190 (nd–778)	135 (26–1872)	54 (nd–203)	–
	Public places	73 (nd–261)	234 (nd–538)	232 (nd–465)	416 (nd–1616)	68 (nd–1029)	629 (116–2357)	44 (nd–74)	–
Saudi Arabia Jeddah (Ali et al. 2016)	House	35 (LOQ–165)	410 (125–1650)	1650 (200–3700)	500 (150–8700)	205 (LOQ–2750)	230 (65–1200)	220 (55–520)	70 (LOQ–270)
	Car	60 (20–6550)	1200 (30–52,300)	1650 (100–92,000)	2700 (100–45,600)	50 (LOQ–12,500)	470 (40–4150)	180 (60–9000)	150(LOQ–850)
	AC filter	10 (LOQ–35)	820 (120–16,500)	2000 (200–11,500)	7800 (1000–61,234)	50 (LOQ–1300)	600 (120–2500)	350 (80–1600)	130 (30–500)
Jeddah, Makkah, and Madina (Ali et al. 2018)	Masjid	–	420 (270–3470)	2055 (1570–4820)	2490 (970–6945)	–	1135 (675–4960)	–	–
	Hotel	–	750 (250–1750)	3315 (375–12,620)	2360 (1150–9050)	–	1590 (230–3730)	–	–

(continued)

Table 4.1 (continued)

Country	Microenvironment	TnBP	TCEP	TCPP	TDCPP	TBOEP	TPhP	EHDPP	TEHP
Pakistan Faisalabad (Ali et al. 2013)	Urban house	10 (nd – 22)	15 (nd – 175)	10 (nd – 85)	2.5 (nd – 255)	16.5 (nd – 145)	175 (nd – 330)	67 (nd – 360)	20 (nd – 50)
	Car	18 (nd–250)	75 (nd–1520)	100 (nd–2615)	29 (nd–1240)	16 (nd–1525)	245 (2–4800)	37 (2–440)	13 (nd–325)
Gujrat (Ali et al. 2012)	Rural house	–	–	–	–	67 (nd–340)	109 (15–185)	–	–
	Mosque	–	<20 (nd–720)	–	<10 (nd–120)	30 (nd–2850)	95 (nd–630)	–	–
Islamabad and Faisalabad (Ali et al. 2014)	Electronic shops	–	–	75 (6–620)	29 (nd–1475)	<15 (nd–100)	195 (10–5000)	45 (2–830)	35 (nd–1025)
	Clothing shops	–	–	21 (nd–95)	7 (nd–30)	32 (nd–70)	62 (nd–220)	45 (nd–120)	25 (nd–175)
	University office	–	–	11 (nd–320)	20 (nd–185)	115 (nd–11,900)	170 (10–23,450)	88 (5–285)	50 (nd–4100)
Philippines Malate	Residential	19 (nd–79)	34 (nd–1200)	–	–	–	89 (8.5–2100)	110 (8–770)	140 (4–970)
Payatas (Kim et al. 2013)	Dumping site	20 (nd–280)	16 (nd–140)	–	–	–	71 (13–440)	34 (8–560)	41 (nd–370)
Kuwait (Ali et al. 2013)	House	58 (21–800)	710 (275–1800)	1460 (120–7065)	360 (60–1555)	855 (31–140,450)	430 (44–6890)	190 (75–10,990)	65 (<5–340)
	Car	730 (20–9830)	1765 (10–13,650)	30,725 (2490–134,055)	7630 (600–165,550)	4465 (15–12,475)	1760 (<2–7415)	520 (2–4285)	135 (5–1305)

EHDPP with concentrations 48–10,000, nd–141,000, nd–6920, and 28–6900 ng/g, respectively. Likewise, He et al. (2015) also found high concentrations of the aforementioned OPEs in the dust sampled from the e-waste site of Qingyuan, China. They found a high concentration of TCP (520–46,600) than TCPP (110–22,300) and TPhP (90–27,400) in their studied e-waste site. He et al. (2015) and Zheng et al. (2015) suggested that the difference in TPhP, TCP, and TCPP concentrations in dust samples was due to the contrasting e-waste recycling practices carried out among various e-waste sites. Conversely, dust samples collected from the rural, urban, and college indoor environment of Guangzhou and Qingyuan, China, showed an abundance of TCEP with the median concentration 7940, 1930, 3780 ng/g, respectively (He et al. 2015). Another recent research by Wu et al. (2016) indicated the occurrence of TCEP, TPhP, TCPP, and TBOEP in the indoor microenvironment of the daycare centers, households, and offices of Beijing, China. The above studies from China indicated the ubiquitous presence of TPhP, Cl-OPEs, and TBOEP with high concentrations in various indoor microenvironments; this advocates their extensive use in Chinese industries of household items and long-term contamination of these chemicals in the indoor environment of China. A recent study by Cao et al. (2014) in Beijing, China, found the presence of OPEs in various particle sizes of the dust samples collected from student dormitory, hotel, kindergarten, and public offices but found that different particle size did not influence the concentrations of FRs. In this study, the concentrations screened OPEs were higher in hotel samples compared to other indoor environments ($p < 0.05$).

In studies carried out in Gulf countries, researchers found TCPP and TDCPP as the major Cl-OPEs in indoor dust from different microenvironments (households, mosque, public and workplaces, vehicles) of Egypt, Kuwait, and Saudi Arabia (Abdallah and Covaci 2014; Ali et al. 2013, 2016, 2018). TPhP and TBOEP were the other important OPEs reported in these studies and their high concentration in the indoor dust could be linked with the consistent use of waxes or PVC covering as well as floor polishes (Abdallah and Covaci 2014). In most of these studies OPEs were found in higher concentrations in the cars and AC filter dust samples than households; this could possibly be attributed to the use of polypropylene polymers moulded parts and ABS in the casing of AC and vehicles with PUFs in the interior upholstery, instrument panels, textiles, and electronics (Ali et al. 2013, 2016; Abdallah and Covaci 2014). However, it is uncertain that OPEs concentrations and profile in the dust samples of cars is influenced by distance driven, manufacturing year, and seat type (Abdallah and Covaci 2014).

The variation in the profiling of OPEs (Fig. 4.2) reported in different indoor microenvironment dust of several countries can describe the prevailing fire safety regulations as well as implementation of up-to-date fire safety protocols along with phasing out the polybrominated diphenyl ethers (PBDEs) (Ali et al. 2013). China is one of the vital manufacturers of OPEs; other developing countries do not produce them, instead they import these chemicals from the major producers to fulfill their consumption demand. Similarly, the occurrence of OPEs in these developing countries might also show their emission from the import of consumer products treated with them. These studies suggested that levels of OPEs in the environment will be increasing in the future with their higher use in the consumer goods and with increasing recycling activities; thus, chronological data is vital to realize the dynamics of OPEs in the environment.

4.2.2 Air and Tree Bark

Air is a vital compartment of the ecosystem which serves as a sink for different carbon-based contaminants emitting in the surrounding environment. The fingerprints of levels and distribution of OPEs in various areas have been revealed by conducting several studies on ambient air (Yang et al. 2014; Bi et al. 2010). The distribution pattern of OPEs from different surroundings varies depending on the emission patterns of these chemicals from various sources such as industries, waste recycling units, and building materials used for construction (Bi et al. 2010; Carlsson et al. 2000). Both Cl-OPEs and other OPEs are reported in the particulate matter present in the air. According to a recent study on the levels and OPE's size-specific distribution, Yang et al. (2014) observed a high concentration of OPEs in the air-suspended particulate matter present in various public offices in China. They found that total OPEs (ΣOPEs) in these offices were ranged between 5 and 148 ng/m^3, where the highest concentration of TCPP (0.8–81 ng/m^3) was observed followed by TCEP (1–13.5 ng/m^3) and TPhP (0.3–10 ng/m^3) (Yang et al. 2014). In the ΣOPEs present in particulate matter, 77% contribution came from Cl-OPEs, among these, TEHP, TDCPP, and TCP were mostly present in the ultrafine particles ($\leq$1 μm), whereas fine particles ($\leq$2.5 μm) contained EHDPP, TPP, and TBOEP (Yang et al. 2014). Similarly, air sample collected around recycling site of e-waste in south China showed tremendously high levels of TPhP and their concentration ranged between 28,600 and 72,600 ng/m^3 (Bi et al. 2010). These authors attributed such high concentrations to the presence of computer's printed circuit boards that were the main e-waste from recycling plants. In line with the aforementioned study, Carlsson et al. (2000) also proposed that computers can be the main source of TPhP present in their surrounding environment. Similarly in another recent study, Luo et al. (2016) studied OPEs in size-fractionated atmospheric dust particles from electronic waste recycling zone and urban area of Guangzhou, China, and found concentrations of ∑OPEs was 130 $\pm$ 130 and 138 $\pm$ 127 ng/m^3, respectively. In this study, the profiles of Cl-OPEs significantly differed between the two sites, but the pattern of OPEs size distribution did not differ at various heights (Luo et al. 2016). In contradiction with other studies carried out in China, TBOEP was the major contributor of OPEs present in the air and this pollutant was 80% and 85% of the ΣOPEs found in the electronic waste recycling site and urban area of Guangzhou, respectively. According to authors, electronic waste recycling operations, waxes and floor polish, etc., were major causes of releasing TBOEP in the air. The estimated risks calculated to OPEs via inhalation in this study were lower than the respective reference doses (RfD), although this suggests that through inhalation potential health risk is low but this is important considering exposure from other pathways.

4.2.3 Water

Recently a number of studies from developing countries have reported OPEs in water samples. For instance, Khan et al. (2016) found that concentrations of ∑OPEs from the portable water sampled from Pakistan background, rural, and urban areas ranged between nd and 71.05 ng/L with TCPP (64.4%) contributed the most, followed by TCEP (18.5%)> TDCPP (10.5%)> TPP (3.0%)> TEHP (2.6%) > and TBP (0.9%). In a similar study by Wang et al. (2014) from 23 locations in Pearl River estuary found OPEs in seawater during dry and wet seasons. Eight OPEs were detected in seawater samples with the main contribution was from TCEP and TCPP. The respective concentration of these compounds were 43% and 45% of the total ΣOPEs during the dry and wet season. The total concentration of OPEs ranged between 2040 and 3120 ng/L during dry and 1080–2500 ng/L in the wet season. The lower concentrations in winter could be explained by the dilution of OPEs with the large flux of water in the wet season (Wang et al. 2014). Hu et al. (2014) performed a study in Yellow Sea and East China Sea and found 91.87–1392 ng/L of halogenated OPEs in seawater sampled near three districts of China: Lianyungang, Xiamen, and Qingdao. Among the OPEs, TCPP, Tris(2,3-dibromopropyl)phosphate (TDBPP), TDCPP, and TCEP were sensed and their mean concentrations were 84.1, 96.7, 109.3, and 134.4 ng/L, respectively (Hu et al. 2014). The highest pollution was found in the district Lianyungang, where TCEP was 550.5 and 618 ng/L in two samples (Hu et al. 2014). The authors linked this pollution to the WWTPs' effluents in the industrial zones present in the surroundings. It is also suggested that Cl-OPEs can enter into the sea or groundwater through their atmospheric transport and washout by precipitation. In addition to seawater, OPEs were also found in the tap drinking water with ΣOPEs concentrations ranged between 85.1 and 325 ng/L (Li et al. 2014). The major components found in tap water were TBOEP (70.1 ng/L), TPP (40.0 ng/L), TCPP (33.4 ng/L), TCEP (12.5 ng/L), and TnBP (7.48 ng/L). The higher concentration and range of TBOEP (24.1–151 ng/L) can be explained by its use in large quantities in plastics, polishes, and rubber. Interestingly, TBOEP in the water was increased whereas TnBP decreased by boiling, hence showing TnBP breakdown and releasing of TBOEP from the other impurities present in water. These authors also observed 10–25% low concentration of OPEs in bottled water when compared it with tap water; however, authors concluded that bottled water contaminated with TBOEP was the main concern. According to Zeng et al. (2015) wastewater samples also contained OPEs that were sampled from the wastewater treatment plants (WWTPs) situated in the PRD, China. The reported concentrations of different OPEs in the effluent wastewater was 21,272 ng/L of TnBP, 4350 ng/L of TBEP, 60.3 ng/L of TDCPP, 299 ng/L of TCPP, and 438 ng/L of TCEP. Moreover, the concentrations of TPhP and TTP were 149.2 and 99 ng/L, respectively. After treatment of water, TnBP (3105 ng/L) and TBOEP (495 ng/L) remained still dominant and no difference in the concentrations of three Cl-OPEs was found, this might indicate their resistance against current wastewater treatment (WWT) technology. Hence, the OPEs that still remained in final effluent can act as a source of pollution to the water bodies of Pearl River where that effluent will be discharged after WWT. Such observations indicating the need for developing such innovative WWT technology treat all OPEs present in the water to save water bodies from the pollution.

4.2.4 Soil, Sludge, and Sediment

The soil, sediments, and sludge are key components of the environment and offer effective awareness about the fingerprints of various pollutants in the ecosystem. The presence of OPEs in the soil, sediment, and sludge around the globe are reported in very few studies in the literature and only limited studies carried out in the developing countries (Wei et al. 2015). In China, Zeng et al. (2015) found OPEs in the water molecules and dewatered sludge. They observed that OPEs distribution was much different in water than sludge and can be linked to the diverse physicochemical properties and difference in the degradation pattern of OPEs during their treatment process. After treatment, TnBP remained dominant with its concentrations ranged between 271 and 3458 ng/g, followed by 184–674 ng/g of TTP, 166–637 ng/g TPhP, and 205–510 ng/g TBOEP (Zeng et al. 2015). On the other hand, these authors found relatively lower levels of 20–209 ng/g of TCEP, 38–320 ng/g of TCPP, and 38–150 ng/g of TDCPP in sludge samples. The low concentration of Cl-OPEs in sludge is attributed to their high solubility in water and low logKow value as compared to the other OPEs, such properties determined OPEs partition or degradation in the water phase and explain the variation in OPEs concentration in the sludge. Hence, high quantity of TPhP and TTP that are hydrophobic OPEs in nature can make a bond with particles present in sludge especially during wastewater treatment, and hence these OPEs entered into the environment during sludge disposal. In another study on OPEs present in the sediments from Taihu Lake, China, Cao et al. (2012) found that domestic, agricultural, and industrial practices could be the main source of OPEs present in the lake sediments. They found the highest levels of TBOEP in sediments with its concentration ranged between 1.03 and 5.0 ng/g. In addition to this OPEs, the respective concentrations of TCEP, TCPP, and TDCPP were ranged between 0.62 and 3.17, nd and 2.27, nd and 5.54 ng/g. TCEP is carcinogenic and mutagenic in nature which has been steadily replaced with TCPP in the region; however, TCEP is still found with high levels in the lake sediments, thus suggesting two possibilities of their presence in the nature (1) their still high use in various industrial, domestic, or agricultural products and (2) the persistent nature that can slow down its degradation in the environment. In addition to levels of the OPEs, the profile was mainly dominated by TBOEP, TCEP, and TCP in the lake sediments which suggested high use of their products in China. Moreover, these OPEs released from various electronic accessories including electronic recycling sites, since TBOEP is the major manufacturing OPEs used in electronics, followed by TCEP and TCPP (Kawagoshi et al. 1999).

4.2.5 Biological Samples

Like sediments, soil, or sludge, only limited research studies are found in the literature that studies the presence of OPEs in the samples of living organisms. Kim et al. (2011a, b) observed that the dominated OPEs concentration of TEHP, TEP, and TnBP was ranging between nd and 2000, nd and 410, and nd and 590 ng/g lipid weight (lw), respectively, in the fish samples. These authors explained that TEHP high levels in biological samples possibly related to the high hydrophobicity, while TEP due to the intensive use in the Philippines (Kim et al. 2011a, b). Ma et al. (2013) observed extremely high concentrations of TBOEP, TCEP, and TnBP ranged between 1647 and 8840, 83 and 4690, and 44 and 2950 ng/g lw in fish samples taken from the Pearl River Delta (PRD). Accordingly, Wang et al. (2014) found much the same higher concentrations of OPEs present in coastal water of PRD. On the other hand, Ma et al. (2013) observed significant levels of TPhP in the biological samples in domestic birds (nd–209 ng/g lw) of southern China. These levels of OPEs could be attributed to the presence of high concentrations of TPhP (46.5 mg/m^3) in air samples analyzed in south China (Bi et al. 2010). These findings revealed the entry of OPEs into food chain; hence, their prevalence in the environment is a source of great concern since their use and production is increasing in the region with time.

In a recent study by Kim et al. (2014) carried out in Vietnam and the Philippines on human breast milk, authors found the presence of OPEs in these samples with median concentrations of ΣOPEs was 600% (70 vs. 10 ng/g lw) higher in the Philippines than Vietnam milk samples. These results could be explained by the frequent utilization of OPEs-containing products in the households of the Philippines than Vietnam. In another study by the same group found that the OPEs profile present in human breast milk was comparable to the indoor dust samples of the same countries (Kim et al. 2013), which indicate the human exposure to indoor dust could be the main source of OPEs present in human milk. These authors found that breast milk obtained from Vietnam and the Philippines have predominated OPEs profile of TCEP, TPhP, and TBOEP; however, concentrations of these OPEs were much higher in the milk sample of latter than former country. Their presence in human milk designates the exposure of people in these countries to OPEs which have a quite longer half-life in the human body. Qiao et al. (2016) tested the human hair for the presence of OPEs and found high levels (10–604 ng/g) of these chemicals. They found 43.9 ng/g dw median concentration of TCIPP, 24.1 ng/g dw of TEHP, and 20.5 ng/g dw of TPHP, which were the dominant OPEs in human hair samples. Moreover, when the various segment of hair was tested for OPEs, it is found that most of the OPEs present in hair distal segments and their concentration was 1.5–8.6 times higher compared to those present in proximal segments; such observations can be linked to a lengthier exposure duration of the former hair segments to exterior pollution sources. Likewise, in this study, authors also found gender-wise difference in OPEs concentration, with female hair, had a significantly higher concentration (62.7–604 ng/g dw) than male hair (10.1–331 ng/g dw).

4.3 Human Exposure to OPEs via Dust Intake

As shown in this chapter as well as in literature, most of the studies on OPEs were primarily focused on indoor dust which suggests indoor dust as an important source of exposure to OPEs for humans. Although humans are getting exposed to these chemicals via intake of contaminated water, air, and food, a number of studies have suggested the involuntary intake of indoor contaminated dust via ingestion, dermal contact, and inhalation are the major exposure pathways for humans. In modern-day lifestyle, people spend more than 90% of their time indoor (home, office, and school). In the time when people are indoor, they are getting exposed to OPEs all the time to varied extant (Ali et al. 2016, 2018). We estimated daily exposure via dust ingestion and hazardous quotient (HQ) for both adults and toddlers using median concentrations of different OPEs reported in household indoor dust of various countries. Since many of OPEs are suspected carcinogenic, incremental lifetime cancer risk (ILCR) is important to calculate, but SFO, an oral slope factor (ng per kg bw per day), which is theoretically the upper-boundary cancer potency, is missing for most of the OPEs in literature which made it difficult to calculate carcinogenic risk here. Therefore, daily exposure via dust ingestion and HQ was calculated using the following equations (Ali et al. 2018):

$$\text{Average daily dose (ADD) (ng/kg bw/day)} = C_{\text{n}} \times R_{\text{ing}}/\text{BW}$$

$$\text{HQ} = \text{ADD}/\text{RfD}$$

C_{n} represent the median levels of selected individual OPEs in indoor dust, R_{ing} is the rate of dust intake (50 mg for adults and 200 mg for toddlers), BW is body weight (70 kg for an adult and 12 for toddler), and RfD is the reference dose value of selected OPEs (Ali et al. 2016). Bioaccessibility of various OPEs varies from dust, but for consistency and worst-case scenario, we assumed 100% absorption on the intake OPEs from dust; similarly, we assumed 24 h indoor. These are preliminary calculations using some assumptions; therefore, caution is needed when interpreting these results (Ali et al. 2018). The calculated ADD (Table 4.2) was lower than the corresponding RfD for both toddlers and adults from all studied countries, and HQ was <1 which showed that the estimated noncancer risk from the individual OPEs is negligible for people living in these areas. However, as discussed above. large number of information on the OPEs are missing in the literature such as SFO values, updated RfD values for some OPEs, and also the synergetic health impact of various OPEs. Other exposure pathways such as air, dermal contact, and diet need to be considered when calculating the overall impact of OPEs on health.

Table 4.2 Estimated daily exposure (ng/kg bw/day) to OPEs for toddlers and adults from various developing countries

Analytes	RfD	Exposed group	China (Wu et al. 2016)	Egypt (Abdallah and Covaci 2014)	Pakistan (Ali et al. 2013)	Saudi Arabia (Ali et al. 2016)	Kuwait (Ali et al. 2013)
TnBP	24,000	Toddler	2.03	0.28	0.17	0.58	0.97
		Adult	0.09	0.01	0.01	0.03	0.04
TCEP	22,000	Toddler	83.20	0.37	0.25	6.83	11.83
		Adult	3.57	0.02	0.01	0.29	0.51
TCPP	80,000	Toddler	34.13	0.47	0.17	27.50	24.33
		Adult	1.46	0.02	0.01	1.18	1.04
TDCPP	15,000	Toddler	7.98	1.20	0.04	8.33	6.00
		Adult	0.34	0.05	<0.01	0.36	0.26
TBOEP	15,000	Toddler	37.65	0.30	0.28	3.42	14.25
		Adult	1.61	0.01	0.01	0.15	0.61
TPhP	70,000	Toddler	6.27	1.12	2.92	3.83	7.17
		Adult	0.27	0.05	0.13	0.16	0.31
EHDPP	NA	Toddler	<0.01	0.70	1.12	3.67	3.17
		Adult	<0.01	0.03	0.05	0.16	0.14
TEHP	NA	Toddler	0.62	<0.01	0.33	1.17	1.08
		Adult	0.03	<0.01	0.01	0.05	0.05

4.4 Concluding Remarks and Future Perspectives

In this chapter, we arranged available data on the presence of OPEs in the environment of various developing countries. The discussed chemicals were considered safe alternatives for the regulated FRs, but their documented presence in the environment presents few questions about their environmental safety. Most of the gathered data in this chapter are from China and information on the production and dynamics of these chemicals are still scantily known in most of the developing countries, especially from South America and Africa. Developing countries are in the race of rapid industrialization and little concerns are shown on the environmental issues, this is partly due to the lack of funding, research infrastructure, and knowledge among public and policymakers. The profiling of the chemicals may vary greatly in various developing countries due to their geographical position, safety regulations, weather, lifestyle, and socioeconomic situations. Therefore, this warrant detailed efforts to understand the dynamics of OPEs in these regions. For this large-scale, well-planned studies are needed on both abiotic and biota environmental samples. Biomonitoring studies using serum, blood, urine, and breast milk are lacking but are very important in designing mitigation plans to describe exposure pathways. The knowledge is much lacking on the degradation, biotransformation, long-range transportation, phase distribution, and environmental fate of hydrophobic OPEs. For this information, large-scale different studies covering metabolomics, different exposure parameters, for example, bioaccessibility and bioavailability, biomagnification, bioaccumulation, and temporal and spatial variations are needed. Toxicological studies on most of the OPEs are missing or very few available in the literature. Toxicological studies covering acute and chronic exposure to OPEs are missing and are needed to understand their impact on human health. The toxicological data is also needed to establish the updated RfDs for these chemicals and studies are needed to know the combined impact of OPEs. There are scarce data regarding the epidemiological studies on these chemicals and different health endpoints (including clinical parameters). Not much information and data are available in the literature on production volumes and usage of OPEs in most of the developing countries. All such information on these chemicals is much important and required to give priority in future research programmes. The knowledge on these chemicals is increasing in recently with an increasing number of scientific studies annually, but still, there are many data gaps exist and this warrant more efforts in the future.

References

Abdallah MAE, Covaci A (2014) Organophosphate flame retardants in indoor dust from Egypt: implications for human exposure. Environ Sci Technol 48:4782–4789

Ali N, Van den Eede N, Dirtu AC, Neels H, Covaci A (2012) Assessment of human exposure to indoor organic contaminants in Pakistan. Indoor Air 22:200–211

Ali N, Ali L, Mehdi T, Dirtu A, Al–Shammari F, Neels H (2013) Levels and profiles of organochlorines and flame retardants in car and house dust from Kuwait and Pakistan: implication for human exposure via dust ingestion. Environ Int 55:62–70

Ali N, Mehdi T, Malik RN, Eqani SAMAS, Kamal A, Dirtu AC, Neels H, Covaci A (2014) Levels and profile of several classes of organic contaminants in matched indoor dust and serum samples from occupational settings of Pakistan. Environ Pollut 193:269–276

Ali N, Eqani SAMAS, Ismail IMI, Malarvannan G, Kadi MW, Albar HMS, Rehan M, Covaci A (2016) Brominated and organophosphate flame retardants in indoor dust of Jeddah, Kingdom of Saudi Arabia: implications for human exposure. Sci Total Environ 569:269–277

Ali N, Shahzad K, Rashid MI, Shen H, Ismail IM, Eqani SA (2017) Currently used organophosphate and brominated flame retardants in the environment of China and other developing countries (2000–2016). Environ Sci Pollut Res 24:18721–18741

Ali N, Ismail IM, Kadi MW, Albar HM (2018) Currently used organophosphate flame retardants determined in the settled dust of masjids and hotels of Saudi Arabia, a new insight into human health implications of dust exposure. Environ Sci Process Impacts 20:798–805

Andresen J, Bester K (2006) Elimination of organophosphate ester flame retardants and plasticizers in drinking water purification. Water Res 40:621–629

Bi XH, Simoneit BRT, Wang ZZ, Wang XM, Sheng GY, Fu JM (2010) The major components of particles emitted during recycling of waste printed circuit boards in a typical e–waste workshop of South China. Atmos Environ 44:4440–4445

Bollmann UE, Moeler A, Xie ZY, Ebinghaus R, Einax JW (2012) Occurrence and fate of organophosphorus flame retardants and plasticizers in coastal and marine surface waters. Water Res 46:531–538

Butte W, Heinzow B (2002) Pollutants in house dust as indicators of indoor contamination. Rev Environ Contam Toxicol 175:1–46

Cao SX, Zeng XY, Song H, Li HR, Yu ZQ, Sheng GY, Fu JM (2012) Levels and distributions of organophosphate flame retardants and plasticizers in sediment from Taihu Lake, China. Environ Toxicol Chem 31:1478–1484

Cao Z, Xu F, Covaci A, Wu M, Wang H, Yu G, Wang B, Deng S, Huang J, Wang X (2014) Distribution patterns of brominated, chlorinated, and phosphorus flame retardants with particle size in indoor and outdoor dust and implications for human exposure. Environ Sci Technol 48:8839–8846

Carlsson H, Nilsson U, Ostman C (2000) Video display units: an emission source of the contact allergenic flame retardant triphenyl phosphate in the indoor environment. Environ Sci Technol 34:3885–3889

Dirtu AC, Ali N, Van den Eede N, Neels H, Covaci A (2012) Country specific comparison for profile of chlorinated, brominated and phosphate organic contaminants in indoor dust case study for Eastern Romania, 2010. Environ Int 49:1–8

EFRA (2007) The European flame retardants association, market statistics. http://wwwceficefracom

He CT, Zheng J, Qiao L, Chen SJ, Yang JZ, Yuan JG, Yang ZY, Mai BX (2015) Occurrence of organophosphorus flame retardants in indoor dust in multiple microenvironments of southern China and implications for human exposure. Chemosphere 133:47–52

Hu M, Li J, Zhang B, Cui Q, Wei S, Yu H (2014) Regional distribution of halogenated organophosphate flame retardants in seawater samples from three coastal cities in China. Mar Pollut Bull 86:569–574

Kawagoshi Y, Fukunaga I, Itoh H (1999) Distribution of organophosphoric acid trimesters between water and sediment at a sea-based solid waste disposal site. J Mater Cycles Waste Manag 1:53–61

Khan MU, Li J, Zhang G, Malik RN (2016) First insight into the levels and distribution of flame retardants in potable water in Pakistan: an underestimated problem with an associated health risk diagnosis. Sci Total Environ 565:346–359

Kim JW, Isobe T, Chang KH, Amano A, Maneja RH, Zamora PB, Siringan FP, Tanabe S (2011a) Levels and distribution of organophosphorus flame retardants and plasticizers in fishes from Manila Bay, the Philippines. Environ Pollut 159:3653–3659

Kim JW, Ramaswamy BR, Chang KH, Isobe T, Tanabe S (2011b) Multiresidue analytical method for the determination of antimicrobials, preservatives, benzotriazole UV stabilizers, flame retardants and plasticizers in fish using ultra high performance liquid chromatography coupled with tandem mass spectrometry. J Chromatogr A 1218:3511–3520

Kim JW, Isobe T, Sudaryanto A, Malarvannan G, Chang KH, Muto M, Prudente M, Tanabe S (2013) Organophosphorus flame retardants in house dust from the Philippines: occurrence and assessment of human exposure. Environ Sci Pollut Res 20:812–822

Kim JW, Isobe T, Muto M, Tue NM, Katsura K, Malarvannan G, Sudaryanto A, Chang KH, Prudente M, Viet PH, Takahashi S (2014) Organophosphorus flame retardants (PFRs) in human breast milk from several Asian countries. Chemosphere 116:91–97

Li J, Yu NY, Zhang BB, Jin L, Li MY, Hu MY, Zhang XW, Wei S, Yu HX (2014) Occurrence of organophosphate flame retardants in drinking water from China. Water Res 54:53–61

Luo P, Bao LJ, Guo Y, Li SM, Zeng EY (2016) Size-dependent atmospheric deposition and inhalation exposure of particle-bound organophosphate flame retardants. J Hazard Mater 301:504–511

Ma YQ, Cui KY, Zeng F, Wen JX, Liu H, Zhu F, Ouyang GF, Luan TG, Zeng ZX (2013) Microwave-assisted extraction combined with gel permeation chromatography and silica gel cleanup followed by gas chromatography–mass spectrometry for the determination of organophosphorus flame retardants and plasticizers in biological samples. Anal Chim Acta 786:47–53

Marklund A, Andersson B, Haglund P (2005) Traffic as a source of organophosphorus flame retardants and plasticizers in snow. Environ Sci Technol 39:3555–3562

Mercier F, Glorennec P, Thomas O, Bot BL (2011) Organic contamination of settled house dust, a review for exposure assessment purposes. Environ Sci Technol 45:6716–6727

Moller A, Sturm R, Xie ZY, Cai MH, He JF, Ebinghaus R (2012) Organophosphorus flame retardants and plasticizers in airborne particles over the Northern Pacific and Indian Ocean toward the polar regions: evidence for global occurrence. Environ Sci Technol 46:3127–3134

Ou Y (2011) Developments of organic phosphorus flame retardant industry in China. Ind Eng Prog 30:210–215

Paustenbach DJ, Finley BL, Long TF (1997) The critical role of house dust in understanding the hazards posed by contaminated soils. Int J Toxicol 16:339–362

Qiao L, Zheng XB, Zheng J, Lei WX, Li HF, Wang MH, He CT, Chen SJ, Yuan JG, Luo XJ, Yu YJ, Yang ZY, Mai BX (2016) Analysis of human hair to assess exposure to organophosphate flame retardants: influence of hair segments and gender differences. Environ Res 148:177–183

Stapleton HM, Klosterhaus S, Eagle S, Fuh J, Meeker JD, Blum A, Webster TF (2009) Detection of organophosphate flame retardants in furniture foam and US house dust. Environ Sci Technol 43:7490–7495

Van der Veen I, de Boer J (2012) Phosphorus flame retardants: properties, production, environmental occurrence, toxicity and analysis. Chemosphere 88:1119–1153

Wang X, Liu J, Yin Y (2010) The pollution status and research progress on organophosphate ester flame retardants. Prog Chem (in Chinese) 22:1983–1992

Wang XW, He YQ, Lin L, Zeng F, Luan TG (2014) Application of fully automatic hollow fiber liquid phase microextraction to assess the distribution of organophosphate esters in the Pearl River Estuaries. Sci Total Environ 470-471:263–269

Wei GL, Li DQ, Zhuo MN, Liao YS, Xie ZY, Guo TL, Li JJ, Zhang SY, Liang ZQ (2015) Organophosphorus flame retardants and plasticizers: sources, occurrence, toxicity and human exposure. Environ Pollut 196:29–46

WHO (1991) EHC 111: tri-n-butyl phosphate. World Health Organization, Geneva, Switzerland

WHO (1997) Flame retardants: a general introduction, 192 environmental health criteria. World Health Organization, Geneva, Switzerland

WHO (1998) Environmental health criteria 209, flame retardants: tris(chloropropyl) phosphate and tris(2-chloroethyl) phosphate. World Health Organization, Geneva, Switzerland

WHO (2000) EHC 218: flame retardants: tris(2-butoxylethyl)phosphate, tris (2-ethyl-exyl)phosphate and tetrakis (hydroxymethyl) phosphonium salts. World Health Organization, Geneva, Switzerland

Wu M, Yu G, Cao Z, Wu D, Liu K, Deng S, Huang J, Wang B, Wang Y (2016) Characterization and human exposure assessment of organophosphate flame retardants in indoor dust from several microenvironments of Beijing, China. Chemosphere 150:465–471

Yang FX, Ding JJ, Huang W, Xie W, Liu WP (2014) Particle size-specific distributions and preliminary exposure assessments of organophosphate flame retardants in office air particulate matter. Environ Sci Technol 48:63–70

Zeng X, Liu Z, He L, Cao S, Song H, Yu Z, Sheng G, Fu J (2015) The occurrence and removal of organophosphate ester flame retardants/plasticizers in a municipal wastewater treatment plant in the Pearl River Delta, China. J Environ Sci Health A 50:1291–1297

Zheng XB, Xu FC, Chen KH, Zeng YH, Luo XJ, Chen SJ, Mai BX, Covaci A (2015) Flame retardants and organochlorines in indoor dust from several e waste recycling sites in South China: composition variations and implications for human exposure. Environ Int 78:1–7

Chapter 5
Global Trends of E-waste Pollution and Its Impact on Environment

Rida Akram, Natasha, Ali Ahmad, Sibgha Noreen, Muhammad Zaffar Hashmi, Syeda Riffat Sultana, Abdul Wahid, Muhammad Mubeen, Ali Zakir, Amjad Farooq, Mazhar Abbas, Khurram Shahzad, Shah Fahad, Shakeel Ahmad, and Wajid Nasim

Abstract At the global scale, electronic waste (e-waste) pollution has been considered as a major threat to human health due to its exposure through numerous pathways. The scientific literature regarding potential adverse health effects of e-waste is mainly related to human exposure through ingestion, occupational exposure, or inhalation. However, data regarding the worldwide production and population exposure is much limited. In this chapter, we have summarized the current

R. Akram · Natasha · M. Mubeen · A. Zakir · A. Farooq
Department of Environmental Sciences, COMSATS University Islamabad, Islamabad, Pakistan

A. Ahmad · S. Noreen
Institute of Pure and Applied Biology, Bahauddin Zakerya University, Multan, Pakistan

M. Z. Hashmi
Department of Chemistry, COMSATS University Islamabad, Islamabad, Pakistan

S. R. Sultana
Department of Agronomy, University of Agriculture Faisalabad, Faisalabad, Pakistan

A. Wahid
Department of Environmental Sciences, Bahauddin Zakerya University, Multan, Pakistan

M. Abbas
Department of Management Sciences, COMSATS University Islamabad, Islamabad, Pakistan

K. Shahzad
Central Cotton Research Institute (CCRI), Multan, Pakistan

S. Fahad
Department of Agriculture, University of Swabi, Swabi, KPK, Pakistan

S. Ahmad
Department of Agronomy, Bahauddin Zakariya University, Multan, Pakistan

W. Nasim (✉)
Department of Environmental Sciences, COMSATS University Islamabad, Islamabad, Pakistan

Department of Agronomy, University College of Agriculture and Environmental Sciences, The Islamia University of Bahawalpur (IUB), Bahawalpur, Pakistan

M. Z. Hashmi, A. Varma (eds.), *Electronic Waste Pollution*, Soil Biology 57,
https://doi.org/10.1007/978-3-030-26615-8_5

scientific information regarding the categories of e-waste. The chapter also outlines the handling of e-waste in developed and developing countries and human exposure to e-waste through the various possible routes. Moreover, the contaminations of different environmental media with respect to toxic elements, particularly heavy metals are also discussed. These given studies indicate that e-waste exposure may cause potential human health hazards. Based on this current state of knowledge, strategies can then be proposed to reduce human exposure to e-waste.

Keywords Global trends · E-waste pollution · Impact · Environment

5.1 Introduction

The antagonistic results for health and the environmental exposure to waste materials from human utilization have been observed. Generally, observed toxic waste produced from discarded electrical and electronic equipment (EEE) (Brune et al. 2013). Such materials have expensive parts that have financial esteem when reused. Nonetheless, EEE likewise holds possibly risky elements that might be directly discharged or produced during recycling practices and produce electronic waste, commonly known as e-waste. The manufacture and discharge of toxic products mainly in the "informal" area of e-waste reprocessing where current mechanical procedures are not utilized and high-quality safety measures are frequently insufficient.

Exposure to e-waste is not prudent for any person. Kids are especially defenseless against a considerable amount of e-waste. The EEE contain things that have either a battery or a power rope. Generally, the e-waste created from disposed EEE is coolers, washers, PCs, screens, laptops, TVs, DVD/mp3 players, cell phones, recreation, and wearing equipment (UNEP 2007). Equipment segments containing batteries, circuit, plastic exteriors, cathode-ray tubes (CRTs), glass, and Pb capacitors additionally are viewed as e-waste (UNEP 2007). There are various estimations with regard to the measure of local, provincial, and worldwide e-waste delivered. As per StEP (Solving the E-waste Problem Initiative), the 2012 worldwide production of e-waste calculated was 45.6 million mt (Duffert et al. 2013). There are various impacts of e-waste recycling during metal recovery process on the environment (Fig. 5.1).

The United Nations Environmental Program (UNEP) predicted that the measure of e-waste generated in 2012 is sufficient to fill hundred Empire State structures and midpoints to more than 6.8 kg for each individual. The worldwide populace is almost 7000 million, however, in spite of the fact that there are just 4500 million toilets around the world, there are evaluated to be somewhere around 6000 million portable phones (UNU 2013). In 2012 alone, China produced 0.0111 billion tons of e-waste and the United States delivered 0.01 billion tons (StEP Initiative 2013). This implies, by and large, every American produces 29.5 kg of e-waste compared with less than

Fig. 5.1 Impacts of e-waste on the environment during metal recovery processes

5 kg by each individual in China. The total volume of e-waste is challenging; however, additional concern is the quick rise of this problem, worldwide waste stream (Lundgren 2012).

The United Nation's report (2012) estimated that global e-waste will build a further 33% from 49.7 to 65.4 million tons annually by 2017 (UNU 2013). The collective volume of e-waste delivered is exponentially growing due to different reasons. Customer interest and a high-quality demand led to the worthless purchase of electronics (Schluep et al. 2014). Likewise, new cell phone models are launched at exceedingly short intervals. Not just cell phone models changed, yet the adornments, like, chargers, regularly modified with new mobile phones. Short advancement phases and low reprocessing ratio increase e-waste production at a higher rate. The life span of electronics has been dropping, making notable intensifications to e-waste (Puckett et al. 2014).

E-waste is a worldwide, inter-regional, and local issue. Of the 0.02–0.05 billion tons produced annually, it is assessed that 75–80% is shipped to Asia and Africa for reprocessing (Diaz-Barriga 2013). Loopholes in the current e-waste directions take into consideration the fare transfer of e-waste from one nation to another under the pretense of "endowment" and "reprocessing." The Basel Convention (Basel Convention, UNEP 2008) and The Partnership for Action on Computing Equipment (PACE) encourage the safe administration of e-waste. Among different responsibilities, PACE has given rules on the usefulness of PCs and PC parts, including batteries, ought to must be viewed as usable PCs and, all things considered, reasonable for donation (UNEP 1992).

5.2 Composition of E-waste

Generally, e-waste contains significant toxic elements. The organization of e-waste depends on components, for example, the kind of EEE, the model, make, production date, and the life span of the electronic piece. E-waste from IT and media transmission frameworks have valuable metals than e-waste from domestic use equipment (Chancerel et al. 2009). For example, a cell phone contains in excess of 40 components, base metals, for example, copper (Cu) and tin (Sn); unique metals, for example, lithium (Li), cobalt (Co), indium (In), and antimony (Sb); and valuable metals, for example, silver (Ag), gold (Au), and lead (Pd) (Liu et al. 2009). The e-waste categorization with respect to their composition is shown in Fig. 5.2.

Metals, for example, Au and Pd can be extracted more successfully from e-waste contrasted with the mining of mineral (Chancerel 2009). E-waste contains polybrominated diphenyl ethers (PBDEs) (commonly fire retardants) blended with plastics and different elements. Circuit sheets present in the majority of the EEE may have arsenic (As), cadmium (Cd), chromium (Cr), Pb, mercury (Hg), and additional poisonous elements. Regular printed circuit sheets treated with Pb in EEE contain roughly 50 g of Sn–Pb (Widmer et al. 2005).

Outdated fridges, chillers, and cooling units contain chlorofluorocarbons (CFCs) exhausting ozone (O_3). The conspicuous constituents, for example, barium (Ba), Cd, Cu, Pb, zinc (Zn), and rare earth metals and have CRTs in PC and TV screens (Chancerel et al. 2009). The normal Pb in CRT screens is 1.6–3.2 kg. In this manner, the USA, few nations in the EU, and Japan have prohibited dumping cathode ray tubes in landfills due to their dangerous impacts on the environment.

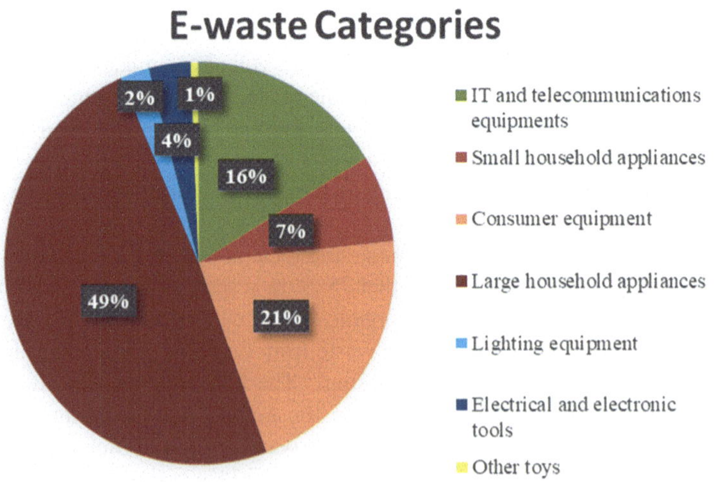

Fig. 5.2 E-waste categories and their contribution

5.3 Global Challenges of E-waste

E-waste has now turned into a worldwide challenge in both developed and developing nations. The total amount of e-waste says a lot about its possible upcoming hazard, on the off chance that it is not overseen legitimately. For instance, as indicated by the UNU (2013), the worldwide production of e-waste was 42 mt, out of this, 6 mt was from Information and Communication Technology (ICT). This additionally implies 6 kilograms of per capita e-waste produced every year. Figure 5.3 shows e-waste generation at the global level and the data shows that Asian continent as the greatest producer of e-waste in 2006 (Balde et al. 2017).

In spite of the fact that the utilization of electronics is most common in developed nations, in the ongoing years, the developing nations also demonstrate a developing pattern of utilization of these advanced types of electronic stuff. However, the greater part of the developed nations' need legitimate systems or foundation or even familiarity with sound treatment and transfer of e-waste. These nations have progressed toward becoming dumping yards of e-waste produced in developing nations. Thus, e-waste in these nations will be treated in problematic courses (UNEP 2007) prompting numerous natural and social concerns/dangers.

Be that as it may, there is a chance if e-waste is appropriately overseen. E-waste contains bunch of profitable materials, for example, Au, Ag, platinum (Pt), and palladium (Pd). On the off chance that appropriately dealt with, these materials can be recouped while lessening the weight on the earth for crude material extraction. Notwithstanding this positive effect on the condition, e-waste has positive social effects as well. In addition, e-waste reusing can open up new enterprises while making practical roads for something that could turn into a social and medical issue.

5.4 E-waste and Developing Countries

E-waste is a critical problem in the waste administration with worldwide interconnections among all developed, developing, and underdeveloped countries. The dependence of society on advanced innovation directs the everyday life in high and middle earning nations where populace uses a lot of electronic items, which more readily progressed toward becoming e-waste. This portion is a quickly rising waste needs proper handling and managing because of the lethal capability toward human health of general and environmental condition (Hammad et al. 2017). Then again, the e-waste has important constituents which might be recovered as valuable metals, reused metals, and plastics by different projects moderating the utilization of natural resources. The e-waste managing framework is to move the worldview from a dangerous contamination cause to a feasible asset with regard to useful advancement. The EU approach advances the round about budget where waste is viewed as assets and sets instructions toward reprocessing of e-waste.

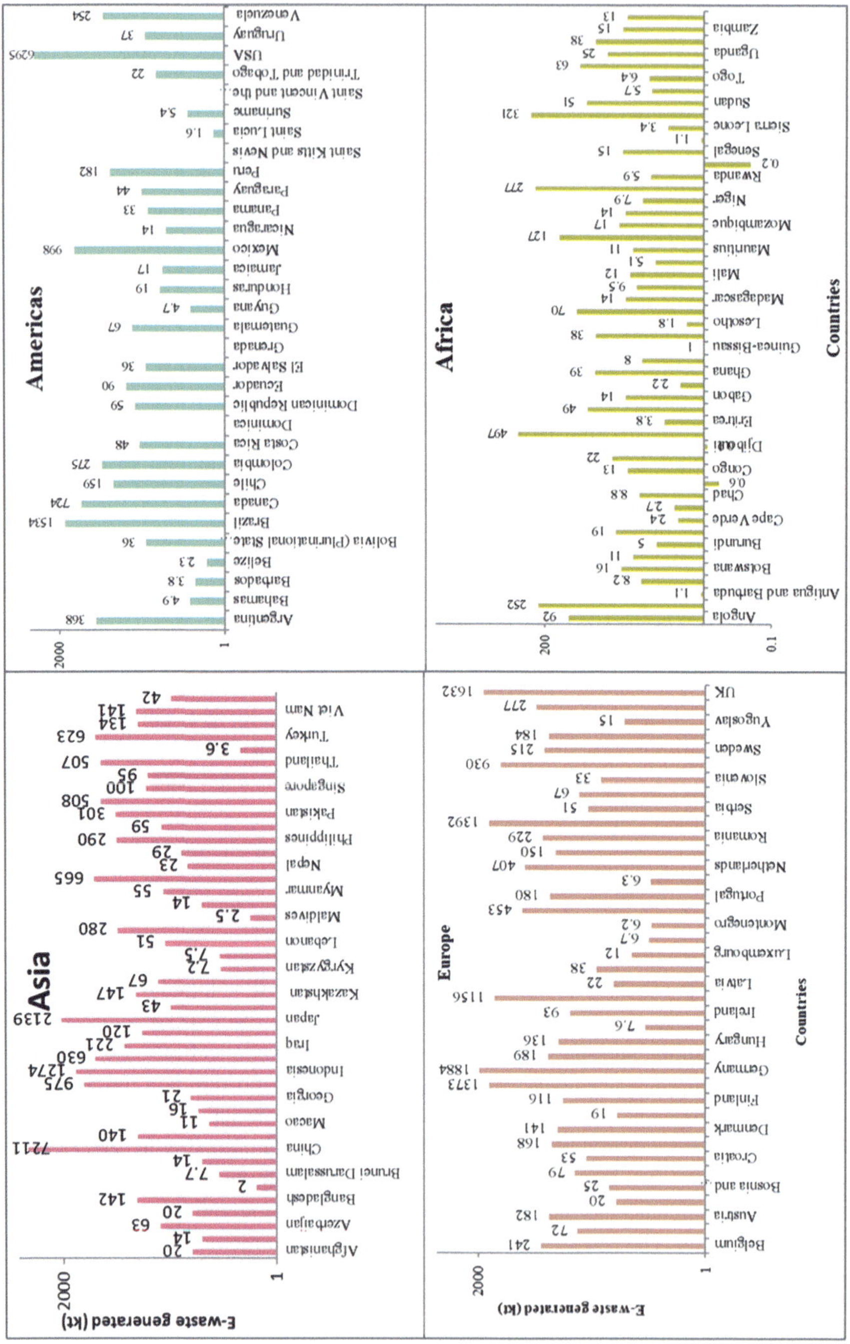

Fig. 5.3 Global e-waste generated (kt) in 2016 (Balde et al. 2017)

Developed nations tend not to reuse e-waste because of the absence of offices, high work expenses, and intense ecological directions and this e-waste is dumped in landfills or sent out to developing nations (Robinson 2009). The Basel Convention Controls the export of harmful e-wastes and prohibits the transfer of poisonous and dangerous e-wastes to developing nations and the directions of developed nations confine the landfill of waste so as to advance the reprocessing and recovery alternatives. National directions that allow, boycott, or overlook the electronic and e-waste trade patterns changed starting with one nation, then onto the next, beside EU that has an increasingly standardized regulation in the countries. Many countries like Nigeria, Cambodia, China, Vietnam, Malaysia, Pakistan prohibited imports of e-waste, some countries like, Benin, Cote D'Ivoire, Kenya, Liberia, Senegal, Uganda, South Africa, and India have not endorsed this issue and but countries like Ghana, Thailand, and the Philippines allowed such imports with exceptional endorsements (Jinhui et al. 2013). India, China, the Philippines, Hong Kong, Indonesia, Sri Lanka, Pakistan, Bangladesh, Malaysia, Vietnam, and Nigeria are the most important regions for e-waste and critical measures of e-waste having dangerous constituents disposed of in open grounds and streams (Heart and Agamuthu 2012).

Such toxic materials are necessary to be evacuated before any dealing or transfer under EU Directives. Categorical directions designed to handle e-waste problems and must be additionally created and implemented by huge stakeholders and companies of e-waste. The absence of enactment, the frail organization, the absence of a formal waste administration framework, poor expectations for everyday comforts expose the developing states to rough recycling practices, open dumping, and burning results in e-waste contamination. China is a noteworthy e-waste reprocessing region with primitive recycling exercises in order to get important metals and valuable resources for industry and to guarantee an ordinary pay regardless of genuine protection dangers (Sepúlveda et al. 2010; Abbas 2011; Nasim et al. 2016, 2018).

Reprocessing organizations compels the poor workers in disassembling zones who complete their work in deprived situations by hand and much of the time with no insurance actions. Such exercises are additionally completed by people at the family unit level. E-waste landfill areas are "problem areas" of intense ecological contamination, generally situated near the housing areas or farming grounds. Such areas release the leachates and lethal fluids into streams, lakes, groundwater, and soil lead to contamination, also have negative impacts on domesticated animals and lastly on its buyers. Table 5.1 shows the occurrence and concentrations of different substances in e-waste and their part in global emissions.

Table 5.1 The occurrence and concentration of different substances in e-waste and their part in global emissions (Morf et al. 2007; Robinson 2009)

Substance and their Occurrence in e-waste	Concentration of substances in e-waste (mg/kg) and Global emissions (tons)
PCB → Transformers PCB → Condensers	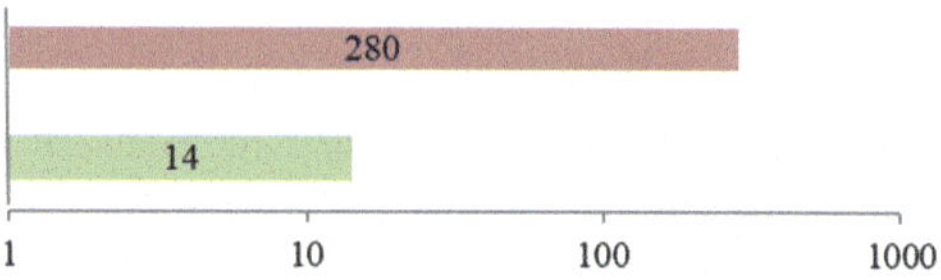
Sb → Fire retardant Sb → Plastics	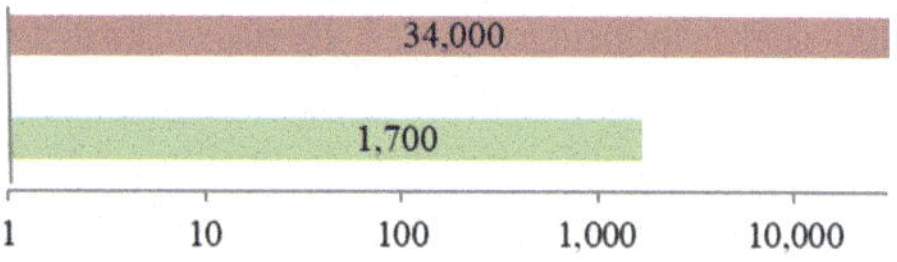
Cd → NiCd-batteries Cd → Photocopying-machines Cd → Fluorescent layer printer inks and toners	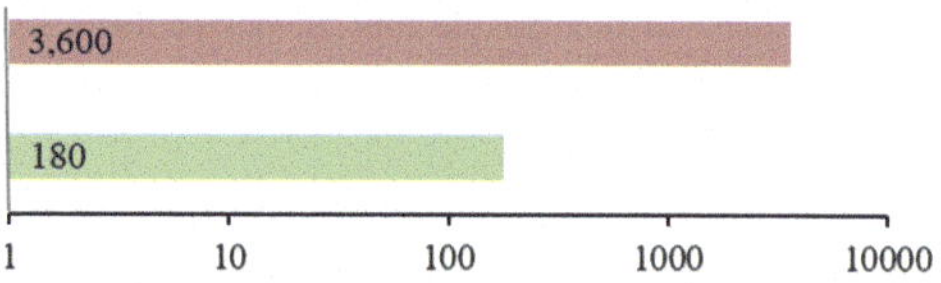
Cr → Data tapes Cr → Floppy-disks	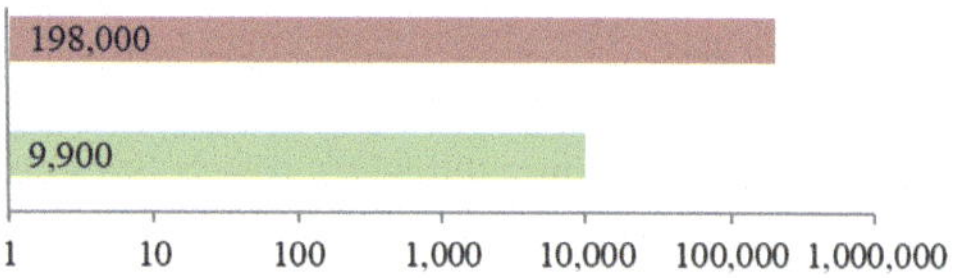
Cu → Cabling	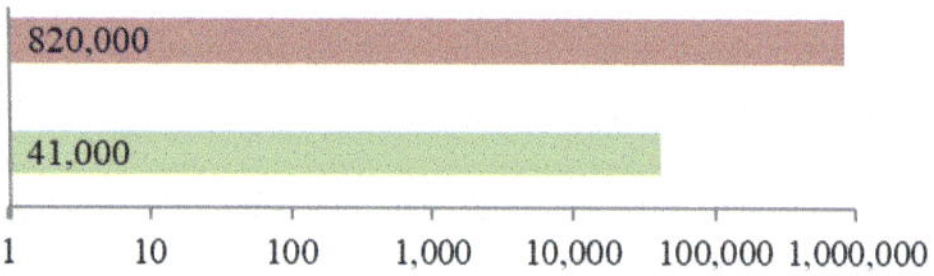

(continued)

Table 5.1 (continued)

Pb → CRT screens; Printed wiring boards; Batteries

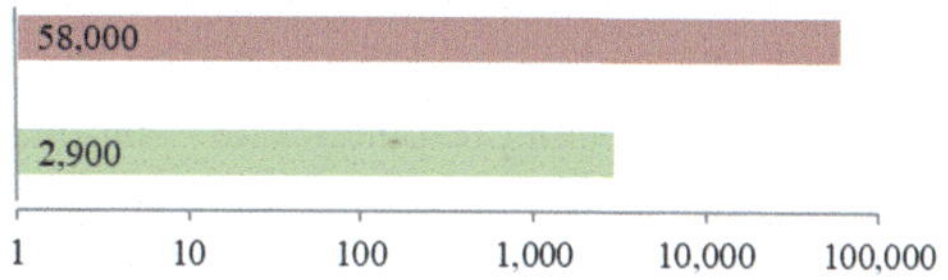

Hg → Mercury-wetted switches; Alkaline batteries; Fluorescent lamps used in LCDs

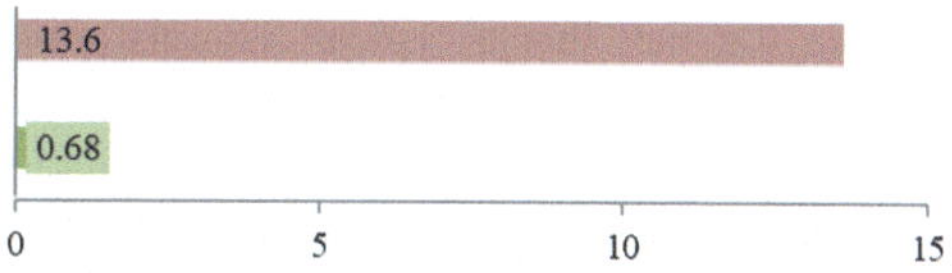

Ni → NiCd-batteries; Ni-MH-batteries; Electron gun in CRT

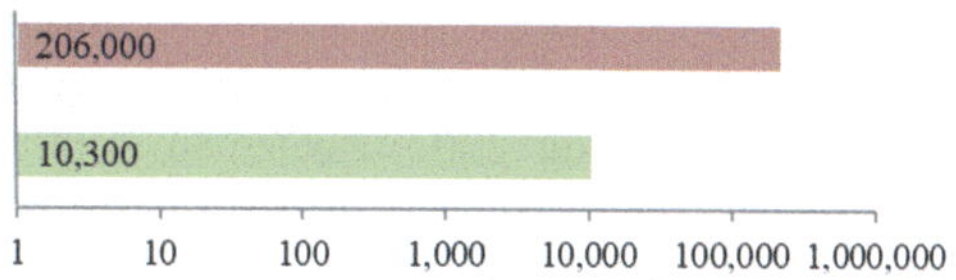

Sn → Solder metal glue; LCD

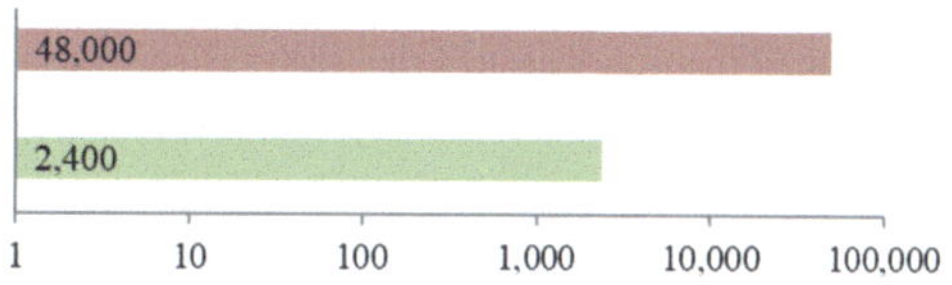

ZnS → Interior of CRT screens; Mixed with rare earth metals

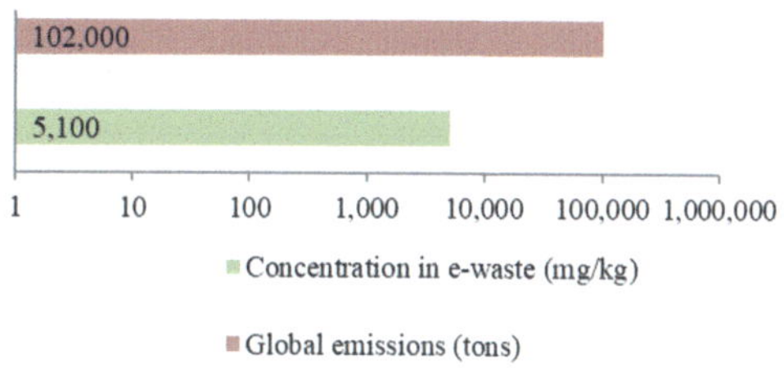

5.5 E-waste and Developed Countries

The WEEE (Waste Electrical and Electronic Equipment) administration is reinforced out globally through various practices and policies. The most comprehensive control framework is the EU order as it influences the entire life cycle beginning from the structure period of electronic equipment to its end of lifetime. There is need to proceed new attention for reusing like reclaim options for this purpose wholesalers need to reclaim an utilized item without a buy of another one as alternative for gathering less WEEE. Japan has built up an administrative framework like the EU (Yoshida and Yoshida 2010; Menikpura et al. 2014).

An alternate administrative methodology is used in the USA, where there is an absence of a typical government enactment about e-waste organizations: each state has characterized its own framework with explicit targets and associations (Kahhat et al. 2008; Kollikkathara et al. 2009; Nasim et al. 2011, 2012). One endeavor toward a new practice has been presented with "National Strategy on Electronics Stewardship" (EPA 2011): it expects to bring up government activities to enhance the structure of electronic items and improve the executives of utilizing or disposed of hardware (Elia and Gnoni 2015; Jabran et al. 2017).

Despite the fact that a typical administrative standard could not be allocated around the world, WEEE the executives frameworks connected have normal highlights just as contrasted dependent on the particular authoritative methodology. One basic fundamental idea is the extended producer responsibility (EPR) rule OECD (2001): the EU enactment is intensely founded on this methodology as group and individual reclaim frameworks will be connected by makers in dealing with all stages in the electronic item's life cycle, including the post-buyer arrangement (Ogushi and Kandlikar 2007; Zoeteman et al. 2010; Toyasaki et al. 2011). The EPR guideline is likewise recognized in Japan: makers and suppliers must compose the reclaim framework for EEE. As of late, the EPR rule has been additionally connected in Canada to characterize advance rules and regulations about electronic equipment (Thompson and Oh 2006) (Fig. 5.4).

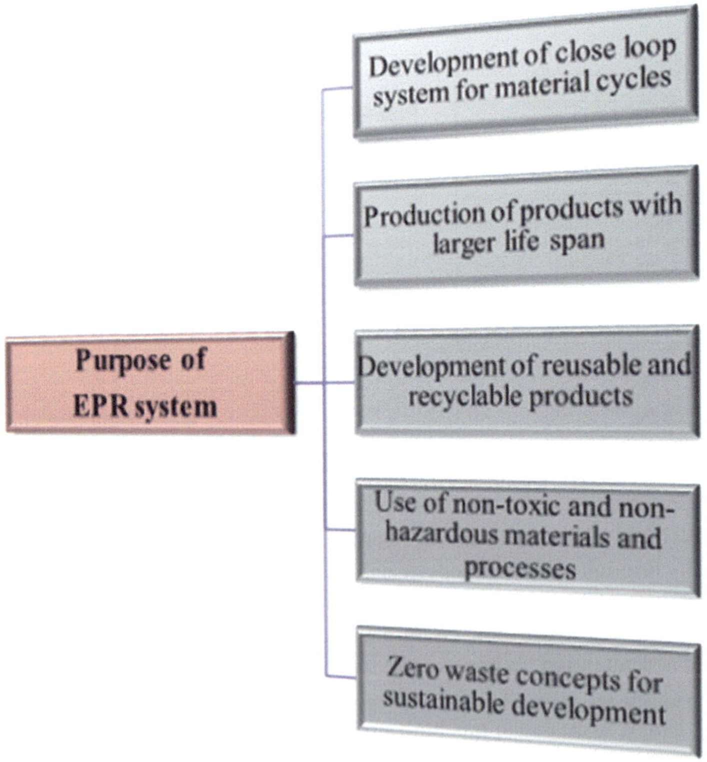

Fig. 5.4 The main purposes of EPR system to control the e-waste generation

5.6 E-waste and Legal Framework

E-waste started having more significance since it was incorporated into the General Law for Prevention and Integral Management of Wastes. This law arranges e-waste managing, controlling, and handling to minimize the hazardous impacts of e-waste on humans health and environment (Karimova and Manrique 2018). This waste contains harmful materials and bioaccumulation can occur. The direction of waste is strengthened by a worldwide organizations that incorporate understandings and arrangements, for example, Basel tradition, Stockholm, and Rotterdam among others.

The consideration of the national and international framework for e-waste administration is essential since it sets a standard in the advancement of legislative issues on waste. The administration of e-waste has picked up significance in view of the existence of a few contaminants found in the waste and when they are transferred not legitimately, can be openly dumped into the earth. Electronic industry developing and mechanical advancement have added to gear process substitution, expanding the quantity of e-waste broadly (Araujo et al. 2012). E-waste subject has increased worldwide significance from its consideration in the motivation of various declarations between nations to advance the activities for decrease natural impacts of e-waste on the environment, for example, the Basel and Stockholm Convention among others (Cano-Robles 2014). These activities chiefly for the distinguishing proof of some determined natural toxins (POPs, e.g., brominated fire retardants just as of heavy metals). These contaminants can be discharged and cause unfavorable ecological and human health impacts at the end of its life cycle (Akram et al. 2017a, b).

5.7 E-waste and Environment

The majority of the toxins and hazardous elements are related to serious ecological and human health effects. Few contaminations can be scattered through the air, groundwater, and soil just as present in the encompassing air in sectors adjacent to the recycling regions. In different circumstances, the waste is dumped directly into the soil or water streams, where the consequent draining of toxins could pollute the earth and impact the natural way of life supplies also. Direct human introduction to these chemicals can, likewise, have negative impacts on human health. The e-waste impacts on soil, vegetation, air quality, water quality, and on human health is given in this segment (Akram et al. 2019). Table 5.2 shows the hazardous impact of e-waste pollutants on the whole environment and their possible route of exposure.

Table 5.2 The hazardous impact of e-waste pollutants on living organisms' health, environment, and their possible route of exposure (BAN and SVTC 2002; Johri 2008; Frazzoli et al. 2010)

E-waste pollutant	Hazardous impact on health	Hazardous impact on environment	Route of exposure
Lead (Pb)	Effects plants, animals and microorganisms, inhibits the functioning of particular enzymes in human body causing severe physiological or neurological effects, damage to the nervous system, kidney, reproductive system in human, blood system and the development of brain in kids	Leaching into groundwater and causes soil acidification	Ingestions, inhalation, and dermal contact
Mercury (Hg)	Stunts fetus growth, contaminants get included in mother's milk. Affects the central nervous system, kidneys, and immune system. Mercury in water bodies can enter the human food chain through aquatic	Contaminate groundwater	Ingestions, inhalation, and dermal contact
Cadmium (Cd)	Causes cancer, softens bones, severe pain in the spine and joints and affects the kidneys	Soil and water	Inhalation and Ingestions
Arsenic (As)	Prolonged exposure to poisonous element leads to skin diseases, lung cancer, and decrease in nerve conduction velocity	Soil and water	Ingestions, inhalation, and dermal contact
Zinc (Zn)	Leads to stomach cramps, skin irritations, nausea, and anemia, can cause severe damage to the pancreas	Soil	Ingestion and inhalation
Lithium (Li)	Affects kidneys	Soil	Ingestion, inhalation, and dermal contact
Beryllium (Be)	Causes lung cancer	Air	Ingestion, inhalation, and trans-placental
Brominated flame retardant (BFR)	Affect thyroid function, have estrogenic effects in animals, contain carcinogens, Dioxins can harm reproductive and immune systems. BFR is present in computer cabinet dust	BFR can leach into landfills. Organic pollutants, source of dioxins and furans in air	Ingestion, inhalation, and trans-placental
Chromium (Cr)	Can cause asthmatic bronchitis, damages liver and kidneys and cause lung cancer	Air	Ingestion and inhalation

(continued)

Table 5.2 (continued)

E-waste pollutant	Hazardous impact on health	Hazardous impact on environment	Route of exposure
Polychlorinated biphenyls (PCBs)	Pollutants accumulate in the crops and vegetation of the sites usually exceeds the permitted level for food	Pollutes soils, sediments, vegetation, and wild aquatic species	Ingestion, inhalation, dermal contact and trans-placental

5.7.1 Soil and Plants

A few types of toxins have been seen in soils and plants close to e-waste handling places. Large amounts of polychlorinated biphenyls (PCBs) were present in the soil and the plant samples of an e-waste reusing sites in China. *Chrysanthemum coronarium* L. from farms and *Bidens pilosa* L. (wild plant) from the e-waste sites were found to have greater concentrations of PCBs than other plants. The soil specimens from the e-waste site displayed a lot higher fixations than other zones; vegetable soils were found to have more elevated amounts of PCBs than paddy soils (Wang et al. 2011; Akram et al. 2018a, b). About 17 kinds of PCDD/Fs, 36 types of PCBs, and 16 types of PAHs were collected from soils near e-waste reprocessing site in Taizhou, China (Shen et al. 2009). Elevated amounts of these three toxins were found in each of the three soils, with the high concentrations of PBDEs and PBBs (polybrominated biphenyls) observed in e-waste dumping sites (Wang et al. 2009).

Leaching of toxic elements apparently occurred in shipyards just as in reprocessing regions. Analysis of toxins discharged from e-waste was completed in a shipyard in Chittagong, Bangladesh. Soil tests demonstrated elevated amounts of Pb, Cd, Cr, Hg, Se, Sn, As, Co, and brominated dioxins (Alam and Bahauddin 2015). An examination of metals (Ag, As, Cd, Co, Cu, Fe, In, Mn, Ni, Pb, and Zn) in surface and soil tests from both formal and informal reprocessing sites of e-waste destinations in Manila and Philippines, and results demonstrated that the concentration of these metals in informal reprocessing destinations were higher than other countries in Asia (Fujimori et al. 2012). Another examination of substantial metal dimensions in the soil surface of a typical e-waste reprocessing site in Manila demonstrated contamination with Cu, Zn, and Pb (Fujimori and Takigami 2014). The difference between informal and formal methods of disposal of e-waste is shown in Fig. 5.5.

5.7.2 Air Quality

Various investigations have been done reporting in real-time contamination brought about by the e-waste handling workouts. Examination of PCDD/Fs, PCBs, and PBDEs was done in surrounding air of an e-waste disassembling region. The concentrations of PCDD/Fs, PCBs, and PBDEs were found to extend from 2.91 to

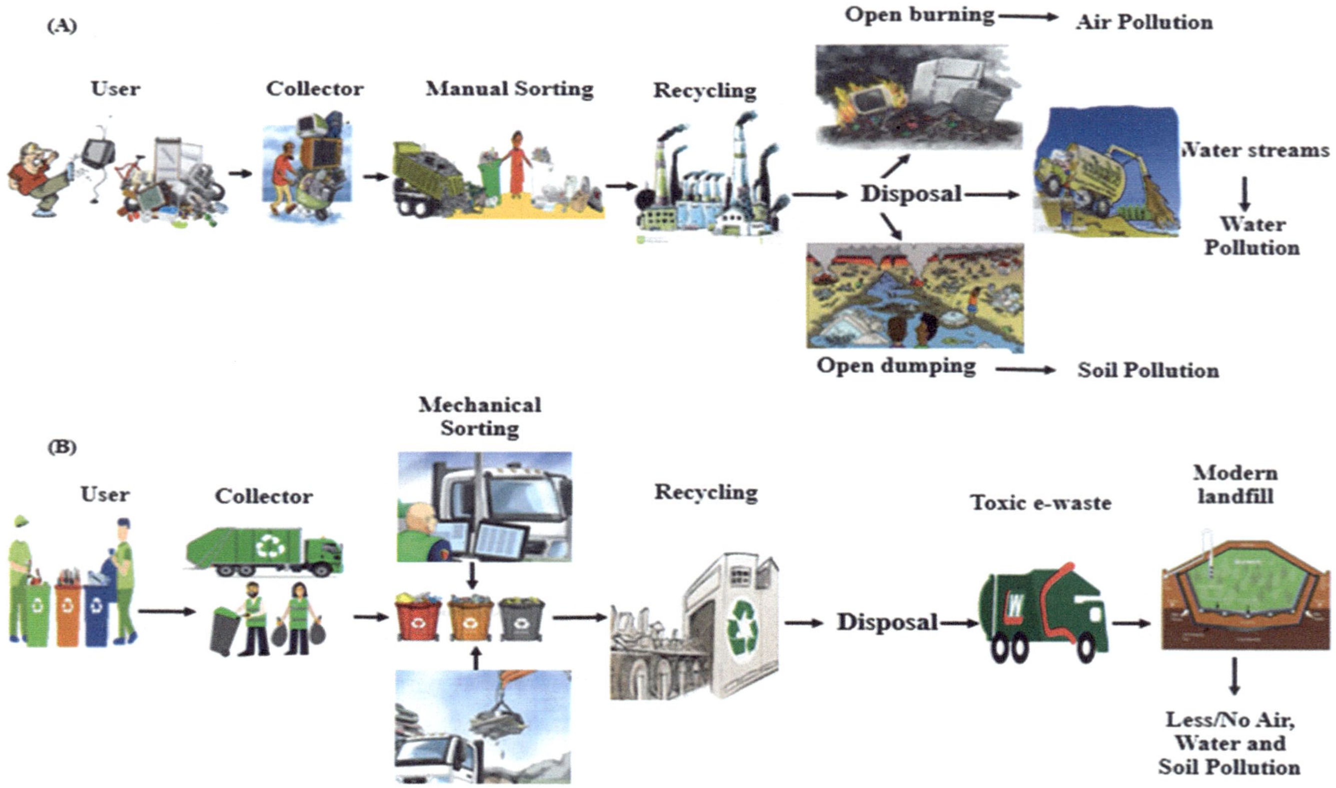

Fig. 5.5 Difference between (**a**) informal and (**b**) formal strategies for the disposal of e-wastes

50.6 pg/m^3, from 4.23 to 1.35 ng/m^3, and from 92 to 3086 pg/m^3, respectively. The dimensions of these three toxins were observed to be directly connected with informal disassembling practices. The chlorinated dioxins and furans were essentially seen in the particulate stage, while PCBs were discovered just in the gaseous phase (Li et al. 2008). Dimensions of PCBs and PBDEs were additionally estimated in the backyard of an e-waste handling territory in Vietnam. The groupings of these two contaminants were seen to be a lot higher (1000–1800 and 620–720 pg/m^3, individually) than in the reference regions (Tue et al. 2013).

Total suspended particles (TSP) and particulate matter ($PM_{2.5}$ μm) were investigated from the atmosphere of Guiyu. PAHs identified with TSP and $PM_{2.5}$ were found to extend from 40.0 to 347 and 22.7 to 263 ng m^{-3}, respectively. The dimensions of Cr, Cu, and Zn in $PM_{2.5}$ were seen to be somewhere in the range of 4 and multiple times of qualities commonly estimated in different countries of Asia. Such an introduction was inescapable for the general population living in the disassembling region (Deng et al. 2006). Another examination in Guiyu demonstrated that all PBDEs broke down in the air were 58 multiple times higher than in different urban communities and were in excess of multiple times higher than recorded in past investigations (Deng et al. 2007).

5.7.3 *Water Quality*

Groundwater has likewise been observed to be polluted by the unrefined e-waste reprocessing practices. An investigation of substantial metals contaminating lakes and well waters was completed in the region of a previous e-waste handling site. Results indicated water contamination with Cd and Cu. The well water was less polluted with heavy metals, however, it was seen that the surface soil demonstrated high concentrations of these toxic metals that were transported to different zones, for example, lake water (Wu et al. 2015).

Lianjiang (China) stream indicated large amounts of As, Cr, Li, Mo, Sb, and Se, while Nanyang (China) waterway had high amounts of Ag, Be, Cd, Co, Cu, Ni, Pb, and Zn. Residues of these waterways had concentrated with Cd, Cu, Ni, Pb, and Zn (Wong et al. 2007).

5.7.4 *Human Health*

Various studies have been conducted to demonstrate the effect of improper handling of e-waste on human health and its related consequences. The breast milk was studied for PCBs and PBDEs in three e-waste reprocessing areas of Vietnam. PBDEs levels were altogether higher in e-waste reprocessing points (20–250 ng/g) than in the reference city, Hanoi. PCBs levels were much lower than PBDEs

(28–59 ng/g). These toxins entered the human body through inhalation and ingestion (Tue et al. 2010).

To check the toxic matals concentration the breast milk were collected from ladies present adjacent to e-waste dumping site in Kolkata, India. The concentration of PCBs achieved is 1700 ng/g lipid weight in e-waste dumping sites, with respect to 60 ng/g lipid weight in reference areas (Devanathan et al. 2012). PBDE investigations of hair samples showed that the levels extended from 22.8 to 1020 ng/g on a dry weight basis, which was multiple times higher than the control tests. PCDD/Fs levels were observed to be 126–5820 pg/g dw, which was higher than control (Ma et al. 2011). The different impacts of e-waste on humans' health is shown in Table 5.2.

5.8 Conclusion and Recommendation

E-waste is a significant worldwide concern as a result of the immense production rates of EEE, furthermore, the absence of legitimate transfer techniques. This expands the concern about asset protection and the negative impacts on human health and the environment. The administration of developing and developed countries has directions identified with e-waste, especially in framework improvement and the board of e-waste yet legitimate usage is not pursued. Awareness and dynamic network cooperation are additionally required in waste isolation, reduction, treatment, and transfer exercises. Reasonable waste administration is needed for the financial development of the developing nations. Reuse, reduce, recycling, and mechanical advantageous interaction can decrease the expanding rate of e-waste and pursue zero waste ideas having natural advantages and building up a simple method to sustain the environment. Following the issues and impacts of e-waste, there is a dire requirement for the definite research on e-waste, including the amount of e-waste production, the gathering strategies, the reprocessing organizations, the current transfer practices, and the different ecological effects for the reasonable administration of e-waste.

References

Abbas Z (2011) WEEE/E waste management in Pakistan WEEE/E-waste management workshop on take-back system. United Nations Environment Programme (UNEP)/International Environment Technology Centre (ITEC), Osaka, Japan

Akram R, Hashmi MZ, Nasim W (2017a) Role of antibiotics in climate change. In: Hashmi MZ, Strezov V, Varma A (eds) Antibiotics and antibiotics resistance genes in soils. Springer, Cham, pp 91–98

Akram R, Amin A, Hashmi MZ, Wahid A, Mubeen M, Hammad HM, Fahad S, Nasim W (2017b) Fate of antibiotics in soil. In: Antibiotics and antibiotics resistance genes in soils. Springer, Cham, pp 201–214

Akram R, Turan V, Wahid A, Ijaz M, Shahid MA, Kaleem S, Hafeez A, Maqbool MM, Chaudhary HJ, Munis MFH, Mubeen M (2018a) Paddy land pollutants and their role in climate change. In: Environmental pollution of paddy soils. Springer, Cham, pp 113–124

Akram R, Turan V, Hammad HM, Ahmad S, Hussain S, Hasnain A, Maqbool MM, Rehmani MIA, Rasool A, Masood N, Mahmood F (2018b) Fate of organic and inorganic pollutants in paddy soils. In: Environmental pollution of paddy soils. Springer, Cham, pp 197–214

Akram R, Fahad S, Hashmi MZ, Wahid A, Adnan M, Mubeen M, Khan N, Rehmani MI, Awais M, Abbas M, Shahzad K (2019) Trends of electronic waste pollution and its impact on the global environment and ecosystem. Environ Sci Pollut Res 26(17):16923–16938

Alam M, Bahauddin KM (2015) Electronic waste in Bangladesh: evaluating the situation, legislation and policy and way forward with strategy and approach. Present Environ Sust Dev 9:81–101

Araujo MG, Magrini A, Mahler CF, Bilitewski B (2012) A model for estimation of potential generation of waste electrical and electronic equipment in Brazil. Waste Manag 32:335–342

Balde CP, Forti V, Gray V, Kuehr R, Stegmann P (2017) The global e-waste monitor 2017: quantities, flows and resources. United Nations University, International Telecommunication Union, and International Solid Waste Association, pp 1–116

Basel Convention, UNEP (2008) The partnership for action on computing equipment (PACE). http://www.basel.int/Portals/4/Basel%20Convention/docs/pub/leaflets/leafP CE.pdf

Brune MN, Goldizen F, Neira M et al (2013) Health effects of exposure to e-waste. Lancet Glob Health 1(2):1–70

Cano-Robles FK (2014) Electronic waste in the international context. International background: basel and Stockholm. In: Bracho LR, García AG, Concepción VA, Robles FC (eds) Electronic waste in Mexico and the world, 1st edn. Secretariat of Environment and Natural Resources. National Institute of Ecology and Climate Change, Mexico, DF, pp 12–15

Chancerel P (2009) Substance flow analysis of the recycling of small waste electrical and electronic equipment: an assessment of the recovery of gold and palladium. Papierflieger. p 1–162

Chancerel P, Meskers CE, Hageluken C, Potter VS (2009) Assessment of precious metal flows during preprocessing of waste electrical and electronic equipment. J Ind Ecol 13:791–810

Deng W, Louie P, Liu W, Bi X, Fu J, Wong M (2006) Atmospheric levels and cytotoxicity of PAHs and heavy metals in TSP and PM2.5 at an electronic waste recycling site in southeast China. Atmos Environ 40:6945–6955

Deng WJ, Zheng JS, Bi XH, Fu JM, Wong MH (2007) Distribution of PBDEs in air particles from an electronic waste recycling site compared with Guangzhou and Hong Kong, South China. Environ Int 33:1063–1069

Devanathan G, Subramanian A, Sudaryanto A, Takahashi S, Isobe T, Tanabe S (2012) Brominated flame retardants and polychlorinated biphenyls in human breast milk from several locations in India: potential contaminant sources in a municipal dumping site. Environ Int 39:87–95

Diaz-Barriga F (2013) Evidence-based intervention programs to reduce children's exposure to chemicals in e-waste sites. Discussion paper for WHO working meeting on e-waste and children's health, pp 1–90

Duffert C, Brune MN, Prout K (2013) Background document on exposures to e-waste. Geneva, Switzerland: World Health Organization. United Nations (UN) Department of Economic and Social Affairs, Population Division. Word population prospects the 2012 Revision

Elia V, Gnoni MG (2015) How to design and manage WEEE systems: a multi-level analysis. Int J Environ Waste Manag 15:271–294

EPA (2011) National strategy for electronics stewardship. http://www.epa.gov/epawaste/conserve/materials/recycling/taskforce/docs/strategy.pdf

Frazzoli C, Orisakwe OE, Dragone R, Mantovani A (2010) Diagnostic health risk assessment of electronic waste on the general population in developing countries scenarios. Environ Impact Assess Rev 30:388–399

Fujimori T, Takigami H (2014) Pollution distribution of heavy metals in surface soil at an informal electronic-waste recycling site. Environ Geochem Health 36:159–168

Fujimori T, Takigami H, Agusa T, Eguchi A, Bekki K, Yoshida A, Terazono A, Ballesteros FC (2012) Impact of metals in surface matrices from formal and informal electronic waste recycling around Metro Manila, the Philippines, and intra-Asian comparison. J Hazard Mater 221:139–146

Hammad HM, Farhad W, Abbas F, Fahad S, Saeed S, Nasim W, Bakhat HF (2017) Maize plant nitrogen uptake dynamics at limited irrigation water and nitrogen. Environ Sci Pollut Res 24:2549–2557

Heart S, Agamuthu P (2012) E-waste: a problem or an opportunity? Review of issues, challenges and solutions in Asian countries. Waste Manag Res 30:1113–1129

Jabran K, Ullah E, Akbar N, Yasin M, Zaman U, Nasim W, Riaz M, Arjumend T, Azhar MF, Hussain M (2017) Growth and physiology of basmati rice under conventional and water-saving production systems. Arch Agron Soil Sci 63:1465–1476

Jinhui L, Brenda NLN, Lili L, Nana Z, Keli Y, Lixia Z (2013) Regional or global WEEE recycling. Where to go? Waste Manag 3:923–934

Johri R (2008) E-waste: implications, regulations and management in India and current global best practices. TERI, New Delhi, p 4

Kahhat R, Kim J, Xu M, Allenby B, Williams E, Zhang P (2008) Exploring e-waste management systems in the United States. Resour Conserv Recycl 52:955–964

Karimova T, Manrique EE (2018) Regulatory frameworks: integration, partnerships and dialogue. World Employ Soc Outlook 2018(2):71–101

Kollikkathara N, Feng H, Stern E (2009) A purview of waste management evolution: special emphasis on USA. Waste Manag 29:974–985

Li Y, Jiang G, Wang Y, Wang P, Zhang Q (2008) Concentrations, profiles and gas-particle partitioning of PCDD/Fs, PCBs and PBDEs in the ambient air of an e-waste dismantling area, southeast China. Chin Sci Bull 53:521–528

Liu Q, Li KQ, Zhao H, Li G, Fan FY (2009) The global challenge of electronic waste management. Environ Sci Pollut Res 16:248–249

Lundgren K (2012) International Labor Office (ILO). The global impact of ewaste: addressing the challenge. 2012. http://www.ilo.org/wcmsp5/groups/public/ed_dialogue/—sector/documents/publication/wcms_196105.pdf

Ma J, Cheng J, Wang W, Kunisue T, Wu M, Kannan K (2011) Elevated concentrations of polychlorinated dibenzo-p-dioxins and polychlorinated dibenzofurans and polybrominated diphenyl ethers in hair from workers at an electronic waste recycling facility in Eastern China. J Hazard Mater 186:1966–1971

Menikpura SNM, Santo A, Hotta Y (2014) Assessing the climate co-benefits from Waste Electrical and Electronic Equipment (WEEE) recycling in Japan. J Clean Prod 74:183–190

Morf LS, Tremp J, Gloor R, Schuppisser F, Stengele M, Taverna R (2007) Metals, non-metals and PCB in electrical and electronic waste – actual levels in Switzerland. Waste Manag 27:1306–1316

Nasim W, Ahmad A, Wajid A, Akhtar J, Muhammad D (2011) Nitrogen effects on growth and development of sunflower hybrids under agro-climatic conditions of Multan. Pak J Bot 43:2083–2092

Nasim W, Ahmad A, Hammad HM, Chaudhary HJ, Munis MFH (2012) Effect of nitrogen on growth and yield of sunflower under semiarid conditions of Pakistan. Pak J Bot 44:639–648

Nasim W, Belhouchette H, Tariq M, Fahad S, Hammad HM, Mubeen M, Munis MFH, Chaudhary HJ, Khan I, Mahmood F, Abbas T (2016) Correlation studies on nitrogen for sunflower crop across the agroclimatic variability. Environ Sci Pollut Res 23:3658–3670

Nasim W, Ahmad A, Amin A, Tariq M, Awais M, Saqib M, Jabran K, Shah GM, Sultana SR, Hammad HM, Rehmani MIA (2018) Radiation efficiency and nitrogen fertilizer impacts on sunflower crop in contrasting environments of Punjab-Pakistan. Environ Sci Pollut Res 25:1822–1836

OECD (2001) Extended producer responsibility. A guidance manual for governments [online]. http://www.oecdilibrary.org/docserver/download/fulltext/9701041e.pdf?expires=340015892and id=idandaccname=oid009239andchecksum=EA57EDD01263DC7C70DD40F70D4756

Ogushi Y, Kandlikar M (2007) Assessing extended producer responsibility laws in Japan. Environ Sci Technol 41:4502–4508

Puckett J, Byster L, Westervelt S, et al (2014) Exporting harm: the high-tech trashing of Asia. Basel Action Network, Silicon Valley Toxics Coalition. http://www.ban.org/Ewaste/technotrashfinalcomp.pdf

Robinson BH (2009) E-waste: an assessment of global production and environmental impacts. Sci Total Environ 408:183–191

Schluep M, Hageluken C, Kuehr R (2014) Sustainable innovation & technology transfer industrial sector studies: recycling from e-waste to resources. United Nations Environment Programme UNEP and StEPe solving the e-waste problem, 2009

Sepulveda A, Schluep M, Renaud FG, Streicher M, Kuehr R, Hageluken C, Gerecke AC (2010) A review of the environmental fate and effects of hazardous substances released from electrical and electronic equipments during recycling: examples from China and India. Environ Impact Assess Rev 30:28–41

Shen C, Chen Y, Huang S, Wang Z, Yu C, Qiao M, Xu Y, Setty K, Zhang J, Zhu Y, Lin Q (2009) Dioxin-like compounds in agricultural soils near e-waste recycling sites from Taizhou area, China: chemical and bioanalytical characterization. Environ Int 35:50–55

StEP Initiative. StEP Annual Report (2013). http://stepinitiative.org/tl_files/step/StEP_AR/StEP_AR.html

The Basel Action Network (BAN) and Silicon Valley Toxics Coalition (SVTC) (2002). Exporting Harm: The High-Tech Thrashing of Asia

Thompson S, Oh S (2006) Do sustainable computers result from design for environment and extended producer responsibility? Analyzing e-waste programs in Europe and Canada. In Proceedings of the International Solid Waste Association's 2006 annual congress, Copenhagen, Denmark

Toyasaki F, Boyacı T, Verter V (2011) An analysis of monopolistic and competitive take back schemes for WEEE recycling. Prod Oper Manag 20:805–823

Tue NM, Sudaryanto A, Minh TB, Isobe T, Takahashi S, Viet PH, Tanabe S (2010) Accumulation of polychlorinated biphenyls and brominated flame retardants in breast milk from women living in Vietnamese e-waste recycling sites. Sci Total Environ 408:2155–2162

Tue NM, Takahashi S, Suzuki G, Isobe T, Viet PH, Kobara Y, Seike N, Zhang G, Sudaryanto A, Tanabe S (2013) Contamination of indoor dust and air by polychlorinated biphenyls and brominated flame retardants and relevance of non-dietary exposure in Vietnamese informal e-waste recycling sites. Environ Int 51:160–167

UNEP (United Nations Environment Program) (1992) Basel Convention on the control of transboundary movements of hazardous wastes and their disposal. http://www.basel.int/Portals/4/Basel%20Convention/docs/text/

UNEP (United Nations Environment Program) (2007) Division of technology, industry, and economics, international environmental technology center. E-waste, vol I: Inventory assessment manual, 2007. http://www.unep.org/ietc/Portals/136/Publications/Waste

UNU (United Nations University) (2013) Solve the E-waste Problem (StEP), Massachusetts Institute of Technology (MIT), National Center for Electronics Recycling (NCER). World e-waste map reveals national volumes, international flows

Wang HM, Yu YJ, Han M, Yang SW, Yang Y (2009) Estimated PBDE and PBB congeners in soil from an electronics waste disposal site. Bull Environ Contam Toxicol 83:789–793

Wang Y, Luo C, Li J, Yin H, Li X, Zhang G (2011) Characterization of PBDEs in soils and vegetations near an e-waste recycling site in South China. Environ Pollut 159:2443–2448

Widmer R, Oswald-Krapf H, Sinha-Khetriwal D, Böni H, Schnellmann M (2005) Global perspectives on e-waste. Environ Impact Assess Rev 25:436–458

Wong CS, Duzgoren-Aydin NS, Aydin A, Wong MH (2007) Evidence of excessive releases of metals from primitive e-waste processing in Guiyu, China. Environ Pollut 148:62–72

Wu Q, Leung JYS, Geng X, Chen S, Huang X, Li H, Huang Z, Zhu L, Chen J, Lu Y (2015) Heavy metal contamination of soil and water in the vicinity of an abandoned e-waste recycling site: implications for dissemination of heavy metals. Sci Total Environ 506–507:217–225

Yoshida F, Yoshida H (2010) Japan, the European Union, and waste electronic and electrical equipment recycling: key lessons learned. Environ Eng Sci 27:21–28

Zoeteman BC, Krikke HR, Venselaar J (2010) Handling WEEE waste flows: on the effectiveness of producer responsibility in a globalizing world. Int J Adv Manuf Technol 47:415–436

Chapter 6
E-waste: Global Scenario, Constituents, and Biological Strategies for Remediation

Srujana Kathi and Anbarashan Padmavathy

Abstract Global technology development and industrialization have led to the increased usage of electronic gadgets. Electronic waste or e-waste is one of the emerging environmental issues in the developing countries. Much of the e-waste globally generated is recycled in the unregulated informal sector and results in significant risk to environmental health. These wastes also consist of economically valuable minerals such as copper, silver, and gold. The multitude of toxic heavy metals present in the components of discarded electrical and electronic equipment such as cadmium, arsenic, antimony, chromium, lead, mercury, selenium, beryllium, brominated flame retardants, PAHs, and PCBs pose threats to the environment. The usage of microbes and plants in minimizing the toxicity of chemicals and metals in the environment is eco-friendly and cost effective. This chapter provides a concise overview of the volume of e-waste generated globally, disposal and reuse/recycle practices; forecasts e-waste production, and discusses environmentally sustainable remediation strategies. The principles, advantages, and disadvantages of bioleaching, biosorption, bioaccumulation, bioprecipitation, biomineralization, and phytoremediation techniques, which are recognized as biological strategies for remediation of contaminants released into different environmental matrices are presented.

Keywords Bioaccumulation · Bioleaching · Biomineralization · Bioprecipitation · Biosorption · Electronic waste · Phytoremediation

6.1 Introduction

Electronic wastes (e-wastes) assume growing pollution concerns globally, due to the occurrence of a range of toxic substances (Kiddee et al. 2013). The pollution of soil ensuing from offensive e-waste recycling might appreciably alter the soil microbiota

S. Kathi (✉) · A. Padmavathy
Department of Ecology and Environmental Sciences, Pondicherry University, Puducherry, Kalapet, Puducherry, India

M. Z. Hashmi, A. Varma (eds.), *Electronic Waste Pollution*, Soil Biology 57,
https://doi.org/10.1007/978-3-030-26615-8_6

(Liu et al. 2015). Unrestrained e-waste recycling releases xenobiotic compounds as well as heavy metals into the environment. These objectionable compounds may be disseminated, bioaccumulated/biomagnified resulting in deleterious effects on human health (Man et al. 2013). Generally, many of the persistent materials are capable of undergoing long-range transport via air/water and pressurize the ecosystems and living organisms located at far of distances from the recycling areas (Wang et al. 2011). Farmland soils collected from nearby roads and the dismantling workshops of e-waste recycling areas in South China showed that the concentrations of polybrominated diphenyl ether (PBDE) were subject to complex environmental processes (Luo et al. 2009).

Conventional physical dismantling methods such as crushing, jigging, shaping, and electrostatic separation are in wide usage for handling PCBs. Acid washing is a hydrometallurgical method. Heating and smelting are the pyrometallurgical technique (Awasthi et al. 2016a). E-waste recycling done in workshops include removing the outer coverings, shredding into pieces, changing their state by melting, flaming, discarding the non-usable components, and trading. These activities might enhance contamination of the ambient environment by heavy metals. Moreover, contamination of the surface soil may be limited by the extent of recycling area, type of soil, and physicochemical properties of soil (Fujimori and Takigami 2014). Unsophisticated recycling e-wastes by physical separation and incineration lead to accumulation of heavy metals in the environment (Kyere et al. 2017). Open burning and open storage of e-wastes should be barred and sustainable wastewater treatment strategies should be implemented at each workshop to decrease contamination by dioxin-like compounds from e-waste (Suzuki et al. 2016). Urban biomining implies the use of microorganisms for extracting and recovering metals from terminated products, electronic and electrical wastes, and exhausted batteries (Nancharaiah et al. 2016).

Inadequate legislations coupled with unemployment paved the way for quick rise of informal e-waste handling centers in developing countries. Contaminants move through the food chain by means of root plant translocation system and to the human body thereby threatening human health (Awasthi et al. 2016a, b). Toxic heavy metals that get accumulated in the water and sediments can have hazardous implications on associated life forms. We understand that surface as well as bottom sediments offer habitat and food for all kinds of associated organisms. There are chances for either direct or indirect contamination of aquatic ecosystems by heavy metals leading to bioaccumulation. Moreover, atmospheric deposition of heavy metals by sedimentation could impact human health as well (Kyere et al. 2017). Abandoned e-waste recycling areas pose serious ecological risk. One of the best possible solutions to address e-waste contamination might be remediation of polluted soil and water (Wu et al. 2015). Risks factors associated with unsustainable waste management practices offers scientific recycling as a viable alternative practice (Echegaray and Hansstein 2017). EPA has estimated globally an increase of 5–10% in e-waste generation per year. Our challenge lies in identifying and developing most innovative, cost-effective, and eco-friendly solutions; this method can be used to decontaminate the polluted environments and convert it into proper functioning ecosystems, which can support life for present and future generations. The present article summarizes the role of microbes in innovative alternative clean-up technologies for the removal of e-waste. Currently, the world is in the quest for novel approaches that could address escalating forms of anthropogenic wastes. Extended Producer Responsibility (EPR) is one such move toward e-waste management. As defined by OECD "EPR is an environmental policy approach in which a producer's responsibility for a product is extended to the post-consumer stage of the product's life cycle, including its final disposal" (OECD 2001). Subsequent to the "Polluter-pays Principle," it might be comprehended according to the EPR policy, considering the environmental impacts of the products manufactured, the expenditures on treatment and disposal of the end products should be incorporated into the market price of the product.

6.2 State-of-the Art Picture of E-waste Management

Today, globally waste produced from Electric and Electronic Equipment's (WEEEs) is reported to be growing annually greater than ever from 3% to 5% (Cucchiella et al. 2015). It is estimated that globally approximately 49.8 million tons of e-waste is generated in 2018 (Cui and Anderson 2016). In India, it was jointly estimated by UNEP and United Nations University (UNU) by the year 2020, e-waste from old computers might record a growth of 500% (Borthakur and Govind 2017). About 6.8 kg of e-waste per person was generated in 2012 according to the estimates of UNEP. The transboundary movement of e-waste from the Northern hemisphere to the Southern hemisphere poses a serious threat to the environment. Several urban areas in Africa have become containers for rejected e-waste from developed countries. As a matter of fact, the industry surrounding e-waste dismantling is known to ruin the quality of ambient environment (Daum et al. 2017). Here, it is apparent to understand global e-waste trading. Therefore, significant regulations on transboundary movement of e-waste from OECD countries to non-OECD countries were imposed by the Basel convention (Wath et al. 2011). To enable environmentally friendly management of used and end-of-life computing equipment, the Partnership for Action on Computing Equipment (PACE) has been initiated by the Parties to the Basel Convention on the Control of Transboundary Movements of Hazardous Wastes and Their Disposal. PACE has framed guidelines on deciding the basis of the functionality of end-use computers and computer components (Perkins et al. 2014; UNEP 1992). However, many independent studies suggest higher concentrations of Persistent Organic Pollutants (POPs), such as polychlorinated biphenyls (PCBs), in subtropical and tropical regions of the world (Breivik et al. 2015). One of the first countries to implement formal e-waste management regimen was Switzerland that recycled 11 kg/capita of WEEE against the 4 kg/capita goal set by EU (Wath et al. 2010). As a matter of fact, the discarded end-of-life computers from Japan are reused in China, while home appliances from Japan are reused in Southeast Asia. Furthermore, we understand that the e-waste scrap produced in Asia

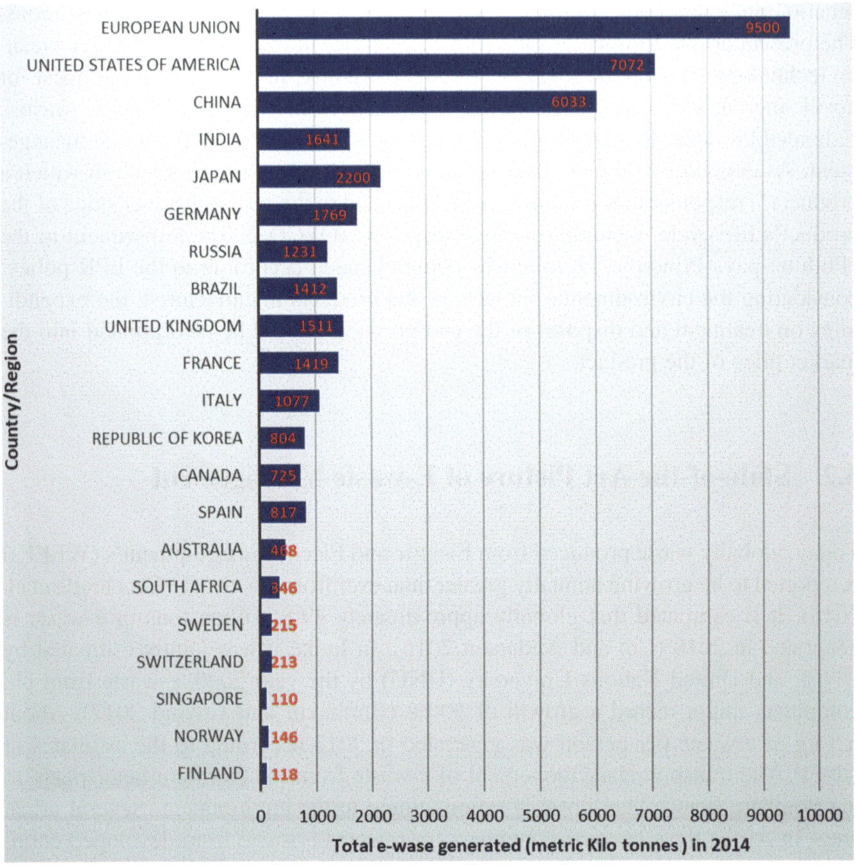

Fig. 6.1 Global e-waste generation based on UNU 2015

is recycled in Guangdong Province of China (Shinkuma and Nguyen 2009). According to an estimate of the European Environment Agency, annually about 1.3 million tons of rejected EEE are shipped from the EU to Asia and Africa (Lundgren 2012). Instances of water contamination have been reported in areas close to e-waste recycling towns in China. Furthermore, escalated levels of dissolved metals were found contaminating the sediments in the rivers nearby e-waste recycling town of Guiyu (Chan and Wong 2013). Mixed contamination of organic and inorganic contaminants in the soil was identified in several studies owing to unscientific recycling of emerging anthropogenic wastes. Presently, there is an urgent need for transformation of land use patterns likely to pose threat to the human health as well as wildlife. Consequently, novel soil remediation techniques that can address these issues are essential (Ye et al. 2015). Subsequently, it is imperative to consider the assorted global experiences and advance comprehensive schedules to discourse the e-waste catastrophe pertaining to any growing economy with prime importance (Borthakur and Govind 2017) (Fig. 6.1).

6.3 E-waste Generation in India

In the Indian context, formulating a single widely accepted e-waste management strategy is quite challenging owing to the presence of a vast diaspora of sociocultural, economic, political, technological, infrastructural, and environmental considerations. However, instead of learning from extraneous experiences in e-waste management we can study from our own proficiencies and frame an inclusive strategy for sustainable e-waste supervision. The highest contribution to waste electrical and electronic equipment (WEEE) in the country is given by West Bengal, Uttar Pradesh, Tamil Nadu, Delhi, Andhra Pradesh, Maharashtra, Karnataka, Punjab, Gujarat, and Madhya Pradesh. Among the total e-waste generated in the country, southern, northern, and eastern regions account for 30%, 21% and 14%, respectively, while western India accounts for the largest volume at 35% (Needhidasan et al. 2014). However, the existing practices of storage, processing, recycling, and disposal of e-waste handling are likely to harm human health and the environment (Borthakur 2015). In India, approximately an 18-fold increase in the usage of mobile phones is expected by 2020 (Perkins et al. 2014). Therefore, framing strategies and implementation in the area of solid waste management is supposed to be one of the most demanding environmental glitches facing local governments in urban India (Cornea et al. 2017). The sale of secondhand electronics items operated from the outskirts of the megacities has a profitable market in the country (Ongondo et al. 2011). Though presently almost 138 formal e-waste recycling setups have come up to mitigate the e-wastes, an appreciable quantity of generated e-wastes is not reaching the recycling centers (CPCB 2014). No validated data regarding e-waste generation in the country is available till date. Consequently, a research gap is visualized on the extrapolation of e-waste generation and possibilities of projected waste disposal patterns in country (Dasgupta et al. 2017).

Enormous quantities of e-waste from developed countries is imported to developing countries like India. Among the total e-waste generated in India 65 cities generate more than 60%. Therefore, the Indian government plans to legislate new regulations that make a producer/manufacturer of equipment accountable for the gathering and proper disposal of e-waste generated when the product is discarded. As per the reports of Confederation of Indian Industries, the total out-of-date electronic and electrical equipment waste generated in India has been valued to be 1,46,000 tons per year. In this case, 22% PCs e-waste generated by households, 78% accounts business sector, approximately 1050 tonnes per year of computer wastes comes from manufacturing as well as retailing sectors. The management and handling of waste in India is currently governed by legislation such as the Environment (Protection) Act of 1986, and Hazardous Material (Management, Handling and Transboundary Movement) Rules of 2008 (Chauhan and Upadhyay 2015). Usage patterns of various e-items, e-waste generation, and disposal, jointly designed a composite concern where buyers desire to reap maximum profit from the clearance arrangement such as reusing and recycling (Dasgupta et al. 2017).

6.4 Constituents of E-wastes

The main components of e-wastes include wood, glass, plastics, ferrous and non-ferrous metals, rubber, concrete and ceramics, plywood, printed circuit boards (PCB), and other items. The presence of elements like arsenic, cadmium, lead, mercury, selenium, and hexavalent chromium and flame retardants beyond threshold quantities of e-waste classifies them as hazardous wastes. Among the several metals associated with discarded equipment, iron and steel constitute about 50% of the total e-waste generated followed by plastics (21%), non-ferrous metals (13%) including other constituents. Non-ferrous metals consist of metals like aluminum (Al), copper (Cu), and precious metals, for example, gold (Au), silver (Ag), palladium, and platinum (Needhidasan et al. 2014). Hazardous heavy metals and organic compounds, mainly polycyclic aromatic hydrocarbons (PAHs), polychlorinated and polybrominated dibenzo-*p*-dioxins (PCDDs and PBDDs), hexabromo cyclododecanes (HBCDs), polybrominated diphenyl ethers (PBDEs), polychlorinated and polybrominated biphenyls (PCBs and PBBs), dechlorane plus (DP), and polychlorinated and polybrominated dibenzofurans (PCDFs and PBDFs) are also found more than permissible levels at the e-waste dumping sites (Pramila et al. 2012). Printed circuit boards (PCB) of mobile phones and computers are mostly copper and other metal-containing materials. The existence of pollutants in mixtures in the nature renders pollution abatement a complex task, especially when a gradient of contamination reaches to the agricultural soils from the hot spots of contamination (Zhang et al. 2012) (Fig. 6.2).

PCBs can associate with more than 40 elements, including some hazardous ones: Co, In, Zn, Al, Pb, Ag, B, C, H, K, N, O, F, S, Au, Ti, Pd, Fe, Mn, Cu, Ni, Sn, and Sb as well as precious metals such as gold, silver, and palladium (Arshadi and Mousavi

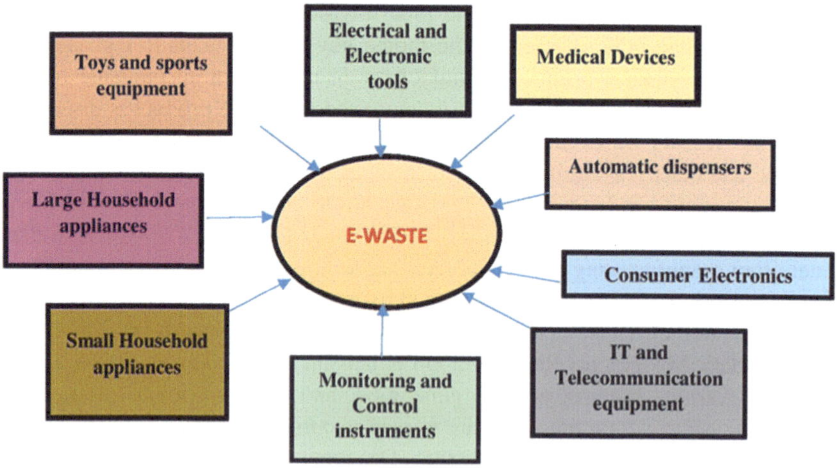

Fig. 6.2 Sources of e-waste generation

Table 6.1 Components of e-wastes

Components	Constituents
Flat screen monitors and switches	Mercury
Printed circuit boards	Lead oxide and cadmium
Cathode ray tubes	Lead and cadmium
Plastics from printers, key boards, monitors	Phthalates
Capacitors and electrical transformers, printing inks	Polychlorinated biphenyls
Plastic cable casings, printed circuit boards	Brominated flame retardants
Cable insulation/coating	Poly Vinyl Chlorides (PVC)
Batteries used in computers	Cadmium

2015). Plastics present in cell phones naturally contain a blend of polycarbonate and polyamide. PBDEs belong to brominated flame retardants (BFRs) that can be utilized in numerous polymeric materials like circuit boards, casings, wires, cable coatings, textiles, resins, and additional substrates to decrease their fire-related hazards. The fourth conference of the parties conducted during May 2009, amended the Stockholm Convention on persistent organic pollutants (POPs) by adding two groups of PBDEs namely Hexa- and Hepta-BDEs, and Tetra- and Penta-BDEs to Annex A (Elimination) of the Convention (Anh et al. 2017). E-waste comprehends 60% of metals, about 30% plastics, and the rest hazardous pollutants of 2.70% (Widmer et al. 2005). Since the lifetime of a computer is around 2–5 years, it is projected that around 17 million computers are thrown away annually. Supposing that the mass the central processing unit (CPU) of computers is around 3 kg, we can evaluate that approximately 50 million kilograms of CPUs get rejected annually. The amount of gold in 1 ton of PCBs from computer amounts to ~17 tons of ores where return outcomes display ~40 wt % of the output was copper (Arshadi and Mousavi 2015) (Table 6.1).

6.5 Effects on Soil Microbial Communities

Soil microbes perform dynamic functions in the ecosystems. These communities may be strongly organized by land-use patterns related to e-waste recycling, which can surge the heavy metal concentration in soils (Wu et al. 2017). Basic e-waste recycling activities discharge enormous quantities of persistent organic pollutants (POPs) and heavy metals into ambient soils, affectating considerable peril to the ecosystems and human health. Microbes capable of metabolizing POPs in soils play essential roles in POPs remediation (Liu et al. 2015). Bioremediation can expand the e-waste management horizons in a sustainable manner. All such strategies should focus on the organic and inorganic portions of the e-wastes. Organic fraction composed of a variety of thermo- and thermosetting plastics. These plastics might be halogenated where soil microbes get involved in the process of dehalogenation. Leaching of inorganic portions of both metallic and non-metallic components can be managed by microbes (Pant et al. 2018). High persistence, toxicity, and bioaccumulation resulted in higher attention for compounds such as PCBs. PBDE patterns from an e-waste recycling site in South China implied that the PBDEs in farmland soils have been subjected to complex environmental processes (Luo et al. 2009). *Microbes such as Solibacter*, *Nitrososphaera*, and *Nitrospira* dominated in e-waste recycling sites (Wu et al. 2017). High-throughput 16S rRNA gene sequencing showed that *Acidobacteria, Bacteroidetes, Chloroflexi, Deltaproteobacteria*, and *Firmicutes* dominated the river sediment microbial assemblages consistently polluted by e-waste (Liu et al. 2018). Enhanced PCB degradation was due to the amplified PCB biostimulation of microbial communities subsequent to the accumulation of plantation and β-cyclodextrin. Besides, experimental results suggested that PCB elimination was mainly subsidized by microbial degradation rather than plant uptake or abiotic dissipation (Luo et al. 2009). The diversity of bacterial communities in the soil around an e-waste recycling workshop in the Taizhou in China was reported gradual change in soil bacterial diversity along the polychlorinated biphenyls pollution gradient (Tang et al. 2013). Concentrations of most of the heavy metals such as cadmium (Cd), chromium (Cr), lead (Pb), nickel (Ni), and zinc (Zn) in soil, water, and plant samples during the wet and dry seasons in and around the largest e-waste dumping site in Nigeria, Alaba International Market in Lagos exceeded maximum permissible levels (Olafisoye et al. 2013).

6.6 Biological Remediation Strategies

Chemical methods assure advanced recovery of metal within a short duration. Requirement of huge quantities of chemical reagents, risks involved in operation and handling including chances for secondary contamination can be understood as the drawbacks in chemical methods. Though physical or mechanical method can rapidly recycle electronic wastes they are costly and it is tough to detach impurities from metals. Even from depleted low-grade wastes, we can recover metals using relatively simple operational technology, energy efficiency, low cost, environmental compatibility, and efficiency in recovery of metals (Priya and Hait 2017; Awasthi et al. 2016a). The occurrence of organic pollutants and epoxy layer coatings on metals should be detached to render metals accessible for microbial attack to ensure quicker extraction and recovery. Microorganisms such as bacteria, yeast, and algae are identified to persist in waste-holding environs due to its aptitude to decrease, amass, confiscate, captivate, and oxidize dissimilar kinds of waste materials making it meagerly soluble and effortlessly precipitated so that their toxicity gets reduced (Varjani et al. 2018) (Fig. 6.3).

Biological methods such as bioleaching, biosorption, bioaccumulation, biotransformation, bio-oxidation, biomineralization, and microbially improved chemisorption of metals can be used in the removal of toxic metals from e-wastes (Patel and Kasture 2014; Dixit et al. 2015). Though bacteria are competent in leaching metals from electronic wastes, the capability of fungus and algae is not much explored in this field. Since the biosorption capacity of fungus is quite efficient for heavy metals,

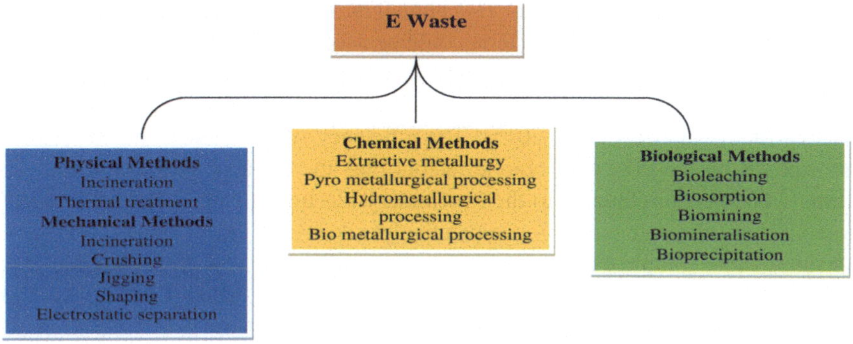

Fig. 6.3 Strategies applicable for the handling e-wastes

bioleaching studies can be performed with fungus, which has efficient enzymes to enhance utilization (Chatterjee and Abraham 2017). Bioaccumulation is the active mechanism of metal uptake, while biosorption is the passive mode in which the dead biomasses of the microorganisms are utilized for sorption (Volesky 1990). Bio-oxidation is the process of extraction of metals, mainly gold from ores, by oxidizing the matrix in which the metals are fixed. Basically, the metals are made available for extraction in this process. In large-scale stirred tanks bio-oxidation process recycled to discharge gold (Nancharaiah et al. 2016).

6.7 Bioleaching

Microbial leaching is a pragmatic method for extracting metals in solid state from e-waste. The natural capability of microbes to oxidize or utilize organic and inorganic substrates is exploited for melting the metals. Autotrophic bacteria, heterotrophic bacteria, and heterotrophic fungi are the three key microbial groups involved in this mechanism (Awasthi et al. 2016a; Karwowska et al. 2014). We can understand bioleaching as a process in which microbes interact with the metals for extraction and purification (Gerayeli et al. 2013). Chemical and biological leaching has its own pros and cons. Technical, economic, and environmental reasons are important for choosing one process over the other. Hybrid methodology helps to overcome the glitches associated with chemical and biological extraction techniques for the metals present in e-waste (Needhidasan et al. 2014). Acidophilic microorganisms are vital for leaching heavy metals from wastes (Hong and Valix 2014). They thrive in acidic pH (2.0–4.0) and help in dissolving the metals. Among the bacteria, *Acidithiobacillus ferrooxidans, Acidithiobacillus thiooxidans, Leptospirillum ferrooxidans*, and *Sulfolobus* sp., are well-known consortia for the bioleaching activity while fungal genera such as *Penicillium* and *Aspergillus niger* help in metal leaching process (Mishra and Rhee 2010). Bioleaching, involving microorganisms such as *Chromobacterium violaceum*, may allow metal recycling (Tay et al.

2013). The bacteria isolated from *Hymeniacidon heliophila* sponge cells showed bioleaching activity and these bacteria could be used as a copper recovery tool from e-waste (Rozas et al. 2017). Au tolerance of *Aspergillus* species could be a suitable indicator for selecting filamentous fungi able to cause bioleaching of gold from WEEE (Madrigal-Arias et al. 2015). The efficiency of the bioleaching process is predisposed by the effects of abiotic factors plus the metallic composition of the waste, with special reference to their secondary reactions (Valix 2017). Biological leaching, on the other hand, is comparatively a low-cost method while it is time consuming and complete recovery of the metal may not be practically feasible sans biological leaching (Pant et al. 2012). Biochar with redox activity can be used to enhance the bioleaching efficiency of metals from a basic e-waste in order to avoid the disadvantages of being time consuming or having a moderately low efficiency. Iron-mediated bioleaching was significantly promoted by biochar and its leaching time was decreased by one-third compared with that of a biochar-free system (Wang et al. 2016). Fungi interact with gold by mobilizing it through a mechanical attack as well as through biochemical leaching by the production of cyanides. Moreover, fungi are also able to release Au through the degradation of cyanide from aurocyanide complexes. Subsequently, fungi can localize gold through biosorption, bioaccumulation, and biomineralization as nanoparticles. The diversity of mechanisms of gold recycling using fungi combined with their filamentous lifestyle, which allows them to thrive in heterogeneous and solid environments such as e-waste, makes fungi an important bioresource to be harnessed for the recovery of gold (Bindschedler et al. 2017) (Table 6.2).

6.8 Biosorption

Biosorption is a physicochemical and metabolically independent process. Mechanisms include absorption, adsorption, ion exchange, surface complexation, and precipitation (Fomina and Gadd 2014). It is a biotechnological revolution, low-cost tool for recovery of precious metals from aqueous solutions. Diverse natural biomaterials such as algae, fungi, bacteria, actinomycetes, yeast; biopolymers, and biowaste materials are acknowledged to bind to the metals (Das and Das 2013). The benefits of an ideal biosorption process include low cost, short operation time, absence of toxicity limitations, absence of requirements for nutrients, avoidance of sudden death of biomass, and easy mathematical modeling of metal uptake by reactors. Passive biosorption also has some disadvantages, such as early saturation and limitation in biological process (Hansda and Kumar 2016). Biomaterials such as rice husk, coconut shell, plant barks, leaves, sawdust, sugarcane bagasse, and peat moss have received much interest (Michalak et al. 2013). The mechanism of biosorption is complex. Biological ligands, biosorbents, characteristics of the targeted metals and characteristics of the metal solution drive these mechanisms (Deng and Wang 2012). Selective biosorption of gold from printed circuit board was achieved by using the combination of ammonium thiosulfate (AT) and *Lactobacillus*

Table 6.2 List of major microbes and metal extraction mechanisms

Metal chelating microbial groups	Name of organisms	Metal extracted	Mechanism	References
Iron oxidizers	*Acidithiobacillus ferrooxidans* *Leptospirillum ferrooxidans* *Leptospirillum ferriphillum*	Cu, Zn, Ni, Pb, Cd	$2Fe^{2+} + \frac{1}{2}O_2 + H^+ \rightarrow 2Fe^{3+} + H_2O$ $Cu^0 + 2Fe^{3+} \rightarrow Cu^{2+} + 2Fe^{2+}$	Dave et al. (2018), Awasthi et al. (2016a), Bas et al. (2013)
Sulfur oxidizers	*Acidithiobacillus thiooxidans* *Sulpholobus* spp.	Cu, Zn, Ni, Al	$2S^0 + 3O_2 + 2H_2O \rightarrow 4H^+ + 2SO_4^{-2}$	Dave et al. (2018)
	Leptospirillum ferriphilum *Sulfobacillus thermosulfidooxidans*	Cu		Wu et al. (2018)
Cyanogenic organisms	*Chromobacterium violaceum Pseudomonas aeruginosa* *Pseudomonas florescence* *Marasmius oreades*	Au, Ag, Pd, Pt	$NH_2CH_2COOH \rightarrow CN^- + CO_2 + 4H^+$	Ilyas et al. (2014)
	Bacillus megaterium	Au, Cu		Arshadi and Mousavi (2015)
	Chromobacterium violaceum	Au		
Organic acids producers	*Aspergillus niger* *Penicillium simplicissimum*	Cu, Zn, Ni, Pb, V, Mo, Al, Co, Li		Dave et al. (2018)
	Acidithiobacillus thiooxidans and Acidithiobacillus ferrooxidans	Cu		Liang et al. (2010)

acidophilus. This process might include a unique leaching-sorption method for gold recovery from AT leachant by the proposed combination (Sheel and Pant 2018). Gold biosorption from leached solutions using chitin as biosorbent enabled about 80% of the gold recovery. This study reported that precipitation followed by biosorption resulted in the best gold recovery, because other species were removed from the leachate solution in the precipitation step (Cortes et al. 2015). Green algae are highly abundant biomaterials employed as useful biosorbents in many studies. *Cladophora hutchinsiae* the biosorbent used for removal of highly toxic chemical such as uranium (Bagda et al. 2017). Novel bacterial strain *Halomonas* BVR 1 isolated from electronic industry effluent is very effective in adsorbing cadmium (Rajesh et al. 2014). Biosorption technique using agricultural or industrial wastes can be employed to recover heavy metals from leachate followed by their precipitation as hydroxides. Combinatorial approach of these two processes provides solutions to the shortcomings of metal leaching. Amended cellulosic materials with specific functional groups, such as carboxyl, amino, and sulfo, are considered conductive to the sustainable development and long-term economic interests (Varshney et al. 2017).

6.9 Biomining

Biomining is an industrial-scale biotechnological process meant for extracting base metals from sulfidic ores. Presently, biomining accounts for about 20% and 5% of the world production of Cu and Au, respectively. Industrial biomining uses oxic conditions and acidophilic iron- or sulfur-oxidizing microorganisms (e.g., *Acidithiobacillus ferrooxidancs*, *At. thiooxidans*). These microbes produce ferric iron which is an oxidant and sulfuric acid, the promoter of dissolution (Nancharaiah et al. 2016). Biomining is now used chiefly to leach copper sulfides and as an oxidative pretreatment for refractory gold ores, though it is also used to recover other base metals, such as cobalt, nickel, and zinc. Current developments have included using acidophiles to process e-wastes, to extract metals from oxidized ores and to selectively recover metals from process waters and waste streams (Johnson 2014). Chemolithotrophs such as *Acidithiobacillus and Leptospirillum* are the most commonly used genera for the biomining of metals from dust. *Acidithiobacillus ferrooxidans* extracted up to 70% of Zn from Fe–Mn alloy industrial dust by oxidative bioleaching (Sethurajan et al. 2018). The microbial solubilization of metals by biomining is positively used in industrial processes to extract numerous metals such as copper, gold, and uranium (Jerez 2017). Biomining organisms are able to grow lithotrophically by oxidizing ferrous iron and/or elemental sulfur as electron donors to produce ferric iron and sulfuric acid, which attack sulfide minerals. Some biomining microbes can also grow autotrophically using CO_2 for growth, whereas others are heterotrophic and thus need an organic carbon source (Schippers 2007). The most frequently used tactics for commercial-scale biomining are based on bioreactors, heaps, and dumps (Kaksonen et al. 2017). Two other engineering designs are used in biomining operations. Stirred tank bioreactors are used almost exclusively for bio-oxidation of refractory gold concentrates in which the fine gold particles are enshrouded by sulfide minerals. Most current commercial operations use the BIOX® process and operate as continuous feed systems, processing between 40 and >8000 tons of concentrate/day (Brierley and Brierley 2013; Rawlings 2002). All biomining operations work under non-sterile conditions which preclude or restrict the use of genetically engineered microorganisms. While pure cultures of bacteria or archaea can degrade sulfide minerals, it is now well recognized that bioleaching and biooxidation are mediated by consortia of acidophilic prokaryotes. These have been categorized as: firstly, ferric iron-generating prokaryotes which produce the mineral oxidant; secondly, sulfuric acid-generating autotrophs which maintain the low pH environment essential; and lastly, heterotrophic and mixotrophic prokaryotes, which degrade organic compounds seeped from autotrophic iron oxidizers and sulfur oxidizers, thereby escaping potential toxicity issues (Johnson and Hallberg 2008). Erüst et al. (2013) reviewed the possible applications of biohydrometallurgy to recover metals from spent batteries and catalysts.

6.10 Bioaccumulation

Bioaccumulation is a biotechnological approach for the removal of heavy metals where living biomass takes up and transports heavy metals. It relies on intrinsic biochemical and structural properties, physiological and genetic adaptation, environmental modification of metal specification, availability, and toxicity (Jaafar et al. 2015). It is a toxicokinetic process that affects the sensitivity of living organisms to chemicals. Organisms can normally resist concentrations of chemicals up to certain levels, beyond which these chemicals become toxic and endanger the organism (Mishra and Malik 2013). Efforts on engineering metal-resistant microorganisms have been conducted mainly on mesophilic strains for bioremediation purposes, such as genetically engineered *Escherichia coli* JM109 for the accumulation of mercury, nickel, and cadmium by overexpression of a metal transport system and a metallothionein or a polyphosphate kinase and metallothionein, a phytochelatin synthase gene, and a heavy metal ATPase gene. Genetically modified strains of *Corynebacterium glutamicum* for arsenite accumulation *E. coli* BL21 for cadmium accumulation by incorporating glutathione synthesis genes, a serine acetyltransferase gene. *Pseudomonas putida X4* for cadmium biosorption by surface display of metallothionein (Gumulya et al. 2018). During augmented metal exposure, bacteria produce metal-binding proteins such as metallothioneins that can bind with metals and transport them into the cell. Several organisms have been reported for the study of bioaccumulation for increased levels of pollutants, including plants, fungi, fish, algae, mussels, oysters, and bacteria (Mosa et al. 2016). The exceptional ability of the GeoChip microarray technique coseals 424,000 genes in over 4000 functional groups elaborated in various significant biological processes. Further, this technique can be applied to search the bioaccumulation capability of several microbial groups to uranium (Van Nostrand et al. 2009).

6.11 Bioprecipitation

Alteration in the ionic equilibrium of metals owing to the addition of certain chemicals. This process is unlike coagulation and flocculation. Concentrations of the metals in the solution and pH determine the removal efficiency of metals. Precipitation includes stages like nucleation, growth of nucleus, aggregation, and crystallization. The chief drawbacks of these processes are sludge generation and pH adjustment (Sethurajan et al. 2018). Ultimately, this process results from the excretion of special proteins like thiol groups rich special proteins/proteins with less molecular weight namely metallothioneins and phytochelatins from bacteria that reacts with metal ions present in solution to yield insoluble metal compounds (Hansda and Kumar 2016). The possibility of selective recovery of metals from multimetallic leachate can be understood as a significant merit of bioprecipitation systems. By altering, the system pH, metals can be selectively precipitated as metal sulfides, since metal sulfide formation is dependent on the solution pH (Sethurajan et al. 2018).

6.12 Biomineralization

This is a process in which harmful metal ions associate with anions or ligands formed from the microbes to form precipitation (Tabak et al. 2005). Applications suggest the potential to remove these metals from contaminated water as well as sequester them in a reusable form. Biomineralization can be studied with respect to phylogenetically diverse microorganisms isolated from pristine and contaminated environments (Lens 2016). Studies including fungal genera reported that gold can be immobilized by biosorption, bioaccumulation, and biomineralization (Bindschedler et al. 2017) (Table 6.3).

6.13 Phytoremediation for Electronic Waste

Green technology utilizes the natural potential of plants to remediate contaminated surface water, soil, and sediments (Kathi and Khan 2011). Utilizing plants to eliminate toxic metals and metalloids is found to be limited by time-consuming remediation process as well as phytotoxicity of heavy metals (Dhankher et al. 2012).

Table 6.3 List of microorganisms identified for various biological extraction of metals

Name of the microorganism	Name of the toxic metal removed	References
Bioleaching		
Acidithiobacillus thiooxidans *Micrococcus roseus* *Thiobacillus ferrooxidans* *Aspergillus fumigates* *Aspergillus niger*	Arsenic, Lead Cadmium Arsenic, Lead Arsenic Cadmium, Lead	Patel and Kasture (2014)
Acidithiobacillus ferrooxidans *Acidithiobacillus thiooxidans* *Sulfobacillus thermosulfidooxidans* *Sulfobacillus thermotolerans* *Leptospirillum ferrooxidans*	Copper	Makinen et al. (2015)
Sulfobacillus thermosulfidooxidans *Thermoplasma acidophilum*	Al, Cd, Cr, Cu, Pb, Zn	Ilyas et al. (2014)
Acidithiobacillus ferrivorans and *Acidithiobacillus thiooxidans*	Cu	Isildar et al. (2015)
Biosorption		
Bacillus sphaericus *Pseudomonas aeruginosa* *Myxococcus xanthus* *Rhizopus arrhizus* *Streptoverticillium cinnamoneum* *Saccharomyces cerevisiae* *Pseudomonas putida* X4 *Bacillus subtilis* *Magnetospirillum gryphiswaldense* *Rhizopus arrhizus (F)* *Acidithiobacillus ferrooxidans*	Chromium Uranium Cadmium, Uranium Lead Cadmium Cadmium	Patel and Kasture (2014) He et al. (2012)
Ecklonia maxima (Algae) *Synechocystis* sp. *Chlorella vulgaris,* *Scenedesmus obliquus*	Zinc Cadmium, Copper, Lead Copper, Nickel, Chromium	Chojnacka et al. (2005) Gajendiran and Abraham (2015)
Bioaccumulation		
Deinococcus radiodurans *E. coli* JM109 *Corynebacterium glutamicum* *Plectonema boryanum UTEX 485 (Cyanobacteria)*	Lead, Cadmium Mercury, Nickel, and Cadmium Arsenic Gold	Jaafar et al. (2015) Gumulya et al. (2018) Villadangos et al. (2014) Al-Homaidan et al. (2015)
Biomineralization		
Bacillus fusiformis *Cupriavidus metallidurans* *Desulfotomaculum auripigmentum* *Sporosarcina ginsengisoli* *Aspergillus flavus* *Cupriavidus metallidurans*	Lead Cadmium Arsenic Arsenic Lead Gold	Patel and Kasture (2014) Lengke et al. (2006)
Bioprecipitation		
Thauera selenatis	Selenium	Francis (1998)

The technique of phytoremediation includes numerous processes such as phytoextraction, phytofiltration, phytovolatilization, phytostabilization, and phytodegradation (Alkorta et al. 2004). Biotechnological approaches are currently being used for the phytoremediation of heavy metals and metalloids, such as Hg, Cd, Pb, Se, Cu, and As. Approaches to engineer plants for phytoremediation of heavy metals and metalloids include: (1) manipulating metal/metalloid transporter genes and uptake systems; (2) enhancing metal and metalloid ligand production; and (3) conversion of metals and metalloids to less toxic and volatile forms (Kotrba et al. 2009). This process has been utilized successfully in areas contaminated with PCBs and other organic pollutants (Pramila et al. 2012). Toxic elements can be mobilized and transported (influx) into roots through plasma membrane transporters. They can then be effluxed out of the roots into the xylem and translocated into the shoots. At this stage, plant tolerance to toxic elements may be boosted through manipulation of influx/efflux transporters or by enhancing the levels of chelators. Volatilization of the toxic elements can be achieved through enzymes that modify these toxic elements. Chelators or efflux transporters can also be used to export the toxic elements out of the cytosol and into vacuoles or the cell wall (Mosa et al. 2016; Dhankher et al. 2012). Next generation sequencing was used to study the whole genomes and transcriptomes of several heavy metal-tolerant organisms (Peña-Montenegro and Dussán 2013). Genomics, proteomics, metabolomics, transcriptomics, and phenomics could support the process of identifying the candidate genes that can be later extrapolated for designing plants for phytoremediation. To achieve this end, several approaches have been put forward including transgenic, cisgenic, gene stacking, metabolic engineering, and genome editing (Mosa et al. 2016).

6.14 Conclusions

Toxic constituents present in e-waste needs to be efficiently reduced at the source of generation. This issue can be sustainably addressed using hybrid microbiological processes. However, the efficiency of biological treatments would be enhanced if it is used with a combination of different methods such as with nanoparticles or with some non-polluting biodegraded agents. Therefore, biological methods can be suggested to curtail the e-waste pollution. In the process of reusing wastes generated from fermentation industry, it can be applied for the removal and extraction of metals from the electronic equipment. e-biologics, either produced using microbes or designed with microbial components can be considered as a significant green solution. Reductive bioleaching of oxidized ores and urban biomining of electronic wastes are the recent innovations in the field of biomining. Commercial phytoremediation of heavy metals and metalloids can be fostered by initiating breeding programs to improve the biomass and growth habits of natural hyperaccumulators and breed those traits into non-food, high biomass, fast growing plants for sustainable removal of contaminants. Moreover, a combination of phytoremediation approach with bioenergy ensures twin use of plants for phytoremediation and biofuel production can be considered as a significant option for remediation of contaminated lands.

References

Al-Homaidan AA, Alabdullatif JA, Al-Hazzani AA, Al-Ghanayem AA, Alabbad AF (2015) Adsorptive removal of cadmium ions by *Spirulina platensis* dry biomass. Saudi J Biol Sci 22:795–800

Alkorta I, Hernández-Allica J, Becerril JM, Amezaga I, Albizu I, Garbisu C (2004) Recent findings on the phytoremediation of soils contaminated with environmentally toxic heavy metals and metalloids such as zinc, cadmium, lead, and arsenic. Rev Environ Sci Biotechnol 3:71–90

Anh HQ, Nam VD, Tri TM, Ha NM, Ngoc NT, Mai PTN, Minh TB (2017) Polybrominated diphenyl ethers in plastic products, indoor dust, sediment and fish from informal e-waste recycling sites in Vietnam: a comprehensive assessment of contamination, accumulation pattern, emissions, and human exposure. Environ Geochem Health 39:935–954

Arshadi M, Mousavi SM (2015) Enhancement of simultaneous gold and copper extraction from computer printed circuit boards using *Bacillus megaterium*. Bioresour Technol 175:315–324

Awasthi AK, Zeng X, Li J (2016a) Integrated bioleaching of copper metal from waste printed circuit board—a comprehensive review of approaches and challenges. Environ Sci Pollut Res 23:21141–21156

Awasthi AK, Zeng X, Li J (2016b) Environmental pollution of electronic waste recycling in India: a critical review. Environ Pollut 211:259–270

Bagda E, Tuzen M, Sarı A (2017) Equilibrium, thermodynamic and kinetic investigations for biosorption of uranium with green algae (*Cladophora hutchinsiae*). J Environ Radioact 175:7–14

Bas AD, Deveci H, Yazici EY (2013) Bioleaching of copper from low grade scrap TV circuit boards using mesophilic bacteria. Hydrometallurgy 138:65–70

Bindschedler S, Bouquet TQTV, Job D, Joseph E, Junier P (2017) Fungal biorecovery of gold from e-waste. In: Advances in applied microbiology, vol 99. Academic Press, pp 53–81

Borthakur A (2015) Generation and management of electronic waste in India: an assessment from stakeholders' perspective. J Dev Soc 31:220–248

Borthakur A, Govind M (2017) Emerging trends in consumers' E-waste disposal behaviour and awareness: a worldwide overview with special focus on India. Resour Conserv Recycl 117:102–113

Breivik K, Armitage JM, Wania F, Sweetman AJ, Jones KC (2015) Tracking the global distribution of persistent organic pollutants accounting for e-waste exports to developing regions. Environ Sci Technol 50:798–805

Brierley CL, Brierley JA (2013) Progress in bioleaching: Part B: Applications of microbial processes by the minerals industries. Appl Microbiol Biotechnol 97:7543–7552

Chan JKY, Wong MH (2013) A review of environmental fate, body burdens, and human health risk assessment of PCDD/Fs at two typical electronic waste recycling sites in China. Sci Total Environ 463:1111–1123

Chatterjee A, Abraham J (2017) Efficient management of e-wastes. Int J Environ Sci Technol 14:211–222

Chauhan R, Upadhyay K (2015) Removal of heavy metal from e-waste: a review. Int J Chem Sci 3:15–21

Chojnacka K, Chojnacki A, Gorecka H (2005) Biosorption of Cr3+, Cd2+ and Cu2+ ions by blue-green algae *Spirulina* sp.: kinetics, equilibrium and the mechanism of the process. Chemosphere 59:75–84

Cornea N, Véron R, Zimmer A (2017) Clean city politics: an urban political ecology of solid waste in West Bengal, India. Environ Plan 49:728–744
Cortes LN, Tanabe EH, Bertuol DA, Dotto GL (2015) Biosorption of gold from computer microprocessor leachate solutions using chitin. Waste Manag 45:272–279
CPCB (2014) Central Pollution Control Board, New Delhi, India. List of registered e-waste dismantler/recycler in the country. http://cpcb.nic.in/Ewaste_Registration_List.pdf. Accessed 14 Mar 2016
Cucchiella F, D'Adamo I, Koh SL, Rosa P (2015) Recycling of WEEEs: an economic assessment of present and future e-waste streams. Renew Sustain Energ Rev 51:263–272
Cui H, Anderson CG (2016) Literature review of hydrometallurgical recycling of printed circuit boards (PCBs). J Adv Chem Eng 6:1–11
Das N, Das D (2013) Recovery of rare earth metals through biosorption: an overview. J Rare Earths 31:933–943
Dasgupta D, Debsarkar A, Hazra T, Bala BK, Gangopadhyay A, Chatterjee D (2017) Scenario of future e-waste generation and recycle-reuse-landfill-based disposal pattern in India: a system dynamics approach. Environ Dev Sustain 19:1473–1487
Daum K, Stoler J, Grant RJ (2017) Toward a more sustainable trajectory for e-waste policy: a review of a decade of e-waste research in Accra, Ghana. Int J Environ Res Public Health 14:135
Dave SR, Asha BS, Devayani RT (2018) Microbial technology for metal recovery from e-waste printed circuit boards. J Bacteriol Mycol Open Access 6:241–247
Deng X, Wang P (2012) Isolation of marine bacteria highly resistant to mercury and their bioaccumulation process. Bioresour Technol 121:342–347
Dhankher OP, Pilon-Smits EA, Meagher RB, Doty S (2012) Biotechnological approaches for phytoremediation. In Plant biotechnology and agriculture, pp 309–328
Dixit R, Malaviya D, Pandiyan K, Singh UB, Sahu A, Shukla R, Paul D (2015) Bioremediation of heavy metals from soil and aquatic environment: an overview of principles and criteria of fundamental processes. Sustainability 7:2189–2212
Echegaray F, Hansstein FV (2017) Assessing the intention-behavior gap in electronic waste recycling: the case of Brazil. J Clean Prod 142:180–190
Erüst C, Akcil A, Gahan CS, Tuncuk A, Deveci H (2013) Biohydrometallurgy of secondary metal resources: a potential alternative approach for metal recovery. J Chem Technol Biotechnol 88:2115–2132
Fomina M, Gadd GM (2014) Biosorption: current perspectives on concept, definition and application. Bioresour Technol 160:3–14
Francis AJ (1998) Biotransformation of uranium and other actinides in radioactive wastes. J Alloys Compd 271:78–84
Fujimori T, Takigami H (2014) Pollution distribution of heavy metals in surface soil at an informal electronic-waste recycling site. Environ Geochem Health 36:159–168
Gajendiran A, Abraham J (2015) Mycoadsorption of mercury isolated from mercury contaminated site. Pollut Res J 34:535–538
Gerayeli F, Ghojavand F, Mousavi SM, Yaghmaei S, Amiri F (2013) Screening and optimization of effective parameters in biological extraction of heavy metals from refinery spent catalysts using a thermophilic bacterium. Sep Purif Technol 118:151–161
Gumulya Y, Boxall NJ, Khaleque HN, Santala V, Carlson RP, Kaksonen AH (2018) In a quest for engineering acidophiles for biomining applications: challenges and opportunities. Genes 9:116
Hansda A, Kumar V (2016) A comparative review towards potential of microbial cells for heavy metal removal with emphasis on biosorption and bioaccumulation. World J Microbiol Biotechnol 32:170
He X, Chen W, Huang Q (2012) Surface display of monkey metallothionein α tandem repeats and EGFP fusion protein on *Pseudomonas putida* X4 for biosorption and detection of cadmium. Appl Microbiol Biotechnol 95:1605–1613
Hong Y, Valix M (2014) Bioleaching of electronic waste using acidophilic sulfur oxidising bacteria. J Clean Prod 65:465–472

Ilyas S, Lee JC, Kim BS (2014) Bioremoval of heavy metals from recycling industry electronic waste by a consortium of moderate thermophiles: process development and optimization. J Clean Prod 70:194–202

Isildar A, van de Vossenberg J, Rene ER, van Hullebusch ED, Lens PNL (2015) Two-step bioleaching of copper and gold from discarded printed circuit boards (PCB). Waste Manag 57:149–157

Jaafar R, Al-Sulami A, Al-Taee A, Aldoghachi F, Napes S (2015) Biosorption and bioaccumulation of some heavy metals by *Deinococcus radiodurans* isolated from soil in Basra governorate—Iraq. J Biotechnol Biomater 5:2

Jerez CA (2017) Biomining of metals: how to access and exploit natural resource sustainably. Microbial Biotechnol 10:1191–1193

Johnson DB (2014) Biomining—biotechnologies for extracting and recovering metals from ores and waste materials. Curr Opin Biotechnol 30:24–31

Johnson DB, Hallberg KB (2008) Carbon, iron and sulfur metabolism in acidophilic micro-organisms. Adv Microb Physiol 54:201–255

Kaksonen AH, Morris C, Wylie J, Li J, Usher K, Hilario F, du Plessis CA (2017) Continuous flow 70 C archaeal bioreactor for iron oxidation and jarosite precipitation. Hydrometallurgy 168:40–48

Karwowska E, Andrzejewska-Morzuch D, Lebkowska M, Tabernacka A, Wojtkowska M, Telepko A, Konarzewska A (2014) Bioleaching of metals from printed circuit boards supported with surfactant-producing bacteria. J Hazard Mater 264:203–210

Kathi S, Khan AB (2011) Phytoremediation approaches to PAH contaminated soil. Indian J Sci Technol 4:56–63

Kiddee P, Naidu R, Wong MH (2013) Electronic waste management approaches: an overview. Waste Manag 33:1237–1250

Kotrba P, Najmanova J, Macek T, Ruml T, Mackova M (2009) Genetically modified plants in phytoremediation of heavy metal and metalloid soil and sediment pollution. Biotechnol Adv 27:799–810

Kyere VN, Greve K, Atiemo SM, Ephraim J (2017) Spatial assessment of potential ecological risk of heavy metals in soils from informal e-waste recycling in Ghana. Environ Health Toxicol 32:1–7

Lengke MF, Ravel B, Fleet ME, Wanger G, Gordon RA, Southam G (2006) Mechanisms of gold bioaccumulation by filamentous cyanobacteria from gold (III)-chloride complex. Environ Sci Technol 40:6304–6309

Lens PNL (2016) Biotechnological applications for electronic waste water processing. Linnaeus Eco-Tech, 103

Liang G, Mob Y, Zhou Q (2010) Novel strategies of bioleaching metals from printed circuit boards (PCBs) in mixed cultivation of two acidophiles. Enzyme Microb Technol 47:322–326

Liu J, He XX, Lin XR, Chen WC, Zhou QX, Shu WS, Huang LN (2015) Ecological effects of combined pollution associated with e-waste recycling on the composition and diversity of soil microbial communities. Environ Sci Technol 49:6438–6447

Liu J, Chen X, Shu HY, Lin XR, Zhou QX, Bramryd T, Huang LN (2018) Microbial community structure and function in sediments from e-waste contaminated rivers at Guiyu area of China. Environ Pollut 235:171–179

Lundgren K, International Labor Office (ILO). The global impact of e-waste: addressing the challenge (2012). Available from http://www.ilo.org/wcmsp5/groups/public/%2D%2D-ed_dialogue/%2D%2D-sector/documents/publication/wcms_196105.pdf

Luo Y, Luo XJ, Lin Z, Chen SJ, Liu J, Mai BX, Yang ZY (2009) Polybrominated diphenyl ethers in road and farmland soils from an e-waste recycling region in Southern China: concentrations, source profiles, and potential dispersion and deposition. Sci Total Environ 407:1105–1113

Madrigal-Arias JE, Argumedo-Delira R, Alarcón A, Mendoza-López M, García-Barradas O, Cruz-Sánchez JS, Jiménez-Fernández M (2015) Bioleaching of gold, copper and nickel from waste cellular phone PCBs and computer goldfinger motherboards by two *Aspergillus niger* strains. Braz J Microbiol 46:707–713

Makinen J, Bacher J, Kaartinen T, Wahlstrom M, Salminen J (2015) The effect of flotation and parameters for bioleaching of printed circuit boards. Miner Eng 75:26–31

Man M, Naidu R, Wong MH (2013) Persistent toxic substances released from uncontrolled e-waste recycling and actions for the future. Sci Total Environ 463:1133–1137

Michalak I, Chojnacka K, Witek-Krowiak A (2013) State of the art for the biosorption process—a review. Appl Biochem Biotechnol 170:1389–1416

Mishra A, Malik A (2013) Recent advances in microbial metal bioaccumulation. Crit Rev Environ Sci Technol 43:1162–1222

Mishra D, Rhee YH (2010) Current research trends of microbiological leaching for metal recovery from industrial wastes. Curr Res Technol Educ Top Appl Microbiol Microb Biotechnol 2:1289–1292

Mosa KA, Saadoun I, Kumar K, Helmy M, Dhankher OP (2016) Potential biotechnological strategies for the cleanup of heavy metals and metalloids. Front Plant Sci 7:303

Nancharaiah YV, Mohan SV, Lens PNL (2016) Biological and bioelectrochemical recovery of critical and scarce metals. Trends Biotechnol 34:137–155

Needhidasan S, Samuel M, Chidambaram R (2014) Electronic waste–an emerging threat to the environment of urban India. J Environ Health Sci Eng 12(1):36

Olafisoye OB, Adefioye T, Osibote OA (2013) Heavy metals contamination of water, soil, and plants around an electronic waste dumpsite. Pol J Environ Stud 22:1431–1439

Ongondo FO, Williams ID, Cherrett TJ (2011) How are WEEE doing? A global review of the management of electrical and electronic wastes. Waste Manag 31:714–730

Pant D, Joshi D, Upreti MK, Kotnala RK (2012) Chemical and biological extraction of metals present in E waste: a hybrid technology. Waste Manag 32:979–990

Pant D, Giri A, Dhiman V (2018) Bioremediation techniques for E-waste Management. In: Waste bioremediation. Springer, Singapore, pp 105–125

Patel S, Kasture A (2014) E (electronic) waste management using biological systems-overview. Int J Curr Microbiol Appl Sci 3:495–504

Peña-Montenegro TD, Dussán J (2013) Genome sequence and description of the heavy metal tolerant bacterium *Lysinibacillus sphaericus* strain OT4b.31. Stand Genomic Sci 9:42–56

Perkins DN, Drisse MNB, Nxele T, Sly PD (2014) E-waste: a global hazard. Ann Glob Health 80:286–295

Pramila S, Fulekar MH, Bhawana P (2012) E-waste-A challenge for tomorrow. Res J Recent Sci 1:86–93

Priya A, Hait S (2017) Comparative assessment of metallurgical recovery of metals from electronic waste with special emphasis on bioleaching. Environ Sci Pollut Res 24:6989–7008

Rajesh V, Kumar ASK, Rajesh N (2014) Biosorption of cadmium using a novel bacterium isolated from an electronic industry effluent. Chem Eng J 235:176–185

Rawlings DE (2002) Heavy metal mining using microbes. Annu Rev Microbiol 56:65–91

Rozas EE, Mendes MA, Nascimento CA, Espinosa DC, Oliveira R, Oliveira G, Custodio MR (2017) Bioleaching of electronic waste using bacteria isolated from the marine sponge *Hymeniacidon heliophila* (Porifera). J Hazard Mater 329:120–130

Schippers A (2007) Microorganisms involved in bioleaching and nucleic acid-based molecular methods for their identification and quantification. In: Microbial processing of metal sulfides. Springer, Dordrecht, pp 3–33

Sethurajan M, van Hullebusch ED, Nancharaiah YV (2018) Biotechnology in the management and resource recovery from metal bearing solid wastes: recent advances. J Environ Manag 211:138–153

Sheel A, Pant D (2018) Recovery of gold from electronic waste using chemical assisted microbial biosorption (hybrid) technique. Bioresour Technol 247:1189–1192

Shinkuma T, Nguyen TMH (2009) The flow of E-waste material in the Asian region and a reconsideration of international trade policies on e-waste. Environ Impact Assess Rev 29:25–31

Suzuki G, Someya M, Matsukami H, Tue NM, Uchida N, Viet PH, Takigami H (2016) Comprehensive evaluation of dioxins and dioxin-like compounds in surface soils and river sediments from e-waste-processing sites in a village in northern Vietnam: heading towards the environmentally sound management of e-waste. Emerg Contam 2:98–108

Tabak HH, Lens P, van Hullebusch ED, Dejonghe W (2005) Developments in bioremediation of soils and sediments polluted with metals and radionuclides–1. Microbial processes and mechanisms affecting bioremediation of metal contamination and influencing metal toxicity and transport. Rev Environ Sci Biotechnol 4:115–156

Tang X, Qiao J, Chen C, Chen L, Yu C, Shen C, Chen Y (2013) Bacterial communities of polychlorinated biphenyls polluted soil around an e-waste recycling workshop. Soil Sediment Contam Int J 22:562–573

Tay SB, Natarajan G, bin Abdul Rahim MN, Tan HT, Chung MCM, Ting YP, Yew WS (2013) Enhancing gold recovery from electronic waste via lixiviant metabolic engineering in *Chromobacterium violaceum*. Sci Rep 3:2236

UNEP (1992) Basel Convention on the control of transboundary movements of hazardous wastes and their disposal. Available from http://www.basel.int/Portals/4/Basel%20Convention/docs/text/BaselConventionText-e.pdf

UNU (2015) E-waste World Map: Update to quantitative data and legal texts – STEP. United Nations University (UNU). http://step-initiative.org/index.php/newsdetails/items/e-waste-world-map/

Valix M (2017) Bioleaching of electronic waste: milestones and challenges. In: Current developments in biotechnology and bioengineering (pp. 407–442)

Van Nostrand JD, Wu WM, Wu L, Deng Y, Carley J, Carroll S (2009) GeoChip-based analysis of functional microbial communities during the reoxidation of a bioreduced uranium-contaminated aquifer. Environ Microbiol 11:2611–2626

Varjani SJ, Gnansounou E, Baskar G, Pant D, Zakaria ZA (2018) Introduction to waste bioremediation. In: Waste bioremediation. Springer, Singapore, pp 1–5

Varshney S, Jain P, Srivastava S (2017) Application of ameliorated wood pulp to recover Cd (II), Pb (II), and Ni (II) from e-waste. J Mater Cycles Waste Manage 19:1446–1456

Villadangos AF, Ordóñez E, Pedre B, Messens J, Gil JA, Mateos LM (2014) Engineered coryneform bacteria as a bio-tool for arsenic remediation. Appl Microbiol Biotechnol 98:10143–10152

Volesky B (1990) Removal and recovery of heavy metals by biosorption. CRC Press, Boca Raton

Wang Y, Luo C, Li J, Yin H, Li X, Zhang G (2011) Characterization of PBDEs in soils and vegetations near an e-waste recycling site in South China. Environ Pollut 152:2443–2448

Wang S, Zheng Y, Yan W, Chen L, Mahadevan GD, Zhao F (2016) Enhanced bioleaching efficiency of metals from E-wastes driven by biochar. J Hazard Mater 320:393–400

Wath SB, Vaidya AN, Dutt PS, Chakrabarti T (2010) A roadmap for development of sustainable e-waste management system in India. Sci Total Environ 409:19–32

Wath SB, Dutt PS, Chakrabarti T (2011) E-waste scenario in India, its management and implications. Environ Monit Assess 172:249–262

Widmer R, Oswald-Krapf H, Sinha-Khetriwal D, Schnellmann M, Böni H (2005) Global perspectives on e-waste. Environ Impact Assess Rev 25:436–458

Wu Q, Leung JY, Geng X, Chen S, Huang X, Li H, Lu Y (2015) Heavy metal contamination of soil and water in the vicinity of an abandoned e-waste recycling site: implications for dissemination of heavy metals. Sci Total Environ 506:217–225

Wu W, Dong C, Wu J, Liu X, Wu Y, Chen X, Yu S (2017) Ecological effects of soil properties and metal concentrations on the composition and diversity of microbial communities associated with land use patterns in an electronic waste recycling region. Sci Total Environ 601:57–65

Wu W, Liu X, Zhang X, Zhu M, Tan W (2018) Bioleaching of copper from waste printed circuit boards by bacteria-free cultural supernatant of iron–sulfur-oxidizing bacteria. Bioresour Bioprocess 5:10

Ye M, Sun M, Wan J, Fang G, Li H, Hu F, Orori Kengara F (2015) Evaluation of enhanced soil washing process with tea saponin in a peanut oil–water solvent system for the extraction of PBDEs/PCBs/PAHs and heavy metals from an electronic waste site followed by vetiver grass phytoremediation. J Chem Technol Biotechnol 90:2027–2035
Zhang WH, Ying-Xin WU, Simonnot MO (2012) Soil contamination due to e-waste disposal and recycling activities: a review with special focus on China. Pedosphere 22:434–455

Chapter 7
Organic and Inorganic Contaminants from E-waste and Their Effects on Environment

Deedar Nabi, Ayesha Ashfaq, and Hamza Bin Riaz

Abstract The environmental footprint of electronic waste is expanding in concert with the rise of electronics industry. There is a need to quantify the area of environmental footprint of e-waste by carefully assessing the types and quantities of contaminants emanating from the e-waste. Proper risk assessment should consider the fate, behavior, and transport of contaminants from e-waste to other environmental compartments such as air, water, and soil. The holistic risk assessment should include the survey of all possible exposure pathways to humans and the environment. A wide variety of electrical and electronic devices contribute to e-waste, and become the source of diverse chemicals leaching into the environment, which cause damage to the environment and to human beings. Persistence, bioaccumulation, and toxicity (PBT) assessment of the chemical released into the environment shows that the pollutants have the attributes of persistence (non-biodegradable with longer environmental half-lives), bioaccumulative, and toxic in nature. The developed world exports a colossal amount of e-waste to developing countries. Pakistan is not an exception to the inward traffic of e-waste. However, the impact assessment studies of e-waste in Pakistan are scarce. Strong measures need to be taken in this direction to tackle this ever-increasing hazard of e-waste to protect humans and the environment from its disastrous effects.

Keywords Exposure assessment · Organic chemicals · E-waste · Handling

7.1 Introduction

With new technology being introduced on a daily basis, amount of electrical and electronic equipment (e-waste) or e-scrap is increasing due to the waste material that is thrown away and is not reused further (Problem and Paper 2014).

D. Nabi (✉) · A. Ashfaq · H. B. Riaz
Institute of Environmental Sciences and Engineering (IESE), National University of Sciences and Technology (NUST), Islamabad, Pakistan
e-mail: deedar.nabi@iese.nust.edu.pk

M. Z. Hashmi, A. Varma (eds.), *Electronic Waste Pollution*, Soil Biology 57,
https://doi.org/10.1007/978-3-030-26615-8_7

7.1.1 E-waste Disposal, Bird's Eye View

In the late twentieth century, precarious e-waste shipment started from developed states to the developing states, located in the regions of Asia, Africa, Central America, and Europe, majorly including countries like China, the Philippines, Pakistan, Ghana, India and Nigeria, resulted in the destructing the environment and human health (Nations, Programme and UNEP 2010; Sajid 2016). Figure 7.1 is showing the pathways of e-waste trafficking across the countries. The environmental damage is happening due to improper reprocessing techniques, no proper disposing site (landfills), incineration of materials, and lack of regulatory rules and educational awareness about environment to the concerned people.

Basel Convention mainly focuses on managing waste for the protection of the environment and human health, along with strict control on the e-waste exports and restriction of e-waste movement like other ordinary export products to developing countries (UNEP 2015). Even though Basel Convention (1989) emphasized controlled trafficking of hazardous waste within nations already having agreement between the regions and statutory writings by different nations implemented in the year 1992, it is resulting in minimal limitation for e-waste trade across regions of the world having less resources for handling such waste (Khan 2016). Despite of the above mentioned steps, e-waste disposal to other countries continued, so further amendments were done for minimal possible trading to developing countries in 1995 and 1998 and then again in 2006 for wide-ranging management of e-waste, making it possible through Nairobi Declaration, also called STEP, i.e., solving the e-waste problems, implemented in 2007 for sharing of the issues related to e-waste along with the information exchange of e-waste disposal and recycling (Umair et al. 2016).

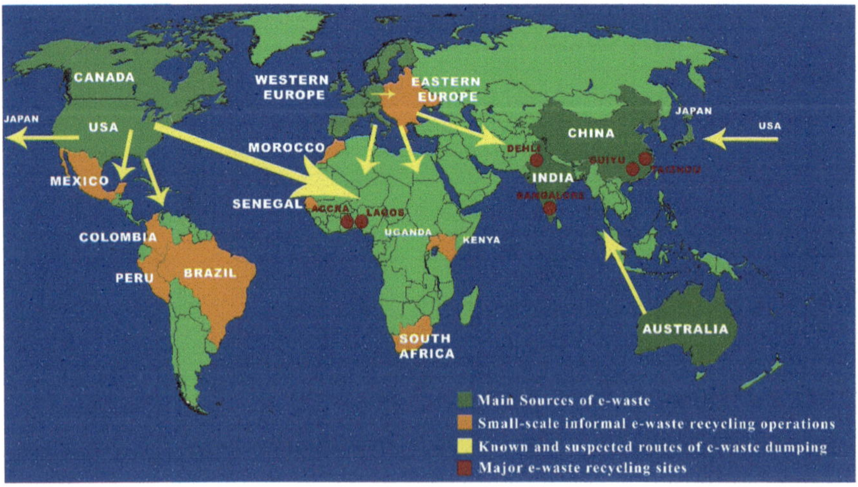

Fig. 7.1 Main sources of e-waste and routes of e-waste dumping (Adapted from: Chen et al. 2011)

7.2 E-waste Classification

E-waste material is classified into three main categories, i.e., consumer, commercial, and industrial.

(a) Consumer/household appliances are further classified into two categories: large household appliances (television, refrigerators, deep freezers, washing machines, printers, etc.) and small household appliances (hairdryer, toaster irons, heaters, mobile phones, etc.).
(b) Commercial appliances include large gears like photocopy machines (Morf et al. 2007).
(c) Industrial material comprises of electrical monitoring and electronic control instruments.

7.3 Material Used for Different Gears of E-waste

Different hazardous compounds and metals are associated with the several components of e-waste given as below:

Telecommunication, Industrial, Consumer/Household Electrical, and Electronic Devices These devices are mainly composed of mercury (Hg), which is used in the electromagnetic induction devices in various parts of the equipment, like electric ovens and heaters, cameras, security systems, discharge tubes or lamps, televisions, LCDs, etc. (Tsydenova and Bengtsson 2011).

Batteries Nowadays, there are different varieties of batteries used in cars, computers, mobiles, UPS systems, etc. These batteries are mainly comprised of lead, cadmium, lithium, and nickel. All of these metals have environmental and human health concerns (Ceballos and Dong 2016).

Printed Circuit Boards (PCBs) PCBs are used as connecting device in electrical and electronic equipment. These devices comprise of circuit boards, mainly of copper (20%) along with other elements like iron (8%), tin (4%), nickel (2%), zinc (1%), lead (2%), cadmium and silver (0.2%), gold (0.1%), palladium (0.005%), etc. The amount of these elements in electronic equipment is found to be 10% more than that in ores in earth's crust (Huang et al. 2009; Guo et al. 2008).

Cathode Ray Tubes (CRTs) Cathode ray tubes found in old version computers, monitors, and television sets. Amount of lead content in them varies depending on the age of the device; older versions contain 2–3 kilogram, while the later versions of CRTs contain 1 kg of lead. Frame and bear present in electron gun consist of 1–2 gram of barium and its compounds (Robinson 2009; Lau et al. 2014).

Liquid-Crystal Displays (LCDs) and Organic Light-Emitting Diodes (OLEDs) LCD screens are used in modern computer monitors, mobile display screens, and TV screens. Nowadays, being replaced by OLEDs, having good light emitting properties, have less mass and better resolution and better color disparity, and use less electricity consumption than LCDs. In spite of all the above advantages, OLEDs have very small life as compared to LCDs generating the bulk of e-waste (Thejokalyani and Dhoble 2014). However, the amount of different metals are greater in OLEDs than in LCDs Metals includes silver, beryllium, chromium, copper, nickel, and tin, copper and titanium. Copper and titanium metals are reported to be in greater concentartions than the threshold limits in California (Kiddee et al. 2013).

Plastics Plastics are used to form the body of the equipment, wire covering and protecting materials. Chemically composed of polyvinyl chloride (PVC). Cadmium, brominated flame retardant (BFR, an organic brominated compound), and phthalate are used for improving the strength and stability of the PVC plastics (Birnbaum and Staskal 2004).

7.4 Sources of Exposure

As discussed in Sect. 7.3, e-waste materials gears comprisesmajorly of hazardous metals like cadmium, lead, copper, tin, mercury, etc. and non-metals like chlorine, bromine, and phosphorous (Herat and Agamuthu 2012). Exposure of these elements to humans causes damage to major organs like the heart, lungs, kidneys, and reproductive systems. Due to the improper disposal and informal recycling of e-waste, risks of damage to human health and the environment are becoming the main highlighted issue in the developing countries (Morf et al. 2007).

There are three different means of exposure of human to e-waste:

1. Connection with the familiar reusing tasks
2. Functioning in the official (formal) and unofficial (informal) recycling factories (occupational exposure)
3. E-waste in the environment even after disposal (environmental exposure)

In informal sectors of e-waste handling, electronics are ripped into pieces to make valuable things from discarded electronic material with the help of very little use of equipment, workers are being exposed to the hazardous chemicals by direct contact with e-waste. Moreover, in these factories preventive measures and examination of the working area are not carried out properly to avoid exposure to toxic chemicals (Wong et al. 2007a, b). Hand utilizing sledges, electric drills, screwdrivers, and cutters are used for disassembling the disposed e-waste into its parts/components like circuit boards, CD drivers, chargers, cables, monitors, wires, etc. These segments are then sold to different markets for reprocessing different items (Huo et al. 2007). Exposure of these elements to the humans causes major organ damage. Due to improper disposal and informal recycling of e-waste, risk for human and environment is increasing and taking a serious form (Morf et al. 2007).

7.5 Toxicants and Chemicals

Reuse or recycling of e-waste is done in uncontrolled conditions, and various toxic chemicals are released in the atmosphere, its reprocessing results in releasing toxic substances into the air, causing airborne diseases (Frazzoli et al. 2010; Sepúlveda et al. 2010; Perkins et al. 2014). Human exposure to the mixture of chemicals causes various types of diseases and disorders like calcium homeostasis, epigenetic modifications, neurotransmission, endocrine disruption, etc. (Chen et al. 2011). burning processes and destroying activities releases toxic substances in particulate matter. (Sepúlveda et al. 2010). Disassembling and chopping processes incorporate hazardous chemicals to wastewater facilities, through evaporation, re-volatilization, and leaching, these chemicals get to surface and groundwater.

7.5.1 Occupational Hazards of Inorganic Contaminants

There are different toxic components present in e-waste materials, its recycling results in severe health damage to workers and their families (Ikhlayel and Nguyen 2017). Several formal recycling methods follow proper processing like separation, renewing, and fixing of different e-waste materials, followed by the stripping and grinding of e-waste using manpower or different machinery having proper air passages; personal protection for employees is taken strictly into consideration; the waste is systematically recycled by other parties to recover the component materials, whereas in non-formal recycling, there is no proper secure procedure adapted, no personal protecting facilities, and no proper air passages. Material in informal units is manually treated in all the steps, i.e., stripping, grinding, and renewing of PCBs. In the easygoing sections of e-misuse, many youths and teenagers are working. Studies have shown that young children are involved in picking and sorting of electronic waste by hand. Also, they get involved in burning of electronic waste and separate them into parts by hand. Youths are a wellspring of disgraceful work in making countries, their small hands give them use in pulverizing things (ILO 2004). Regardless, paying little notice to whether youths are not especially managing e-misuse things in the reusing system, there are a couple of observations as to how children may get in contact with hazardous mixes. Youngsters are familiar with the effects of EEE-misuse when they live in, or near to homes where reusing practices happen as depicted by studes on exposure of e-waste at different recycling sites in Table 7.1. Toxins remaining on parents' clothes who are exposed to e-waste during burning process have serious impact on children's health. Often, recycled e-waste

Table 7.1 Studies done on the e-waste recycling sites

Recycling place	Samples and sites under consideration	Experimental findings
Informal sites		
Guiyu (Zheng et al. 2016)	Hair and blood samples of children near burning site	High tin levels in people living in the vicinity of site Higher lead, nickel, copper, selenium, and cadmium levels as compared to reference site population
Ghanna (Carlson 2016)	Blood and urine samples	Cadmium and lead higher in concentration
Formal sites		
USA (Kim et al. 2015)	Blood, urine, personal air, and skin wipes	Workers and population near dismantling site have high blood lead level
Sweden (Julander et al. 2014)	Samples of blood, urine, and plasma	Chromium, cobalt, indium, mercury high levels
France (Lecler et al. 2015)	Air quality w.r.t metals	Higher lead concentrations

materials are used for making toys and other children-related products, which can also lead them to getting exposed to hazardous chemicals (Pascale et al. 2016; Heacock et al. 2015).

Copper metal is obtained from electric wires having plastic covering, the plastic is removed by burning, leaving behind a lot of burnt waste, as summarized in Table 7.2 which when dumped in open air and into water bodies causes severe contaminating toxicants which can damage human health very badly, In developed countries, e.g., the USA, no proper dust handling amenities have been reported. Lead and cadmium are reported to be higher than the threshold values for processing of CRT scrubbing and grinding procedure. Two employees working for cleaning process have been reported to have high levels of lead in their blood samples. Recycling of alkaline batteries releases high levels of mercury in the atmosphere, which can disturb central nervous system and can cause other respiratory diseases in humans.

7.5.2 *Environmental Hazards*

7.5.2.1 Soil Fertility

Contamination of heavy elements and acidic percolations causes land infertility due to less availability of total organic matter, resulting in metal accumulation such as copper, lead, and cadmium. Studies have shown that in China, the concentrations of copper have found to be 10 times more on burning sites, 40 times more in dumping sites, and 60 times more in acid leaching soils than the prescribed threshold limits—grade II guidelines (Li et al. 2011; Wu et al. 2015).

Table 7.2 Toxic products released by the combustion of e-wastes

Organic compounds' emissions during e-waste combustion	
Class of chemicals	Compounds
Furan	Dibenzofurans
Dioxins	Polychlorinated dibenzodioxins
Aromatic hydrocarbons and Derivatives	Polychlorinated biphenyl
	Naphthalene, acenaphthene, acenaphthylene, phenanthrene, fluorene, fluoranthene, anthracene, benzo(a)anthracene, chrysene, pyrene, benzo(a)pyrene, benzo(b)fluoranthene, benzo(k) fluoranthene, dibenz(a,h)anthracene, benzo(ghi)perylene, indeno [1,2,3-cd]pyrene

Adapted from: Frazzoli et al. (2010), Grant et al. (2013)

7.5.2.2 Water Contamination

Water bodies become highly contaminated with metal, along with lowering of pH due to e-waste recycling and incineration. Washing of e-waste is usually done with sulfuric acid (H_2SO_4), which is then released into nearby water bodies. Moreover, discarded e-waste parts are buried in dumping sites located in the vicinity of water bodies, causing leachate to travel into the streams through rainfall and hence making them contaminated. Studies have shown that metal like cadmium remains in the upper 200 cm of the soil, where the percolation chances are very low, but during rainfall it get added to the water through runoff, and thus waters are reported to have high metal concentrations by then. Such contaminated water if used for irrigating vegetables and rice fields can lead to accumulation of these heavy metals in plants. Studies have reported that in order to reduce the number of illegal e-waste dumping sites in the developed countries, legal e-waste dumping sites have been made in the developed countries, thereby reducing the amount of metal contamination and acidic strength (Wu et al. 2015). Lead contamination is 0.4 mg/L in a study done at Guiyu, which is eight times higher when compared with permissible drinking water levels (0.05 mg/L). Furthermore, higher water concentrations of silver, chromium, lithium, molybdenum, bismuth, and selenium have been reported (Li et al. 2011).

7.5.2.3 Air Contamination

Air contamination is mainly caused by the soot created by e-waste renewing, sorting, grinding, combusting, and recycling activities. Thus, heavy metals like cadmium, copper, chromium, nickel, lead, zinc, manganese, and arsenic are emitted into air which affects the environment and also causes damage to human health through different pathways like breathing, ingestion, and sponging up through the skin (Grant et al. 2013).

7.5.2.4 Effects on Health

To study the harmful effects of the EEE, research is being done for more than last 30 years; results of such research show that combustion of e-waste causes damaging effects. Reason of toxic and harmful impacts is release of almost over 1000 chemicals from e-waste. Abnormalities like thyroid-stimulating hormones imbalance and serum PBDEs may occur and aggravate by exposure to e-waste toxicants and are found to be common in people working in the dismantling process of e-waste and have a direct contact with e-waste-released chemicals and dust. In such dismantling sites, heavy metals like nickel, manganese, and chromium are most commonly found (Zheng et al. 2013). It may cause spontaneous abortion in pregnant women (Wu et al. 2016; Xu et al. 2016) and premature births and also affects babies by causing reduced birth weight. In the early age period, reduced height of children (Zheng et al. 2013), changes in cell level (Li et al. 2008b), and reduction in social inconveniences (Li et al. 2008a) are caused by e-waste exposure. Incineration is done very frequently in e-waste-related processes, which releases a number of toxic heavy metals, e.g., mercury, lead, and cadmium (Duffert et al. n.d.; Frazzoli et al. 2010; Chan and Wong 2013). The emissions from poor burning of e-waste at low temperatures usually contain organic poly-halogenated dioxins and furans (Frazzoli et al. 2010). Toxicants exposure may also happen when polystyrene frame, elastic, tires, trim deposit, biomass, etc., are used as fuel. Furthermore, the working materials used in reusing procedures are proven to be harmful to human health (Duffert et al. n. d.). These working materials may include cleaning solvents and reagents, such as cyanide and other solid draining acids. Indirect exposures of human to e-waste-related hazards may also occur due to the contamination of soil, air, and water, major environmental contamination. Studies done in China have shown that water and sediments near e-waste recycling sites have been found to be contaminated with metal impurity; rivers have also been reported to have excess amounts of dissolved metals (Duffert et al. n.d.; Wong et al. 2007a, b). Direct contact including skin interaction with damaging substances and inhalation of particulate matter are the major pathways by which humans come in contact with such contaminants. Workers who deal with e-waste without proper protective measures are reported to have frequent work-related health problems. Dangerous reusing frameworks that are put to practice for recovering valuable materials from e-waste create a lot of risk for exposure to hazardous materials. Singing (light burning) of plastic is frequently done at low temperatures to either discard PC housings or to recover metals from the electronic chips and other different parts, which also releases a number of toxicants. Release of hazardous and dangerous engineered chemicals into the environment can lead to bioaccumulation of impurities for a longer period of time, increasing the risk of exposure. Issues and concerns related to transplacental and chest deplete presentation are high, but no evident data on levels of introduction are available (Duffert et al. n.d.; Tue et al. 2010; Chen et al. 2011; Chan and Wong 2013). In children, the present effects of e-waste have been addressed (Zheng et al. 2013): decreased stature social inconveniences (Li et al. 2008a) and changes in cell levels and purpose of

Table 7.3 Group-wise summary of inorganic elements along with their health effects

Group I-A	
Lithium	Irritant for lungs and eyes
Group II-A	
Barium	Neurotoxicity, digestive tract irritation, cardiomuscular toxicity
Beryllium	Beryllium poisoning, carcinogen
Cadmium	Carcinogen, cardiotoxic, endocrine disruptor
Group III-A	
Aluminum	Lungs irritation, neuron toxicant
Indium	Probably carcinogenic to humans, neurotoxic, indium lung diseases
Gallium	Skin, eyes, and mucous membrane irritant
Thallium	Neurotoxicity, cardiomuscular toxicity, liver toxicity, reproductive defects
Group IV-A	
Tin	Irritant of lungs, eyes, and skin
Lead	Probability of inducing cancer, cardiotoxicity, neurotoxicity, endocrine gland disruption
Group V-A	
Antimony	Irritation of lungs, eyes, and gastrointestine
Arsenic	Hemolysis, endocrine disorders
Transition elements	
Group VII-B	
Cobalt	Allergic reactions, carcinogenic probability to human beings
Group VI-B	
Chromium (VI)	Sensitivity induction carcinogenic to lungs, irritant to skin
Group VIII-B	
Manganese	Cardiotoxic, neurotoxic, irritant to lungs, endocrine gland disruptor
Group II-B	
Group VII-B	
Platinum	Irritant for eyes and lungs
Group I-B	
Copper	Eyes, lungs, and digestive tract irritant
Silver	Nephron toxicity, reproductive organs defects
Group II-B	
Zinc	Induction of fume fever
Mercury	Endocrine disorders

Adapted from: Bakhiyi et al. (2018)

restriction (Li et al. 2008b) as summarized in Table 7.3. A study on ornaments revealed that in Chinese market, lead obtained from recycled material is used in making jewelry, hence potentially available for human absorption. There is also a perception that toys made in China also have incorporated recycled materials containing brominated flame retardants and lead in higher concentrations.

7.6 E-waste Scenario in Pakistan

Pakistan is a developing country and is located in Southeast Asia. Pakistan also receives a huge chunk of e-waste in several forms; hence e-waste is becoming a major contributor for the health- and environment-related issues, as the material received has reusable content up to 2% only. Huge marketing of e-waste takes place in three major cities, i.e., Karachi, Rawalpindi, and Gujranwala. The unused material is recycled informally by workers including women and children, without using any safety precautions. Also, limited information is available for exposure of workers and environment. A study done on some sites in Karachi on flame retardants reported their concentration to be below the RFDs in the soil as well as air, but this study is based on limited number of samples, so concentration may not depict complete overall scenario (Iqbal et al. 2017). Another study done in 2016 reported there are many loopholes in local as well as international legislations to stop trafficking of e-waste to developing countries. There is a need to develop formal recycling sites in Pakistan, similar to those in China and India, to protect laborers and environment, and health awareness programs should also be conducted (Umair et al. 2016).

7.7 Contaminant Transport, and Effects of Contaminants

Recycling of e-waste can promote immediate or variant introduction to a number of dangerous substances present in electric and electronic equipment, in confined and free form, which are risky reusing practices (Table 7.4).

Table 7.4 Summary of organic and inorganic pollutants, their contact environment, and exposure pathways from WEEE

Class of organic pollutants	Compounds	Human exposure	Transport ways
Polyaromatic compounds and their derivatives	Polybrominated diphenyl ethers and polybrominated biphenyls	Breathing, assimilation, placental transfer	Air, water, and soil
	Polychlorinated biphenyls	Absorption, breathing, or contact with epidermis, and placental transfer	Sea food, air, water, and soil
Dioxins	Polychlorinated dibenzodioxins (PCDBs)	Absorption, skin contact, breathing, and placental transfer	Sea food, air, water, and soil

Adapted from: Frazzoli et al. (2010), Grant et al. (2013)

7.8 Persistence, Bioaccumulation and Transport Assessment of Chemicals

Environmental PBTs of the organic contaminants contamination are summarized in Table 7.5 by using EPI Suite showing that the some of organic species used in the formation of the gears of electronic and electrical equipments when released into the atmosphere pose several threats. The LC (acute toxicity) of compound is <1 ppm, making these compounds very toxic in nature; among the selected pollutants there are many which are acutely toxic in fish. Bioaccumulation potential of chemicals is >2, depicting that chemicals remain in the living organisms for a longer period of time; environmental transport involves interaction of chemicals with the living and non-living matter as well as the half-life period of the pollutants; the half-life estimation of half-life done by using BioWin 3 model showed that certain chemicals have values in between 1 and 2, depicting that contaminants are having longer degradation period in years and months.

Table 7.5 PBTs of organic contaminants in environment

Compounds	Smiles	CAS no	LC_{50} (ppm)	BCFBAF	Biodegradation
Dibenzofuran	o(c(c(c1cccc2)ccc3)c3)c12	000132-64-9	3.985	3.07	2.827
2,3,7,8-Tetrabromodibenzofuran	Brc1cc2c(oc3c2cc(c(c3)Br)Br)cc1Br	067733-57-7	0.007	2.275	1.581
1,2,3,7,8-Pentabromodibenzo-P-dioxin	Brc3c(c(cc2c3Oc1c(cc(c(c1)Br)Br)O2)Br)Br	109333-34-8	0.00378	1.661	1.124
2,3,7,8-Tetrabromodibenzo-p-dioxin	BrC3=CC(O2)=C(C=C3Br)OC1=C2C=C(Br)C(Br)=C1	050585-41-6	0.002	2.317	1.434
2,3,7,8-TCDD	Clc3cc2Oc1cc(Cl)c(Cl)cc1Oc2cc3Cl	001746-01-6	0.010	3.797	1.545
1,2,3,4,7,8-Hexachlorodibenzo-p-dioxin	c12c(Oc3cc(Cl)c(cc3O1)Cl)c(c(Cl)c(c2Cl)Cl)Cl	039227-28-6	0.000852	2.770	0.9796
Decabromodiphenyl ether	O(c(c(c(c(c1Br)Br)Br)Br)c1Br)c(c(c(c(c2Br)Br)Br)Br)c2Br	001163-19-5	$6.56e^{-007}$	−0.039	−0.3386
2,2′,3,4,4′,5′,6-Heptabromodiphenyl ether	BrC1=C(OC2=CC(Br)=C(Br)C=C2Br)C(Br)=C(Br)C(Br)=C1	207122-16-5	0.000124	1.411	0.5925
Acenaphthylene	c1ccc2cccc3c2c1C=C3	000208-96-8	2.281	2.740	2.8628
Phenanthrene	c(c(c(c(c1)ccc2)c2)ccc3)(c1)c3	000085-01-8	1.147	3.080	2.2194
Fluoranthene	c(c(ccc1)ccc2)(c1c(c3ccc4)c4)c23	000206-44-0	0.386	2.586	1.9529

References

Bakhiyi B et al (2018) Has the question of e-waste opened a Pandora's box? An overview of unpredictable issues and challenges. Environ Int 110(August 2017):173–192. https://doi.org/10.1016/j.envint.2017.10.021

Birnbaum LS, Staskal DF (2004) Brominated flame retardants: cause for concern? Environ Health Perspect 112(1):9–17. https://doi.org/10.1289/ehp.6559

Carlson K (2016) Electronic waste worker health. Proceedings of the eighth international conference on information and communication technologies and development – ICTD '16, pp 1–4. https://doi.org/10.1145/2909609.2909618

Ceballos DM, Dong Z (2016) The formal electronic recycling industry: challenges and opportunities in occupational and environmental health research. Environ Int 95:157–166. https://doi.org/10.1016/j.envint.2016.07.010

Chan JKY, Wong MH (2013) A review of environmental fate, body burdens, and human health risk assessment of PCDD/Fs at two typical electronic waste recycling sites in China. Sci Total Environ 463–464:1111–1123. https://doi.org/10.1016/J.SCITOTENV.2012.07.098

Chen A et al (2011) Developmental neurotoxicants in e-waste: an emerging health concern. Environ Health Perspect 119(4):431–438. https://doi.org/10.1289/ehp.1002452

Duffert C, Brune MN, Prout K (n.d.) Background document on exposures to e-waste. World Health Organization, Geneva

Frazzoli C et al (2010) Diagnostic health risk assessment of electronic waste on the general population in developing countries' scenarios. Environ Impact Assess Rev 30(6):388–399. https://doi.org/10.1016/j.eiar.2009.12.004

Grant K et al (2013) Health consequences of exposure to e-waste: a systematic review. The Lancet Global Health 1(6):e350–e361. https://doi.org/10.1016/S2214-109X(13)70101-3

Guo J, Rao Q, Xu Z (2008) Application of glass-nonmetals of waste printed circuit boards to produce phenolic moulding compound. J Hazard Mater 153(1–2):728–734. https://doi.org/10.1016/j.jhazmat.2007.09.029

Heacock M et al (2015) E-waste and harm to vulnerable populations: a growing global problem. Environ Health Perspect 124(5):550–555. https://doi.org/10.1289/ehp.1509699

Heræt S, Agamuthu P (2012) E-waste: a problem or an opportunity? Review of issues, challenges and solutions in Asian countries. Waste Manag Res 30(11):1113–1129. https://doi.org/10.1177/0734242X12453378

Huang K, Guo J, Xu Z (2009) Recycling of waste printed circuit boards: a review of current technologies and treatment status in China. J Hazard Mater 164(2–3):399–408. https://doi.org/10.1016/j.jhazmat.2008.08.051

Huo X et al (2007) Elevated blood lead levels of children in Guiyu, an electronic waste recycling town in China. Environ Health Perspect 115(7):1113–1117. https://doi.org/10.1289/ehp.9697

Ikhlayel M, Nguyen LH (2017) Integrated approaches to water resource and solid waste management for sustainable development. Sustain Dev 25(6):467–481. https://doi.org/10.1002/sd.1683

ILO (2004) Addressing the exploitation of children in scavenging (waste picking): a thematic evaluation of action on child labour. International Labour Office, Geneva

Iqbal M et al (2017) E-waste driven pollution in Pakistan: the first evidence of environmental and human exposure to flame retardants (FRs) in Karachi City. Environ Sci Technol 51 (23):13895–13905. https://doi.org/10.1021/acs.est.7b03159

Julander A et al (2014) Formal recycling of e-waste leads to increased exposure to toxic metals: an occupational exposure study from Sweden. Environ Int 73:243–251. https://doi.org/10.1016/j.envint.2014.07.006

Khan SA (2016) E-products, E-waste and the Basel convention: regulatory challenges and impossibilities of international environmental law. Rev Eur Comp Int Environ Law 25(2):248–260. https://doi.org/10.1111/reel.12163

Kiddee P, Naidu R, Wong MH (2013) Electronic waste management approaches: an overview. Waste Manag 33(5):1237–1250. https://doi.org/10.1016/j.wasman.2013.01.006

Kim KH, Kabir E, Kabir S (2015) A review on the human health impact of airborne particulate matter. Environ Int 74:136–143. https://doi.org/10.1016/j.envint.2014.10.005

Lau WKY et al (2014) Human health risk assessment based on trace metals in suspended air particulates, surface dust, and floor dust from e-waste recycling workshops in Hong Kong, China. Environ Sci Pollut Res 21(5):3813–3825. https://doi.org/10.1007/s11356-013-2372-8

Lecler MT et al (2015) Exposure to hazardous substances in Cathode Ray Tube (CRT) recycling sites in France. Waste Manag 39:226–235. https://doi.org/10.1016/j.wasman.2015.02.027

Li Y, Xu X, Liu J et al (2008a) The hazard of chromium exposure to neonates in Guiyu of China. Sci Total Environ 403(1–3):99–104. https://doi.org/10.1016/j.scitotenv.2008.05.033

Li Y, Xu X, Wu K et al (2008b) Monitoring of lead load and its effect on neonatal behavioral neurological assessment scores in Guiyu, an electronic waste recycling town in China. J Environ Monit 10(10):1233–1238. https://doi.org/10.1039/b804959a

Li Y et al (2011) Assessment of cadmium exposure for neonates in Guiyu, an electronic waste pollution site of China. Environ Monit Assess 177(1–4):343–351. https://doi.org/10.1007/s10661-010-1638-6

Morf LS et al (2007) Metals, non-metals and PCB in electrical and electronic waste – actual levels in Switzerland. Waste Manag 27(10):1306–1316. https://doi.org/10.1016/j.wasman.2006.06.014

Nations U, Programme, E. and UNEP (2010) United Nations Environment Programme, English. https://doi.org/10.1016/0959-6526(93)90036-B

Pascale A et al (2016) E-waste informal recycling: an emerging source of lead exposure in South America. Ann Glob Health 82(1):197–201. https://doi.org/10.1016/j.aogh.2016.01.016

Perkins DN et al (2014) E-waste: a global hazard. Ann Glob Health 80(4):286–295. https://doi.org/10.1016/j.aogh.2014.10.001

Problem E and Paper W (2014) One global definition of E-waste, 3576(June)

Robinson BH (2009) E-waste: an assessment of global production and environmental impacts. Sci Total Environ 408(2):183–191. https://doi.org/10.1016/j.scitotenv.2009.09.044

Sajid M (2016) Assessment of electronic waste generation, recycling and disposal practices in Pakistan by Institute of Environmental Sciences and Engineering (IESE) School of Civil and Environmental Engineering (SCEE) National University of Sciences and Technology

Sepúlveda A et al (2010) A review of the environmental fate and effects of hazardous substances released from electrical and electronic equipments during recycling: examples from China and India. Environ Impact Assess Rev 30(1):28–41. https://doi.org/10.1016/j.eiar.2009.04.001

Thejokalyani N, Dhoble SJ (2014) Novel approaches for energy efficient solid state lighting by RGB organic light emitting diodes – a review. Renew Sustain Energy Rev 32:448–467. https://doi.org/10.1016/j.rser.2014.01.013

Tsydenova O, Bengtsson M (2011) Chemical hazards associated with treatment of waste electrical and electronic equipment. Waste Manag 31(1):45–58. https://doi.org/10.1016/j.wasman.2010.08.014

Tue NM et al (2010) Accumulation of polychlorinated biphenyls and brominated flame retardants in breast milk from women living in Vietnamese e-waste recycling sites. Sci Total Environ 408(9):2155–2162. https://doi.org/10.1016/j.scitotenv.2010.01.012

Umair S, Anderberg S, Potting J (2016) Informal electronic waste recycling in Pakistan. J Solid Waste Technol Manag 42(3):222–235. https://doi.org/10.5276/JSWTM.2016.222

UNEP (2015) Waste crime – waste risks

Wong CSC et al (2007a) Evidence of excessive releases of metals from primitive e-waste processing in Guiyu, China. Environ Pollut 148(1):62–72. https://doi.org/10.1016/j.envpol.2006.11.006

Wong CSC et al (2007b) Trace metal contamination of sediments in an e-waste processing village in China. Environ Pollut 145(2):434–442. https://doi.org/10.1016/j.envpol.2006.05.017

Wu Q et al (2015) Heavy metal contamination of soil and water in the vicinity of an abandoned e-waste recycling site: implications for dissemination of heavy metals. Sci Total Environ 506–507:217–225. https://doi.org/10.1016/j.scitotenv.2014.10.121

Wu PF et al (2016) Maternal transfer of dechloranes and their distribution among tissues in contaminated ducks. Chemosphere 150:514–519. https://doi.org/10.1016/j.chemosphere.2015.11.008

Xu L et al (2016) Differential proteomic expression of human placenta and fetal development following e-waste lead and cadmium exposure in utero. Sci Total Environ 550:1163–1170. https://doi.org/10.1016/j.scitotenv.2015.11.084

Zheng G et al (2013) Association between lung function in school children and exposure to three transition metals from an e-waste recycling area. J Expo Sci Environ Epidemiol 23(1):67–72. https://doi.org/10.1038/jes.2012.84

Zheng X et al (2016) Ambient air heavy metals in PM2.5 and potential human health risk assessment in an informal electronic-waste recycling site of China. Aerosol Air Qual Res 16 (2):388–397. https://doi.org/10.4209/aaqr.2014.11.0292

Chapter 8
Environmental and Health Effects: Exposure to E-waste Pollution

Muhammad Usman Khan, Athanasios Besis, and Riffat Naseem Malik

Abstract E-waste is a growing and emerging global problem because plethora of toxic chemicals are released to the environment during dismantling, burning, and recycling processes. Therefore, for this purpose, the current chapter aims to investigate with special emphasis on e-waste associated impacts on the environmental and human health. Informal e-waste activities with no safety measures exist in the developing countries like China, India, Ghana, Vietnam, Pakistan, etc. The major e-waste producer like USA and Europe have been increased the e-waste transportation into the developing countries for further processing. Besides that, large quantities of e-waste are also produced and received by China. Toxic chemicals, viz., PBDEs, DPs, PCDD/Fs, PAHs, PCBs, and heavy metals are reported to be prevalent in the e-waste sites at environmental compartments (air, soil/dust, drinking water) and human samples (placenta, umbilical cords, blood, hair, urine, and other tissues). Most of the studies showed that an elevated hazardous effect of e-waste-derived toxic chemicals on human population (children, adults, male, and female) are mainly prevailed at e-waste exposure sites as compared to the non-e-waste exposure or reference sites. The current chapter overview suggests improving: novel dismantling and remediation technologies; illegal transportation; lack of proper stringent policies and regulations for its safety measures and implementation; and there is also need of awareness regarding the toxic chemicals in e-waste and its exposure with ill-health impacts on human.

Keywords Environmental · Health effects · Exposure · E-waste · Pollution

---

M. U. Khan (✉)
Department of Botanical and Environmental Sciences, Faculty of Biological Sciences, Kohat University of Science & Technology, Kohat, Pakistan

Department of Marine Sciences and Convergence Engineering, Hanyang University, Ansan, Republic of Korea

A. Besis
Department of Chemistry, Environmental Pollution Control Laboratory, Aristotle University of Thessaloniki, Thessaloniki, Greece

R. N. Malik
Department of Environmental Sciences, Ecotoxicology and Environmental Biology Laboratory, Faculty of Biological Sciences, Quaid I Azam University, Islamabad, Pakistan

M. Z. Hashmi, A. Varma (eds.), *Electronic Waste Pollution*, Soil Biology 57,
https://doi.org/10.1007/978-3-030-26615-8_8

8.1 Introduction

Electronics/or electrical waste (e-waste) are the products that are discarded as waste materials with no reuse (Step 2014). E-waste may also be defined as "any appliances utilizing/or exercising electrical supply that has accomplished to end-of-life during its product lifetimes" (Organization for Economic Cooperation and Development (OECD)). E-waste is a homogenous mixture of different materials from many household products such as mobile phones, computers, television, music recorders, washing machines, and air conditioners that contains various toxic chemicals (Xu et al. 2015a, b, c, d, e). All these consumer products are intensively used on a daily basis worldwide, especially in developed and developing countries. Due to the strict regulation in developed countries e-wastes are sent for processing to developing countries because they have low cost of labors and lacking stringent regulations. The illegal recycling activity provides a potential source of income/job opportunities for poor peoples in developing countries.

The major e-waste producers are USA, Western Europe countries, China, Canada, India, Japan, and Australia (Chen et al. 2011), while the major recipients are Asian and South African countries. E-waste recycling activities are practiced in countries of Asia and Africa like China (Guiyu and Taizhou); in India (Delhi and Bengaluru); in Pakistan (Karachi, Lahore, Faisalabad, Gujranwala, Peshawar, and Rawalpindi); in Vietnam (Dong Mai, Bui Dau and Trang Minh); at Lagos in Nigeria; and in Ghana (Agbogbloshie and Accra) (Huo et al. 2007; Xu et al. 2015a, b, c, d, e; Lundgren 2012; Awasthi et al. 2016; Iqbal et al. 2015; Tue et al. 2010, 2016; Osibanjo and Nnorom 2007; Caravanos et al. 2011; Asante et al. 2012). Some countries like Brazil, Colombia, Morocco, Mexico, Kenya, South Africa, Peru, Senegal, Uganda, etc., have small recycling facilities for e-waste (Chen et al. 2011), while in many other countries recycling activities are yet unexplored. However, informal collection and transboundary movement in developing countries are still not well known (Balde 2015).

United Nations has estimated that about 40 million tons of e-wastes are produced every year worldwide but recently other studies increase this estimation to 42 million tons per year (Balde 2015). E-waste production globally was increased up to 65.4 million tons per year in 2017 (Kandil et al. 2013). About 80% of the e-wastes are transported unlawfully from developed to developing countries (Sthiannopkao and Wong 2013). It has been estimated that 60–70% of the e-waste collected from Europe is shifted to Asian and African countries for dismantling and recycling (Nordbrand 2009).

The high amount of e-waste generated throughout the globe are due to the products least life span, cheap/low prices, lifestyles of consumers, change of consumption patterns or increased consumption, illegal transboundary movement, low durability, which are non-cost effective, and unregulated. This rapid increase in e-waste generation is linked with the time to time upgradation advances in technologies, and further continuous fluctuation or fall in prices (Ni and Zeng 2012). Most of the developing countries lack proper disposal, safety measure, and proper management of the e-waste pollution. Therefore, the processing and dismantling of e-waste were carried out to recover the valuable components of gold, copper, platinum, tin, zinc, iron, silver and other metals (Huo et al. 2007). The extraction of e-waste parts was performed by tools, fire, acid, and other methods (Huo et al. 2007; Ikhlayel 2018; Heacock et al. 2016). In the aforementioned processes associated with e-wastes especially in developing countries, various toxic chemicals are released into the environment that is further linked to problems in human health (Bakhiyi et al. 2018). Several toxic chemicals originated from e-waste are responsible for endocrine disruption, reproductive and developmental abnormalities, cell/DNA damage, and cases (Leung et al. 2006). Therefore, it is of utmost importance for workers or any expose group to protect from e-waste toxic chemicals exposure using special clothes at recycling or dismantling places (Iqbal et al. 2015).

This chapter focuses on the different e-waste-derived toxic chemicals with special emphasis in context of their exposure effects on environment (air, dust, and potable water) and human health.

8.2 Toxic Chemicals Release and Environmental/Ecological Effect Exposure from E-waste

E-waste is a homogenous mixture of different materials like plastic, glass, ceramics, metals, and some other inorganic and organic hazardous chemicals. The inorganic toxic chemicals released from the e-waste are metals (e.g., Fe, Cu, Al, Tn, Pb, Ni, Ag, Au, As, Cd, Cr, Hg, Se, and Zn), while the organic chemicals are polybrominated diphenyl ethers (PBDEs), dechlrane plus (DP), polychlorinated biphenyls (PCB), polycyclic aromatic hydrocarbons (PAHs), dioxin and dioxin-like compounds, etc. (Ni and Zeng 2012; Grant et al. 2013; Song and Li 2014; Zhao et al. 2015; Heacock et al. 2016; Iqbal et al. 2017; Pascale et al. 2018; Awasthi et al. 2018; Luo et al. 2018). Most of the valuable heavy metals used in the electrical and electronic products are recovered unlawfully by different methods, viz., burning, dismantling, and acid extraction. These processes lead to the emissions of potentially toxic metals and organic compounds to local environmental matrices (air, dust, water, soil, food crops, etc.) or long-range atmospheric transport (Deng et al. 2007; Gullett et al. 2007; Sitaramaiah and Kumari 2014). Various studies in India and China show that toxic chemicals emitted from e-waste "travel" to other countries (Zhang et al. 2011).

Grant et al. (2013) further categorized exposure to e-waste pollution into three different sectors, viz. informal and formal recycling and exposure to hazardous compounds from e-waste remain in the environment (environmental exposure). Moreover, the peoples may expose to the occupational (formal or informal) e-waste-derived toxic chemicals through the contaminated soil/dust, air, drinking water, etc. (Norman et al. 2013; Grant et al. 2013). Exposure of a population to e-waste pollution is a very complicated process, where many important variables such as variant routes, exposure sources, diverse exposure time length, and the synergistic and inhibitory effects of different chemical exposure are necessary. Exposure to e-waste may vary with the quantity and type of e-waste, processing activities history, and length (Heacock et al. 2016). The extent of chemical contaminants in e-waste associated with adverse health effects is not well understood. Chen et al. (2011) have observed high predictable exposure at informal recycling station operating more than decade. Heacock et al. (2016) investigation showed that most of the toxic chemicals after release from electronics and electric goods are accumulated or sustain in the environment and further accompanied with adverse health effects at or nearby processing unit.

Different environmental matrices (soil, dust, air) have been polluted with e-waste-derived heavy metals and organic chemicals such as PAHs and PBDEs (Luo et al. 2011; Wang et al. 2011; Jinhui et al. 2011; Li et al. 2011; Zheng et al. 2011; Ni and Zeng 2012; Awasthi et al. 2018). Many researchers have studied the health risk exposure to human health from e-waste derived toxic chemicals via different environmental media, viz., soil/or dust, air, and water (Ni and Zeng 2012). Following are the overview of the e-waste-derived toxic chemicals concentrations in the associated matrices and their exposure to human via different dietary and non-dietary pathways (Fig. 8.1).

8.2.1 E-waste-Derived Toxic Chemicals Concentrations in Soil/Dust and Its Exposure to Human

Dust can act as a sink and can provide us information about the status and fates of toxic chemicals in the environment (Bi et al. 2011; Labunska et al. 2013; Ma et al. 2009). Inorganic and organic chemicals after their release from the e-waste source may get absorbed onto the dust particles that later on could be the part of human body either via ingestion, inhalation and dermal contact. The dust is especially dispersed due to dry weather, high temperature, and blowing air at the e-waste sites. Grant et al. (2013) observed that children are mainly prone to particle bound toxic chemicals while Wang et al. (2016) determined that PCB levels in the indoor dust samples at e-waste workshop sites (14,766 ng/g) were higher than local residents houses (1945 ng/g) at Taizhou and Zhejiang, China. At e-waste sites, the concentrations of heavy metals in indoor dust samples were higher than the control/urban sites (He et al. 2017). At the e-waste recycling sites of Guiyu, China, the surface dust is saturated with heavy metals (Ni (605), Zn (5080), Cu (13,400), and Pb (52,770) mg/kg) (Leung et al. 2008). Also, the Pb levels were found multifold higher than the floor dust sample in Delhi, India. PBDEs (2722 ng/g) were also found in house dust near Qingyuan e-waste sites (Yu et al. 2018). Besides dust, different e-waste-derived toxic chemicals are deposited/concentrated in the soil through dry or wet deposition and dust settlement from the air or directly by dumping e-waste materials in the soil. Researchers conducted a detailed studies on

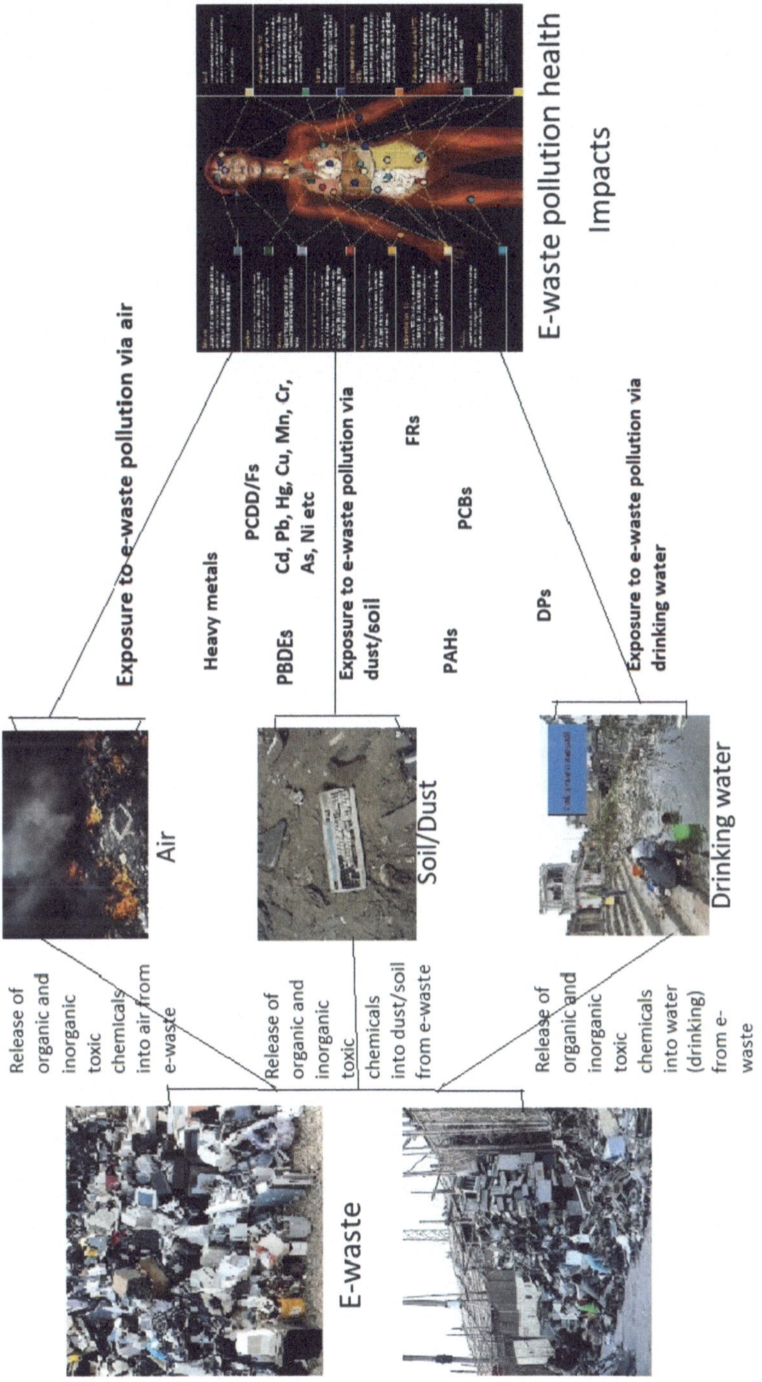

Fig. 8.1 Summary of environmental and human health exposure to e-waste pollution (Images collected from https://www.google.com.pk)

the soils at e-waste or in the vicinity of e-waste soils such as paddy soil, acid leaching area soil, dumping site soil, burning site soil, vegetation field soil, e-waste area soil, and farmyard soil (Zhao et al. 2015; Nie et al. 2015; Wang et al. 2015). Higher heavy metals and PBDEs concentrations were recorded in most of the soil samples at or in the vicinity of the e-waste sites as compared to the reference sites (Zhao et al. 2015; Nie et al. 2015; Iqbal et al. 2017; Luo et al. 2018).

The origin of these metals in soil reflects dumping and burning activities from a plethora of circuit boards and exhibit health hazards, especially to children and adults through dust ingestion and dermal contacts (Wong et al. 2007). A study from South China was conducted in the soils of incineration e-waste sites assessing elevated concentration for Zn (3690), Cu (11,140), Pb (4500), and Cd (17.1) mg/kg attributing their negative impact on the overall environment (Luo et al. 2011). Iqbal et al. (2017) found flame retardants (PBDEs, DPs, NBFRs, and OPFRs) in the surface soil at informal recycling e-waste sites in Karachi, Pakistan. They observed that BDE-209 (726 ng/g), DBDPE (551 ng/g), BTBPE (362 ng/g), and ∑TPP (296 ng/g) were found to be the dominant chemicals in the informal recycling e-waste soils. Ma et al. (2009) observed chlorinated PAHs (26.8 ng/g dry weight) in the e-waste soil samples at Taizhou, while the levels in e-waste garbage and e-waste recycling workshop dust were 59.1 ng/g and 103 ng/g, respectively. Ma et al. (2008) also measured higher levels of PCDD/Fs (854–10,200 pg/g dry weight) in different soil samples at e-waste sites than the two control sites (72.8–456 pg/g) and the cultivated soil (3.44–33.8 pg/g). Another site at Guiyu also showed higher PCDD/Fs concentration at e-waste recycling sites, where they argued that combustion activities are mainly responsible for these pollutants (Leung et al. 2007). Improper e-waste activities result in emissions of dioxin such as PCDD/Fs to the terrestrial environment (Leung et al. 2007). The e-waste after burning practices released compounds, especially PCDD/Fs into the air and dust (Ni and Zeng 2012). They also observed a significant correlation between PCDD/Fs concentration in dust samples and air at the Taizhou e-waste sites.

8.2.2 E-waste-Derived Toxic Chemicals Concentrations in Air and Its Exposure to Human

During the e-waste recycling process, a plethora of toxic chemicals released into the environment (Sepúlveda et al. 2010; Perkins et al. 2014). These toxic substances release made their way to the atmosphere from e-waste dismantling/or recycling activities; as ashes from burning practices; and as a fume from other different activities (Sepúlveda et al. 2010). Sepúlveda et al. (2010) and Awasthi et al. (2016) have investigated higher concentrations of polybrominated diphenyl ether (PBDE), dibenzofurans (PCDFs), polychlorinated dibenzodioxins (PCDDs), and lead (Pb) in the air, dust, and water at e-waste recycling sites of China (Guiyu) and India (Delhi). Higher concentrations of PCDD/Fs (64.9–2765 pg/m^3) and PBDD/Fs (14.3 ng/m^3) in air at Guiyu and around Taizhou e-waste sites were recorded. However, PCDD/Fs and PBDD/Fs atmospheric levels at Chendian, that is, 50 Km away from Guiyu also depicted at considerable high levels and pose negative impacts on the associated environment and health (Ni and Zeng 2012). In Ghana, the e-waste open burning sites of Agbogbloshie were dominated by PCDD/Fs, PBDD/Fs, and PXDD/Fs in air, while in Latin America (Euroguay), Pb concentration was recorded in ground surfaces, where cable burning activities prevailed (Pascale et al. 2016). Higher chlorinated, brominated, and dioxin and dioxin-like substances were also released into the air from burning plastic wastes (Tue et al. 2016). Ding et al. (2018) have investigated PBDEs (883 pg/m^3) concentrations in the air sample (PM 2.5) at Guangzhou, China. Heavy metals, on the other hand, have also close relationship with suspended particles in the atmosphere (Ni and Zeng 2012). Heavy metals were investigated at Guiyu e-waste site in the atmosphere at particulate phase (PM 2.5 μm) and the levels were relatively higher than Asian countries and some other sites in China. Such occurrence of heavy metals in air pose higher risk to the workers and local population via inhalation (Deng et al. 2006). At the e-waste site of Guiyu, elevated heavy metal concentrations (Cr, Cd, Mn, Pb) in particles (PM $<$ 2.5 μm) were observed in comparison with the reference sites (Zheng et al. 2016). Huang et al. (2016) measured heavy metals (As (22), Pb (150), Cr (21), Se (11), Zn (820), Cd (5.8) Cu (80), Sb (13), and Ni (10) (ng/m^3)) concentrations in air and dust at Qingyuan e-waste sites. Iqbal et al. (2017) also investigated flame retardants (PBDEs, DPs, NBFRs, and OPFRs) in the atmosphere (gaseous + particulate phase) at informal e-waste sites in Karachi, Pakistan. They further observed ∑OPFRs, ∑NBFRs, ∑PBDEs, and ∑DP concentrations in air samples (particulate + gaseous) ranged between 5903 and 24,986 ng/m^3, 21.0 and 170 ng/m^3, 40.0 and 275 ng/m^3, and 15.0 and 85.0 ng/m^3, respectively. They also concluded that the most leading congeners were OPFRs (5903–24,986 ng/m^3) in air. The concentration of PCDD/Fs in air at Guiyu e-waste sites was recorded in the range between 64.9 and 2765 pg/m^3 (Li et al. 2007), while at Taizhou it was measured as 14.3 ng/m^3 around e-waste sites (Li et al. 2008a, b).

8.2.3 E-waste-Derived Toxic Chemicals Concentrations in Drinking Water and Its Exposure to Human

In most part of the world, groundwater is mainly consumed by the human population for drinking, but some countries used surface water after purification/filtration for their populations' needs. Surface water is polluted by discharge of effluents from industries, homes, and also by direct dumping of wastes into it or by dry or wet deposition processes from atmosphere. Groundwater is contaminated by leaching activities from the surface water through recharging or dumping/or landfilling practices and e-waste processing may also contaminate the drinking water through the aforementioned processes. In developing countries, the health risk via consumption of potable water is increasing due to lack of filtering facilities, as most of the population is living below the poverty line. Postigo and Barcelo (2015) have highlighted in their study that the groundwater quality is usually deteriorated due to the inclusion of different toxic chemicals, especially from local and diffuse sources. E-waste-derived toxic chemicals can pollute the drinking water as stored in the storage tanks through dust or air deposition processes. Besides drinking purposes, groundwater is widely used for agricultural, industrial, and domestic purposes (Postigo and Barcelo 2015). Wu et al. (2015) measured relatively higher levels of heavy metals (Cu, 112; Cd, 5.60; Zn, 34.0; Mn, 138; Pb, 1.37; Cr, 0.058; Ni, 3.07 mg/L) in water samples from wells and traces of heavy metals in the tap water at Longtang, Guangdong, China. Grant et al. (2013) investigated that human populations are also exposed to the e-waste derived toxic chemicals through intake of water. So, there is a scarcity/lacking of available research in regard to toxic chemicals in drinking water at the e-waste sites.

Many studies investigated the intake of toxic chemicals (PBDEs, PCBs, PCDD/Fs, DP, PAHs, and heavy metals) on a daily basis by the population living at/or nearby e-waste sites (Iqbal et al. 2017; Ni and Zeng 2012; Grant et al. 2013). Various factors, viz., dietary habit differences, consumption pattern, inhalation or ingestion exposure time, lifestyle, and safeguard procedures are some factors responsible to affect or fluctuate the daily intake in human body (Song and Li 2014).

It was concluded from different studies thoroughly that as long as the e-waste-derived toxic chemicals concentrations in the variant environmental matrices increases, then it also enhances the chance of toxicity to human via exposure to dietary (water, food stuff, etc.) and non-dietary pathways (soil/dust, air, etc.) (Fig. 8.1). For the workers in e-waste facilities, the non-dietary exposure pathways are more significant than the dietary exposure pathways. It is difficult to identify the consequences in human health from e-waste-derived toxic chemicals.

8.3 E-waste Pollution Exposure and Body Burden

Prevalence of toxic pollutants in different environmental matrices at e-waste sites could attribute to assess the link with human exposure and risks by investigating directly the toxic chemical concentrations in human tissues (blood and serum, urine, breast milk, placenta, blood of umbilical cord, hair, and other tissues) (Ma et al. 2008). The occurrence of the pollutants in the body depends on its persistency, bioaccumulation, toxicity, lipophilic nature, and long-range transport behavior, exposure time, and other factors. Most of the previous literature highlighted that heavy metals are bioaccumulated at low concentrations in different tissues (blood, hair, urine, kidney, etc.) of human body and further showed a close relationship between heavy metals in the human body and the levels in e-waste recycling processes or e-waste sites (Ni and Zeng 2012). Awasthi et al. (2018) have carried out detailed studies and reviewed that heavy metals, PCBs, PCDD/Fs, PBBs, and PBDEs are detected in different human tissues/or samples (blood, hair, breast milk, and semen) at the same variant as in the e-waste sites (Wenling, Guiyu, Qingyuan) in China. Most of the previous studies conducted evidently depicted the relationship among informal recycling practices at e-waste sites and human body burdens (Song and Li 2014). So, different toxic chemicals after their release from the e-waste materials make their way to human via different exposure pathways (Fig. 8.1).

Researchers have observed the accumulation of toxic chemicals in the placenta of the newborn babies. Heavy metals (Cd, Ni, Cr, Pb), PCDD/Fs, and PBDEs were recorded in higher concentrations in human placenta at e-waste sites (Guiyu, Taizhou) as compared to the reference sites (Chaonan, Shantou) (Guo et al. 2010; Li et al. 2011; Chan et al. 2007; Leung et al. 2010).

Different studies conceded elevated levels of PCB, PBDE, and heavy metals (Cd, Cr, Ni, Pb) in the blood of umbilical cords at e-waste sites as compared to the control sites (Li et al. 2011; Wu et al. 2010, 2011). All the previous results showed that neonates burden to pollutants is mainly concerned with the mother involvement at e-waste sites.

Biomonitoring studies showed that breast milk is considered as an indicator of pollutant burdens in the body, that is, it reflects the transfer of pollutants from mother to their children (Leung et al. 2010). In the e-waste sites (Guiyu and Taizhou), there were also recorded high levels of PCBs, PBDEs, and PCDD/Fs in human breast milk as compared to the control sites (Lin'an) (Xing et al. 2009; Leung et al. 2010; Chan et al. 2007). Elevated concentrations of dioxin in the human breast milk at e-waste sites were reported as compared to the background areas (Chan et al. 2007; Asante et al. 2011). Li et al. (2017) also observed higher PBDEs concentration in human breast milk samples at Wenling e-waste sites. The PCDD/Fs levels in 80% of the human milk samples at Taizhou and at Hangzhou were above of the allowable limit (3 pg WHO-TEQ/g lipid) (Chan et al. 2007).

The blood and serum of human can be used as a biomarker to screen exposure to e-waste pollution. It was concluded from many studies that most of the pollutants originated from e-waste have been concentrated at relatively high levels in the human blood and serum as compared to other tissue or organs (Song and Li 2014). PBDEs levels were comparable to US population but on the other hand, its concentration were recorded as multifold higher than the populations of Japan, Singapore, and Europe (Inoue et al. 2006; Li et al. 2005; Hardell et al. 2007; Meironyte et al. 1999, 2001). PCB concentrations of about 257–399 ng/g lipid were also recorded in human samples in China (Zhao et al. 2009a). Higher BDE-209 (3100 ng/g lipid) concentrations were found in human serum at Guiyu e-waste sites (Bi et al. 2007). At Guiyu, the daily human intake of PCDD/Fs was higher in winter than summer. The daily intake for PCDD/Fs in children was at least two times higher than adults (Li et al. 2007). Higher $\sum$16PAHs concentration were measured in the blood samples (68.53 μg/dL) at Guiyu e-waste site than control groups (26.92 μg/dL) (Xu et al. 2015b, c). Zheng et al. (2017) have also measured PCBs and PBDEs levels in serum samples of male (3173, 690 ng/g lipid) and female (3423, 2309 ng/g lipid) groups, respectively, in South China. Cao et al. (2018) also measured Pb levels in the blood samples of exposed (5.06 μg/dL) group and reference group (3.6 μg/dL) at Guiyu and Haojian, China. A significant correlation among Pb levels in blood and e-waste sites were also observed (Ni and Zeng 2012). Cr in the human blood samples (61.9 μg/dL) at Guiyu e-waste sites was present in higher concentration as compared to the control blood group samples (22.5 μg/dL) (Xu et al. 2015d). Other heavy metals, viz., Cu, Mn, Cd, Pb, Zn, As, Hg, Cr, Ni, and Se were accumulated in higher levels in human blood samples exposed groups as compared to the reference/control groups (Lin et al. 2017). Zhao et al. (2010) estimated relatively high dietary daily intake for heavy metals (Pb, Cr, Cd, Hg, As) at different sites of Zhejiang.

To elucidate the bioindicator approach for e-waste pollution, hair is used as invasive tool in monitoring studies and can be utilized as an indicator for short- and long-term diverse pollutants/chemicals in the human body. Toxic chemicals may enter in human hair samples through different pathways like deposition/absorption from the atmosphere; diffusion from sebum or/sweat secretions; absorption from the bloodstream and skin during hair shaft development (Pragst and Balikova 2006). The Zn, Ni, Cd, Pb, Cu, Cr, Mn, As, PCB, PBDEs, PCDD/Fs, PBB, and Dechlorane plus were also concentrated in the human hair samples from workers at e-waste and residential sites (Zheng et al. 2011; Wang et al. 2009; Chan et al. 2007; Leung et al. 2010; Zhao et al. 2008). PBDEs concentration in human hair samples was recorded as 870 ng/g by lipid at Luqiao e-waste recycling sites, while high PBDEs concentrations in the resident hair samples around e-waste site of Zhejiang in comparison to control sites were also measured (Zhao et al. 2008). Huang et al. (2015) also observed high levels of Sb (160.78 ng/g) in the human hair samples at recycling e-waste sites of Guiyu, China than reference sites (61.74 ng/g).

Urine samples can be used as an indicator tool of pollutants/chemicals burdens in the human body and health risk analysis. Different levels of toxic chemicals, viz., Be, Cu, Zn, Mn, Cd, and Pb were detected at considerable levels in the urine samples (Wang et al. 2011; Zhang et al. 2007). In Ghana e-waste recycling site, high Pb, Fe, Sb, and PAHs concentrations were found in the urine samples of e-waste recycling workers as compared to the other sites (Asante et al. 2012; Feldt et al. 2014).

Zhao et al. (2009b) have carried out detailed study on cancer patients at e-waste dismantling sites (Zhejiang) in China and observed the occurrence of PBDEs (182.3, 174.1, 174.2 ng/g), PCBs (399.4, 455.1, 257.6 ng/g), and PBBs (185.8, 180.6, 191.8 ng/g) in kidney, liver, and lung, respectively. Higher mean concentrations for BDE-28 were recorded in the human semen samples at the e-waste sites of Qingyuan (5.02 pg/g ww) as compared to control groups (1.62 pg/g ww) (Yu et al. 2018). Higher TEQ values at e-waste sites (21.0–33.8 pg WHO-TEQ/g lipid) in human samples prevailed as compared to the control samples (5.59–11.9 pg WHO-TEQ/g lipid) (Ni and Zeng 2012).

From all these studies conducted on different human tissues showed that higher daily intake of pollutants were observed in the workers of e-waste or near sites as compared to the reference/control sites. The toxic chemicals released from e-waste were accumulated in adults during work and then transferred to their children through breastfeeding and placental exposure. It was conceded that there is ultimate need for more in-depth research work in regards to unexplored toxic chemicals screening and health impacts in human and other living organisms at the unidentified e-waste sites.

8.4 Health Effects Exposure from E-waste Pollution

A detailed general appraisal of human exposure effects to e-waste-derived toxic chemicals (organic and inorganic) is portrayed in Table 8.1.

In general view, the workers who are mainly engaged in e-waste recycling processes are more probably exposed to toxic chemicals than the general population (Ni and Zeng 2012). Human exposure to toxic chemicals released from e-waste has their key impacts worldwide, especially in the developing countries. However, different toxic chemicals in e-waste exposure may vary among genders and ages (male, female, and children) (Awasthi et al. 2018). Li et al. (2017) stated that children are more prone to e-waste-derived toxic chemicals than adults due to breastfeeding behaviors that enhance the transfer of these toxic chemicals. Heacock et al. (2016) also highlighted that pregnant women and children are considered to be the most vulnerable populations in regard to e-waste exposure. All the population exposed during working to e-waste-derived toxic chemicals is the reason due to lack of any protective measures. It was concluded that the peoples living near e-waste sites have shown higher concentrations of toxic chemicals as compared to reference sites or sites that are far away from the e-waste sites. Most of the studies highlighted those toxic chemicals (organic and inorganic) in e-waste exposure linked with different health outcomes (Heacock et al. 2016; Pascale et al. 2018; Grant et al. 2013). It was observed from previous study that in Guiyu about 80% of children severely experienced respiratory diseases (Li et al. 2007) and was assumed to be linked with e-waste pollution.

The poor family background of the children generally involved them to work in e-waste by searching and picking the recyclable and reusable parts from the mix of e-waste dump to sell them further. Additionally, they are preferred as cheap labors in developing countries during dismantling and burning activities of e-waste (Pascale et al. 2018). Awasthi et al. (2018) also recorded higher concentration of pollutants in the residents living in the vicinity of informal e-waste dismantling area than the reference sites. Several studies depicted that the populations working in e-waste are also exposed through gestation and among neonates when compared with reference sites (Guo et al. 2010; Li et al. 2008a, b; Zheng et al. 2008). E-waste pollution effects is also concerned with many ill reproductive effects like premature childbirth, stillbirths, abortion cases, and neonatal consequences such as reduction in childbirth weight and height; face difficulties in behavioral changes and cellular level changes and its length at the time of birth (Xu et al. 2012; Guo et al. 2012; Li et al. 2008a, b; Zheng et al. 2013a, b).

8.4.1 Potentially Toxic Elements Health Impacts

China is one of the leading manufacturer, recipient, and dumping ground for e-waste (Song et al. 2012). Besides, e-waste is a mixture of plethora of organic compounds, it

Table 8.1 E-waste-derived toxic chemicals exposure effects on human health (Adapted from Grant et al. 2013)

	Exposure setting	Exposed population	Primary toxicant	Health effect
Thyroid function				
Ju et al. (2008)	Ecological: exposed town vs control town	Mothers and newborn babies ($n = 93$)	Not assessed	Maternal: TSH higher (2.63 vs 2.10 mIU/L) and free T4 lower (16.47 vs 160.76 pmol/L) in the exposed group. Cord blood: TSH (6.35 vs 5.47 mIU/L), free T4 (8.45 vs 9.52 pmol/L)
Yuan et al. (2008)	Formal recycling	Formal and informal workers vs not exposed ($n = 49$)	PBDE	Serum PDBE: 382 vs 158 ng/g lipid weight, $p = 0.045$. Serum TSH: 1.7×10^6 vs 1.1×10^6 mIU/L, $p < 0.01$
Zhang et al. (2010)	Ecological: exposed town vs control town	Pregnant women ($n = 50$)	PCCDs and PCCFs, PCBs and PDBEs	Higher body burdens of PCCD and PCCFs, PCBs, and PDBEs. Serum TSH: 1.15 vs 2.65 nmol/L, $p = 0.015$. Free T4: 112.5 vs 139.0 nmol/L, $p = 0.015$
Wang et al. (2010)	Ecological: workers in exposed town vs those in control town	Population ($n = 442$)	PBDEs	Lower TSH (1.26 vs 1.57 μIU/mL), T3 (1632.4 vs 1817.2 pmol/L), and free T3 (4188.8 vs 4404.4 pmol/L), all $p < 0.001$. No difference in T4.
Han et al. (2011)	Ecological: exposed town vs control town	Population ($n = 369$)	Not assessed	Serum TSH: 1.8×10^6 vs 3.3×10^6 mIU/L. No p value reported
Lung function				
Zheng et al. (2013a, b)	Ecological: exposed town vs control town	School children (aged 8–13 years; $n = 144$)	Chromium, manganese, and nickel	Blood manganese: 374.92 nmol/L vs 271.18 nmol/L, $p < 0.01$. Nickel: 5.3 vs 3.0 mg/L, $p < 0.01$. FVC in boys aged 8–9 years: 1859 vs 2121 mL, $p = 0.03$. Decrease in FVC with increased chromium (11-year-old $\beta = -14.02$, $p = 0.018$, 13-year-old $\beta = -43.23$, $p = 0.027$), decreased FVC with increased nickel (10-year-olds $\beta = -18.47$, $p = 0.035$)

(continued)

Table 8.1 (continued)

	Exposure setting	Exposed population	Primary toxicant	Health effect
Reproductive health				
Guo et al. (2010)	Ecological: exposed town vs control town	Mother–infant pairs ($n = 220$)	Lead, chromium, cadmium, and nickel	No differences in birthweight, birth length, or gestational age. Negative correlation between placental nickel and gestational age (rs = −0.16, $p = 0.017$)
Guo et al. (2012)	Ecological: exposed town vs control town	Mother–infant pairs ($n = 183$)	PAHs	Cord blood total PAH: 108.5 vs 79.63 ppb, $p = 0.003$, chromium: 1.57 vs 1.05 ppb, $p = 0.049$, BaP: 2.14 vs 1.64 ppb, $p = 0.001$. DahA 12.26 vs 11.59 ppb, $p = 0.031$. Increased BaA, chrysene, and BaP in neonates with adverse birth outcomes (prematurity, low birthweight, still birth, and malformations) ($p < 0.05$). BaA negatively associated with neonatal height ($r = -0.23$, $p = 0.006$); chrysene and BaP negatively associated with gestational age ($r = -0.20$, $p = 0.013$; $r = -0.17$, $p = 0.042$, respectively).
Wu et al. (2010)	Ecological: exposed town vs control town	Mother–infant pairs ($n = 153$)	PBDEs	Higher total PBDEs in those with adverse birth outcomes (41.97 vs 9.88 ng/g, $p = 0.004$). No effect on neonatal length, GA, or infant sex
Wu et al. (2011)	Informal recycling	Mother–infant pairs ($n = 167$)	PCBs	Cord blood PCBs: 338.6 vs 140.2 ng/g, correlated with mothers' recycling activity. Higher total PCBs with adverse birth outcomes ($t = -2.26$, $p = 0.03$). Negative associations between individual PCB congeners and neonatal height, neonatal weight, Apgar score, gestational age, and BMI (all $p < 0.05$)
Wu et al. (2012)	Informal recycling	Pregnant women ($n = 167$)	PFOA	Serum PFOA: 17.0 vs 8.7 ng/mL ($p < 0.001$) correlated with recycling activity. Negative association between PFOA and spontaneous abortion ($t = -3.035$,

(continued)

Table 8.1 (continued)

	Exposure setting	Exposed population	Primary toxicant	Health effect
				$p = 0.003$), preterm birth ($t = -2.209$, $p = 0.029$), 1 log increase in PFOA associated with 15.99 (95% CI 2.72–4.25) reduction in GA, 267.3 g (37.2–573.3) reduction in birthweight, 1.91 cm (0.52–3.31) reduction in birth length and 1.37 (0.32–2.42) reduction in Apgar score
Li et al. (2008a, b)	Ecological: exposed town vs control town	Newborn infants (302)	Cr	Chromium Cord blood chromium, 2006: 306.2 vs 1.95 mg/L; 2007: 99.9 vs 32.48. No association with birth weight or birth length
Xu et al. (2012)	Informal recycling	Newborn infants ($n = 531$)	Pb	Lead Cord blood lead: 10.87 vs 2.25 mg/dL ($p < 0.01$), correlated with recycling activity. Higher rates of adverse birth outcomes: stillbirth (4.72% vs 1.03%, $p < 0.05$), preterm birth (5.68% vs 5.24%, $p < 0.05$), lower birthweight (3168 vs 3258 g, $p < 0.05$), and lower Apgar scores (9.6 vs 9.9, $p < 0.05$)
Growth				
Zheng et al. (2013a, b)	Ecological: exposed town vs control town	School children (aged 8–13 years; $n = 144$	Mn, Ni	Height: 126.8 vs 135.0 cm, $p < 0.001$. Weight: 24.7 vs 30.2 kg, $p < 0.01$. BMI: 15.2 vs 16.5, $p < 0.01$. Negative correlations between serum manganese and height (rs $= -0.303$, $p < 0.001$) and weight (rs $= -0.228$, $p = 0.006$) and serum nickel and height (rs $= -0.417$, $p < 0.001$), weight (rs $= -0.399$, $p < 0.001$), and BMI (rs $= -0.213$, $p = 0.011$)
Huo et al. (2007)	Ecological: exposed town vs control town	Children (younger than 6 years; $n = 226$)	Pb	Blood lead: 15.3 vs 9.94 mg/dL ($p < 0.01$). No differences in height, weight, chest circumference, or head circumference

(continued)

Table 8.1 (continued)

	Exposure setting	Exposed population	Primary toxicant	Health effect
Mental health outcomes				
Liu et al. (2011)	Ecological: exposed town vs control town	Children (aged 3–7 years; n = 303)	Pb	Blood lead: 0.64 vs 0.40 μmol/L (13.2 vs 8.3 mg/dL), $p < 0.01$. Temperament scores: activity level (mean ± SD 4.53 ± 0.83 vs 4.18 ± 0.81, $t = 3.377$), adaptability (4.96 ± 0.73 vs 4.67 ± 0.83, $t = 2.96$) and approach withdrawal (4.62 ± 0.85 vs 4.3 ± 0.89, $t = 2.877$)
Li et al. (2008a, b)	Ecological: exposed town vs control town	Newborn babies ($n = 152$)	Pb	Cord blood lead: 11.3 vs 6.0 mg/dL, $p < 0.001$; meconium lead: 2.5 vs 1.2 mg/g, $p < 0.001$. NBNA scores: total (38.45 vs 38.92, $Z = -2.023$, $p = 0.043$), behavior cluster (10.91 vs 11.29, $p = 0.012$). Negative associations between meconium lead and total NMNA ($r = -0.903$, $p < 0.01$), activity tone ($r = -0.637$, $p < 0.01$), and behavioral ($r = -0.826$, $p < 0.01$) scores
DNA Damage				
Chen et al. (2010)	Ecological: exposed town vs control town	Population ($n = 138$)	Not accessed	Micronuclei in binucleated cells: median 4.0% (range 2.0–7.0) vs 1.0% (0.0–2.0), $p < 0.01$
Yuan et al. (2008)	Recycling activity	Recycle workers vs farmers ($n = 49$)	Not accessed	Micronuclei in binucleated cells: median 5.0% (range 0.0–96.0) vs 0.0% (0.0–5.0), $p < 0.01$
Wang et al. (2011)	Recycling activity	Recycle workers vs not exposed ($n = 104$)	Lead, copper, and cadmium	Micronuclei in binucleated cells: median 4.0% (range 2.0–7.0) vs 1.0% (0.0–2.0), $p < 0.01$. Positive correlation between blood lead and micronuclei in binucleated cells ($r = 0.245$, $p < 0.01$). No associations with copper or chromium

(continued)

Table 8.1 (continued)

	Exposure setting	Exposed population	Primary toxicant	Health effect
Liu et al. (2009)	Recycling activity	Recycle workers vs not exposed ($n = 201$)	Not accessed	Micronuclei in binucleated cells: median 16.92% vs 3.47%, $p < 0.01$, chromosomal aberration (5.5% vs 1.7%), and DNA damage (comet assay). Women more affected than men ($p < 0.05$)
Li et al. (2008a, b)	Ecological: exposed town vs control town	Newborn infants ($n = 302$)	Cr	Comet assay: DNA damage (33.2% vs 10.7%, $p < 0.01$), length of tail (4.49 ± 1.92 μM vs 2.09 ± 0.65, $p < 0.01$). Blood chromium correlated with DNA damage (rs = 0.95, $p < 0.01$) and tail length (rs = 0.95, $p < 0.01$)
Gene expression				
Li et al. (2012)	Ecological: exposed town vs control town	Men	Not accessed	miRNA expression profiles in spermatozoa: differential regulation (109 downregulated, 72 upregulated)
Li et al. (2011)	Ecological: exposed town vs control town	Newborn infants ($n = 423$)	Cd	Metallothionein expression in placenta: 67.0% vs 3.7%, $p < 0.01$; and correlated with cord blood ($r^2 = 0.21$, $p = 0.01$) and placenta ($r^2 = 0.76$, $p = 0.01$) cadmium concentrations
Zhang et al. (2011)	Ecological: exposed town vs control town	Pregnant women ($n = 105$	Not accessed	Placental S100P protein (0.026 vs 0.032, $p = 0.045$) and mRNA (0.175 vs 1.462, $p < 0.001$); metallothionein expression (0.051 vs 0.035, $p = 0.003$)

TSH thyroid-stimulating hormone, *T4* tetraiodothyronine, *PDBE* polybrominated diphenyl ethers, *PCCD* polychlorinated dibenzodioxins, *PCCF* polychlorinated dibenzofurans, *PCB* polychlorinated biphenyls, *FVC* forced vital capacity, *PAH* polycyclic aromatic hydrocarbons, *BaP* benzo[a]pyrene, *DahA* dibenz[a,h]anthracene, *BaP* benzo[a]pyrene, *BMI* body-mass index, *Ppb* parts per billion, *NBNA* neonatal behavioral neurological assessment

is also considered as a vast source of potentially toxic metals, specifically at the informal e-waste sites (Luo et al. 2011; Wong et al. 2007). These potentially toxic metals are practiced dominantly in the production of different electronic products (Cd and Pb are used in circuit boards; Cd is used in batteries of computer) and electrical goods (Cu is used in electrical wiring) (Song et al. 2012; Zeng and Li 2013; Zeng et al. 2014). Electronic goods are containing about 60 toxic chemicals (Grant et al. 2013). These metals are mainly used in coating board circuits, cathode rays tubes, chips of semiconductors, and batteries. A mobile phone may include about 40 metals like Cu, Co, Au, platinum, indium, and ruthenium that also used in other electronics (Heacock et al. 2016). Each metal has a specific effect on human health. Pb has the ability to cause behavioral changes and interfere with learning abilities of human; Cu concedes to have adverse effects to liver; and the Cd chronic exposure enhances the chances of kidney damage and lung cancer (Grant et al. 2013; Esteban-Vasallo et al. 2012; Chan and Wong 2013). Chen et al. (2011) carried out a detailed study and concluded that neurotoxicants in e-waste, such as Cd, Pb, Hg, and brominated flame retardants (BFRs), are responsible for irreparable cognitive dearth in behavior and dysfunction of motor ability across the lifespan in children. High Pb, Cd, and Cu concentrations were recorded in rice and dust samples near e-waste homes increasing the risk of neurotoxicity and developmental disorders in children, via food consumption and dust ingestion (Zheng et al. 2013a, b). Zheng et al. (2013a, b) have conducted a study by linking heavy metals (Cr, Mn, and Ni) exposure in e-waste sites to lung cancer. High heavy metals levels in blood samples of children at e-waste sites associated with human health problems like asthma and other respiratory problems and also cause problems to the production of immunoglobulin E (Zeng et al. 2016a, b). Xu et al. (2015a) also stated that about 50% of children who are exposed to e-waste pollution were not able to develop immunity to hepatitis.

The exposures of children to heavy metals are the reason for the increasing gastrointestinal uptake and the permeability of blood–brain barrier (Huo et al. 2007; Guo et al. 2010). Particularly, in comparison to adults, children are more prone to the toxic chemicals due to multiple exposure pathways (i.e., placental and breastfeeding exposure), hand to mouth practices and rapid physiological changes (Garbino 2004). Sitaramaiah and Kumari (2014) argued that Cd and Be have sign of carcinogenicity and Cd were accumulated in the kidney of human body. Pb and Hg are coupled to reproductive, nervous, and genitourinary system problems.

8.4.2 *Organic Chemicals Health Effects*

Persistent pollutants like flame retardants are added in electronics and electrical and other consumer's products like carpet, upholstery, toy, and other various consumer goods to resist or slowdown the impact of fire to burn. Most of the studies correlated e-waste-derived toxic chemicals burdens with different health problems (Chen et al. 2010; Guo et al. 2012; Xu et al. 2012). Song and Li (2014) depicted that the toxic chemicals in e-waste concentrated in different organs of living bodies could cause different diseases (thyroid dysfunction, DNA damage, cancer, mental disorders, etc.). It was deduced from the previous literature survey that concentrations of organic chemicals from e-waste sites were considerably higher in residents living near these sites as compared to the reference sites. Recycling activities at the informal e-waste sites in China, India, and some west African countries have boasted different flame retardants (FRs) (Iqbal et al. 2017).

The organic compounds like PAHs, PBDEs, and PCBs are known to be carcinogenic (IARC 2010, 2011, 2012) and genotoxic (De-Flora et al. 1990; Barber et al. 2006; Lankoff et al. 2006; Danadevi et al. 2004; Hirano et al. 2013). PAHs are the toxic compounds even at a very low concentration can cause cancer and endocrine system dysfunction. It can also cause lower birth weight in the exposed population (Awasthi et al. 2018). Sitaramaiah and Kumari (2014) investigated that PAHs are responsible to cause bladder, skin, and lung cancer.

Most of the PBDEs, PBBs, and PCBs were detected in cancer patients around the e-waste recycling sites of Zhejiang, China (Zhao et al. 2009a). PCBs and PBDEs were reported to have their role in sperm quality (Joensen et al. 2009; McAuliff et al. 2012; Harley et al. 2010) and neurodevelopmental abnormalities. Yuan et al. (2008) have reported elevated concentrations for PBDEs in serum of Chinese individual at the e-waste workplace. It was also observed that extremely high levels of PBDEs exposure to human populations from the recycling e-waste sites can cause genotoxic damage and also affect the thyroid stimulating hormones (Yuan et al. 2008). PBDEs are also responsible to adversely affect the neonatal birth outcomes (Wu et al. 2010, 2011).

The occurrences of dioxin and dioxin-like PCBs and heavy metals (Cr, Cd) have increased the chances of certain diseases, viz., hypertension, obesity, cardiovascular, and type 2 diabetes (Everett et al. 2011, 2008; Halldorsson et al. 2012; Zhang et al. 2012). PAHs, PCBs, PBDEs, and heavy metals (Hg, Cd, Pb) were investigated and concluded that these e-waste-derived chemicals have potential association with cognitive functions impairment and decrease in IQ levels (intelligence) (Li et al. 2012; Chen et al. 2005; Eskenazi et al. 2013; Herbstman et al. 2010; Stewart et al. 2003).

Further detailed studies regarding the e-waste-derived toxic chemicals are needed to further investigate the individual and mixture of chemicals toxic effects on human health.

8.5 Current Policy/Legislation and Convention Regarding E-waste for Environmental Health Exposure

Nowadays, e-waste is the emerging and growing global problem of the current industrialized era. This problem gains their significant importance due to illegal transboundary e-waste shipment and disposal by the industrialized/developed countries to the poor or developing countries. It is quite hard to control and assess the e-waste international trade. There is lacking of any universal policy or legislation regarding e-waste pollution. Basel Convention (1989) exists that controls the transboundary movement of the hazardous wastes or disposal among nations or from developed to developing countries. A total of 181 countries participated in this convention, while only 53 countries signed their decisions. Its amendment in 1995, prevent all of the hazardous materials flow for any reason due to its transboundary movement. Other Rotterdam (1998) and Stockholm (2001) conventions are also involved in prevention/minimize of toxic chemical/persistent organic pollutants and struggle to tackle the e-waste flow in developing countries.

Bali declaration in 2008, have overview the waste management for human health and livelihood to empower political support to promote the public and private investments for the betterment of safer and eco-friendly technology (UNEP 2008). The third international conference at Busan, Korea in 2009 was held regarding the children health interacting with the toxic chemicals emanating from the consumer products at different places (schools, playgrounds, home, health care's, etc.) (WHO 2009). In 2013, Geneva intervention was put forward to state about the health of children and e-waste exploring the issues concerned with exposure and health outcomes. Pacific base consortium also addresses the same socioeconomic health issues of children and e-waste. These conventions have mainly focused on the root cause of sources exposure to e-waste and its impacts. Besides, all these conventions, each country have their own local regulations and implementations for e-waste and transboundary movement. Different international, national, and non-governmental agencies have been actively involved to solve e-waste regarding health problems and rules, regulations concerned with their practical application for its implementation and management (Grant et al. 2013). The United Nations industrial development organization was empowered to encourage eco-friendly industry for e-waste recycling and management in the developing nations. World Health Organization (WHO) have also launched "e-waste and child health initiative" for the protection of children's health to e-waste exposure (WHO 2013), while Geneva declaration was also forwarded to save children's health from e-waste pollution exposure and suggestive measures for its management. It is difficult to limit or locate the e-waste pollution because it is carried out at a small level, that is, local level and its mobile behavior and its illegal movement. Therefore, there is an urgent need to resolve this issue on a priority basis and to locate and identify such ongoing activities and to develop a coordination link with the local populations.

8.6 Concluding Remarks and Future Recommendations

Most of the toxic chemicals studied at e-waste sites pose greater health risk to the residing population as compared to the reference sites. Most of the e-waste pollution and their derived toxic chemicals in different environmental matrices and in human samples showed that they are highly prevalent in China as compared to other developed and developing countries. Mainly there is a lack of proper implementation plan, awareness, and safety measures. It was concluded that the electrical and electrical products with low life span and ever increasing demands for its upgradation/updating increase the e-waste production, so for this purpose potential policies and discourage plans are necessary. However, the lack of legislation exists to probe the illegal e-waste import and export with their production. There is indeed a dire need for proper rules, regulation for its management, and implementation both in developed and developing countries. E-waste inventory for its assessment are even lacking or poorly exist and therefore broader steps are needed in this regard. The combined effect of e-waste-derived toxic chemicals should be determined rather than individual chemical effect investigated at different studies. There is a severe scarcity of proper knowledge to the local populations regarding the health effects of the e-waste treatment processes. There is also a dire need for introducing emerging and novel approaches for improving the dismantling methods, sound eco-friendly management options, and practicing advanced remediation technologies for the removal of toxic chemicals at e-waste sites. Moreover, detailed future monitoring and health-related researches in regard to toxic chemicals derived from e-waste are highly encouraged.

References

Asante KA, Adu-Kumi S, Nakahiro K, Takahashi S, Isobe T, Sudaryanto A et al (2011) Human exposure to PCBs, PBDEs and HBCDs in Ghana: temporal variation, sources of exposure and estimation of daily intakes by infants. Environ Int 37:921–928

Asante KA, Agusa T, Biney CA, Agyekum WA, Bello M, Otsuka M, Itai T, Takahashi S, Tanabe S (2012) Multi-trace element levels and arsenic speciation in urine of e-waste recycling workers from Agbogbloshie, Accra in Ghana. Sci Total Environ 424:63–73

Awasthi AK, Zeng X, Li J (2016) Environmental pollution of electronic waste recycling in India: a critical review. Environ Pollut 211:259–270

Awasthi AK, Wang M, Awasthi MK, Wang Z, Li J (2018) Environmental pollution and human body burden from improper recycling of e-waste in China: a short-review. Environ Pollut 243:1310–1316. https://doi.org/10.1016/j.envpol.2018.08.037

Bakhiyi B, Gravel S, Ceballos D, Flynn MA, Zayed J (2018) Has the question of e waste opened a Pandora' s box? An overview of unpredictable issues and challenges. Environ Int 110:173–192

Balde C (2015) The global e-waste monitor 2014: quantities, flows and resources [Online]. Available from https://i.unu.edu/media/unu.edu/news/52624/UNU-1stGlobal-E-Waste-Monitor-2014-small.pdf. Accessed 20.08.18

Barber JL, Walsh MJ, Hewitt R, Jones KC, Martin FL (2006) Low-dose treatment with polybrominated diphenyl ethers (PBDEs) induce altered characteristics in MCF-7 cells. Mutagenesis 21:351–360

Bi XH, Thomas GO, Jones KC, Qu WY, Sheng GY, Martin FL, Fu JM (2007) Exposure of electronics dismantling workers to polybrominated diphenyl ethers, polychlorinated biphenyls, and organochlorine pesticides in South China. Environ Sci Technol 41(16):5647–5653. https://doi.org/10.1021/es070346a

Bi X, Li Z, Zhuang X, Han Z, Yang W (2011) High levels of antimony in dust from e-waste recycling in southeastern China. Sci Total Environ 409(23):5126–5128

Cao J, Xu X, Zhang Y, Zeng Z, Hylkema MN, Huo X (2018) Increased memory T cell populations in Pb-exposed children from an e-waste-recycling area. Sci Total Environ 616:988–995

Caravanos J, Clark E, Fuller R, Lambertson C (2011) Assessing worker and environmental chemical exposure risks at an e-waste recycling and disposal site in Accra, Ghana. J Health Pollut 1(1):16–25

Chan JK, Wong MH (2013) A review of environmental fate, body burdens, and human health risk assessment of PCDD/Fs at two typical electronic waste recycling sites in China. Sci Total Environ 463–464:1111–1123

Chan JKY, Xing GH, Xu Y, Liang Y, Chen LX, Wu SC, Wong CKC, Leung CKM, Wong MH (2007) Body loadings and health risk assessment of polychlorinated dibenzo-p-dioxins and dibenzofurans at an intensive electronic waste recycling site in China. Environ Sci Technol 41(22):7668–7674. https://doi.org/10.1021/es071492

Chen A, Dietrich KN, Ware JH, Radcliff EJ, Rogan WJ (2005) IQ and blood lead from 2 to 7 years of age: are the effects in older children the residual of high blood lead concentrations in 2-year-olds? Environ Health Perspect 113:597–601

Chen L, Yu C, Shen C, Zhang C, Liu L, Shen K, Tang X, Chen Y (2010) Study on adverse impact of e-waste disassembly on surface sediment in East China by chemical analysis and bioassays. J Soils Sediments 10(3):359–367. https://doi.org/10.1007/s11368-009-0176-8

Chen A, Dietrich KN, Huo X, Ho S-M (2011) Developmental neurotoxicants in e-waste: an emerging health concern. Environ Health Perspect 119(4):431–438

Danadevi K, Rozati R, Banu BS, Grover P (2004) Genotoxic evaluation of welders occupationally exposed to chromium and nickel using the Comet and micronucleus assays. Mutagenesis 19:35–41

De-Flora S, Bagnasco M, Serra D, Zanacchi P (1990) Genotoxicity of chromium compounds. A review. Mutat Res 238:99–172

Deng WJ, Louie PKK, Liu WK, Bi XH, Fu JM, Wong MH (2006) Atmospheric levels and cytotoxicity of PAHs and heavy metals in TSP and PM2.5 at an electronic waste recycling site in southeast China. Atmos Environ 40(36):6945–6955

Deng WJ, Zheng JS, Bi XH, Fu JM, Wong MH (2007) Distribution of PBDEs in air particles from an electronic waste recycling site compared with Guangzhou and Hong Kong, South China. Environ Int 33(8):1063–1069

Ding N, Chen SJ, Wang T, Wang T, Mai BX (2018) Halogenated flame retardants (HFRs) and water-soluble ions (WSIs) in fine particulate matter (PM2.5) in three regions of South China. Environ Pollut 243(238):823–832

Eskenazi B, Chevrier J, Rauch SA et al (2013) In utero and childhood polybrominated diphenyl ether (PBDE) exposures and neurodevelopment in the Chamacos study. Environ Health Perspect 121:257–262

Esteban-Vasallo MD, Aragonés N, Pollan M, López-Abente G, Perez-Gomez B (2012) Mercury, cadmium, and lead levels in human placenta: a systematic review. Environ Health Perspect 120(10):1369–1377

Everett CJ, Mainous AG, Frithsen IL, Player MS, Matheson EM (2008) Association of polychlorinated biphenyls with hypertension in the 1999–2002 National Health and Nutrition Examination Survey. Environ Res 108:94–97

Everett CJ, Frithsen I, Player M (2011) Relationship of polychlorinated biphenyls with type 2 diabetes and hypertension. J Environ Monit 13:241–251

Feldt T, Fobil JN, Wittsiepe J, Wilhelm M, Till H, Zoufaly A, Burchard G, Goen T (2014) High levels of PAH-metabolites in urine of e-waste recycling workers from Agbogbloshie, Ghana. Sci Total Environ 466–467:369–376

Garbino JPD (2004) Children's health and the environment: a global perspective. In: Pronczuk de Garbino J (ed) A resource manual for the health sector. World Health Organization, New York, pp 1–378

Grant K, Goldizen FC, Sly PD, Brune M-N, Neira M, Van Den Berg M, Norman RE (2013) Health consequences of exposure to e-waste: a systematic review. Lancet Glob Health 1(6):350–361

Gullett BK, Linak WP, Touati A, Wasson SJ, Gatica S, King CJ (2007) Characterization of air emissions and residual ash from open burning of electronic wastes during simulated rudimentary recycling operations. J Mater Cycles Waste Manage 9(1):69–79

Guo Y, Huo X, Li Y, Wu K, Liu J, Huang J, Zheng G, Xiao Q, Yang H, Wang Y (2010) Monitoring of lead, cadmium, chromium and nickel in placenta from an e-waste recycling town in China. Sci Total Environ 408:3113–3117

Guo Y, Huo X, Wu K, Liu J, Zhang Y, Xu X (2012) Carcinogenic polycyclic aromatic hydrocarbons in umbilical cord blood of human neonates from Guiyu, China. Sci Total Environ 427:35–40

Halldorsson TI, Rytter D, Haug LS et al (2012) Prenatal exposure to perfluorooctanoate and risk of overweight at 20 years of age: a prospective cohort study. Environ Health Perspect 120:668–673
Han G, Ding G, Lou X et al (2011) Correlations of PCBs, dioxin, and PBDE with TSH in children's blood in areas of computer e-waste recycling. Biomed Environ Sci 24:112–116
Hardell L, Carlberg M, Hardell K, Bjornfoth H, Wickbom G, Ionescu M, van-Bavel B, Lindstrom G (2007) Decreased survival in pancreatic cancer patients with high concentrations of organochlorines in adipose tissue. Biomed Pharmacother 61(10):659–664
Harley KG, Marks AR, Chevrier J, Bradman A, Sjodin A, Eskenazi B (2010) PBDE concentrations in women's serum and fecundability. Environ Health Perspect 118:699–704
He CT, Zheng XB, Yan X, Zheng J, Wang MH, Tan X et al (2017) Organic contaminants and heavy metals in indoor dust from e-waste recycling, rural, and urban areas in South China: spatial characteristics and implications for human exposure. Ecotoxicol Environ Saf 140: 109–115
Heacock M, Kelly CB, Asante KA, Birnbaum LS, Bergman ÅL, Bruné MN, Buka I, Carpenter DO, Chen A, Huo X (2016) E-waste and harm to vulnerable populations: a growing health. Environ Health Perspect 124(5):550–555
Herbstman JB, Sjodin A, Kurzon M et al (2010) Prenatal exposure to PBDEs and neurodevelopment. Environ Health Perspect 118:712–719
Hirano M, Tanaka S, Asami O (2013) Classification of polycyclic aromatic hydrocarbons based on mutagenicity in lung tissue through DNA microarray. Environ Toxicol 28(11):652–659. https://doi.org/10.1002/tox.20761
Huang Y, Ni W, Chen Y, Wang X, Zhang J, Wu K (2015) Levels and risk factors of antimony contamination in human hair from an electronic waste recycling area, Guiyu, China. Environ Sci Pollut Res 22:7112–7119
Huang CL, Bao LJ, Luo P, Wang ZY, Li SM, Zeng EY (2016) Potential health risk for residents around a typical e-waste recycling zone via inhalation of size-fractionated particle-bound heavy metals. J Hazard Mater 317:449–456
Huo X, Peng L, Xu XJ, Zheng LK, Qiu B, Qi ZL, Zhang B, Han D, Piao ZX (2007) Elevated blood lead levels of children in Guiyu, an electronic waste recycling town in China. Environ Health Perspect 115(7):1113–1117. https://doi.org/10.1289/ehp.9697
Ikhlayel M (2018) An integrated approach to establish e-waste management systems for developing countries. J Clean Prod 170:119–130. https://doi.org/10.1016/j.jclepro.2017
Inoue K, Harada K, Takenaka K, Uehara S, Kono M, Shimizu T, Takasuga T, Senthilkumar K, Yamashita F, Koizumi A (2006) Levels and concentration ratios of polychlorinated biphenyls and polybrominated diphenyl ethers in serum and breast milk in Japanese mothers. Environ Health Perspect 114(8):1179–1185. https://doi.org/10.1289/ehp.9032
Iqbal M, Breivik K, Syed JH, Malik RN, Li J, Zhang G, Jones KC (2015) Emerging issue of e-waste in Pakistan: a review of status, research needs and data gaps. Environ Pollut 207:308–318
Iqbal M, Syed JH, Breivik K, Chaudhry JI, Li J, Zhang G, Malik RN (2017) E-waste driven pollution in Pakistan: The First evidence of environmental and human exposure to flame retardants (FRs) in Karachi city. Environ Sci Technol 51(23):13895–13905
Jinhui L, Huabo D, Pixing S (2011) Heavy metal contamination of surface soil in electronic waste dismantling area: site investigation and source-apportionment analysis. Waste Manag Res 29: 727–738
Joensen UN, Bossi R, Leffers H, Jensen AA, Skakkebaek NE, Jorgensen N (2009) Do perfluoroalkyl compounds impair human semen quality? Environ Health Perspect 117:923–927
Ju Y, Xu G, Chen L, Shi J (2008) Effects of the electronic waste pollution on the levels of thyroid hormone and its receptor expression in the parturient woman and the newborn. Wei Sheng Yan Jiu 37:536–539. (in Chinese)
Kandil E, Noureldine SI, Abbas A, Tufano RP (2013) The impact of surgical volume on patient outcomes following thyroid surgery. Surgery 154(6):1346–1353

Labunska I, Harrad S, Santillo D, Johnston P, Brigden K (2013) Levels and distribution of polybrominated diphenyl ethers in soil, sediment and dust samples collected from various electronic waste recycling sites within Guiyu town, southern China. Environ Sci Processes Impacts 15(2):503–511

Lankoff A, Banasik A, Duma A et al (2006) A comet assay study reveals that aluminium induces DNA damage and inhibits the repair of radiation-induced lesions in human peripheral blood lymphocytes. Toxicol Lett 161:27–36

Leung A, Cai ZW, Wong MH (2006) Environmental contamination from electronic waste recycling at Guiyu, southeast China. J Mater Cycles Waste Manage 8(1):21–33. https://doi.org/10.1007/s10163-005-0141-6

Leung AOW, Luksemburg WJ, Wong AS, Wong MH (2007) Spatial distribution of polybrominated diphenyl ethers and polychlorinated dibenzo-p-dioxins and dibenzofurans in soil and combusted residue at Guiyu, an electronic waste recycling site in southeast China. Environ Sci Technol 41(8):2730–2737. https://doi.org/10.1021/es0625935

Leung AOW, Duzgoren-Aydin NS, Cheung KC, Wong MH (2008) Heavy metals concentrations of surface dust from e-waste recycling and its human health implications in southeast China. Environ Sci Technol 42(7):2674–2680. https://doi.org/10.1021/es071873x

Leung AO, Chan JK, Xing GH, Xu Y, Wu SC, Wong CK et al (2010) Body burdens of polybrominated diphenyl ethers in childbearing-aged women at an intensive electronic-waste recycling site in China. Environ Sci Pollut Res Int 17:1300–1313

Li QQ, Loganath A, Chong YS, Obbard JP (2005) Determination and occurrence of polybrominated diphenyl ethers in maternal adipose tissue from inhabitants of Singapore. J Chromatogr B 819(2):253–257

Li HR, Yu LP, Sheng GY, Fu JM, Peng PA (2007) Severe PCDD/F and PBDD/F pollution in air around an electronic waste dismantling area in China. Environ Sci Technol 41(16):5641–5646. https://doi.org/10.1021/es0702925

Li Y, Xu X, Liu J, Wu K, Gu C, Shao G, Chen S, Chen G, Huo X (2008a) The hazard of chromium exposure to neonates in Guiyu of China. Sci Total Environ 403:99–104

Li Y, Xu X, Wu K, Chen G, Liu J, Chen S, Gu C, Zhang B, Zheng L, Zheng M (2008b) Monitoring of lead load and its effect on neonatal behavioral neurological assessment scores in Guiyu, an electronic waste recycling town in China. J Environ Monit 10:1233–1238

Li J, Duan H, Shi P (2011) Heavy metal contamination of surface soil in electronic waste dismantling area: site investigation and source-apportionment analysis. Waste Manag Res 29(7): 727–738. https://doi.org/10.1177/0734242x10397580

Li Y, Li M, Liu Y, Song G, Liu N (2012) A microarray for microRNA profiling in spermatozoa from adult men living in an environmentally polluted site. Bull Environ Contam Toxicol 89: 1111–1114

Li X, Tian Y, Zhang Y, Ben Y, Lv Q (2017) Accumulation of polybrominated diphenyl ethers in breast milk of women from an e-waste recycling center in China. J Environ Sci 52:305–313

Lin X, Xu X, Zeng X, Xu L, Zeng Z, Huo X (2017) Decreased vaccine antibody titers following exposure to multiple metals and metalloids in e-waste-exposed preschool children. Environ Pollut 220:354–363

Liu Q, Cao J, Li KQ et al (2009) Chromosomal aberrations and DNA damage in human populations exposed to the processing of electronics waste. Environ Sci Pollut Res Int 16:329–338

Liu J, Xu X, Wu K et al (2011) Association between lead exposure from electronic waste recycling and child temperament alterations. Neurotoxicology 32:458–464

Lundgren K (2012) The global impact of e-waste: addressing the challenge. Geneva: International Labour Organization. https://www.ilo.org/wcmsp5/groups/public/%2D%2D-ed_dialogue/-sector/documents/publication/wcms_196105.pdf

Luo C, Liu C, Wang Y, Liu X, Li F, Zhang G, Li X (2011) Heavy metal contamination in soils and vegetables near an e-waste processing site, South China. J Hazard Mater 186(1):481–490. https://doi.org/10.1016/j.jhazmat.2010.11.024

Luo J, Wu J, Huo S, Qi S, Gu XS (2018) A real scale phytoremediation of multi-metal contaminated e-waste recycling site with *Eucalyptus globulus* assisted by electrical fields. Chemosphere 201: 262–268

Ma J, Kannan K, Cheng J, Horii Y, Wu Q, Wang W (2008) Concentrations, profiles, and estimated human exposures for polychlorinated dibenzo-p-dioxins and dibenzofurans from electronic waste recycling facilities and a chemical industrial complex in Eastern China. Environ Sci Technol 42(22):8252–8259. https://doi.org/10.1021/es8017573

Ma J, Horii Y, Cheng J, Wang W, Wu Q, Ohura T, Kannan K (2009) Chlorinated and parent polycyclic aromatic hydrocarbons in environmental samples from an electronic waste recycling facility and a chemical industrial complex in China. Environ Sci Technol 43(3):643–649. https://doi.org/10.1021/es802878w

McAuliff ME, Williams PL, Korrick SA, Altshul LM, Perry MJ (2012) Environmental exposure to polychlorinated biphenyls and p,p′-DDE and sperm sex-chromosome disomy. Environ Health Perspect 120:535–540

Meironyte D, Bergman A, Nore NK (1999) Analysis of polybrominated diphenyl ethers in Swedish human milk. A time-related trend study, 1972–1997. J Toxicol Environ Health A 58(6): 329–341. https://doi.org/10.1080/009841099157197

Meironyte GD, Bergman A, Nore NK (2001) Polybrominated diphenyl ethers in Swedish human liver and adipose tissue. Arch Environ Contam Toxicol 40(4):564–570. https://doi.org/10.1007/s002440010211

Ni HG, Zeng EY (2012) Mass emissions of pollutants from e-waste processed in China and human exposure assessment. In: Bilitewski B, Darbra R, Barceló D (eds) Global risk-based management of chemical additives II. The handbook of environmental chemistry, vol 23. Springer, Berlin

Nie X, Fan C, Wang Z, Su T, Liu X, An T (2015) Toxic assessment of the leachates of paddy soil sand river sediments from e-waste dismantling sites to microalga, *Pseudokirchneriella subcapitata*. Ecotoxicol Environ Saf 111:168–176

Nordbrand S (2009) Out of control: e-waste trade flows from the EU to developing countries. SwedWatch, Stockholm, 46

Norman RE, Carpenter DO, Scott J, Bruen MN, Sly PD (2013) Environmental exposures: an underrecognized contribution to noncommunicable diseases. Rev Environ Health 28:59–65

Osibanjo O, Nnorom I (2007) The challenge of electronic waste (e-waste) management in developing countries. Waste Manag Res 25:489–501

Pascale A, Sosa A, Bares CB, Battocletti A, Moll MJ, Pose D, Laborde A, Gonzalez H, Feola G (2016) E-waste informal recycling: an emerging source of lead exposure in South. Ann Glob Health 82(1):197–201

Pascale A, Bares C, Laborde A (2018) E-waste: Environmental and Health Challenges. Encycl Anthropoc 5:269–275. https://doi.org/10.1016/b978-0-12-809665-9.10013-8

Perkins DN, Brune Drisse MN, Nxele T, Sly PD (2014) E-waste: a global hazard. Ann Glob Health 80(4):286–295

Postigo C, Barcelo D (2015) Synthetic organic compounds and their transformation products in groundwater: occurrence, fate and mitigation. Sci Total Environ 503:32–47

Pragst F, Balikova MA (2006) State of the art in hair analysis for detection of drug and alcohol abuse. Clin Chim Acta 370(1–2):17–49

Sepúlveda A, Schluep M, Renaud FG, Streicher M, Kuehr R, Hagelüken C, Gerecke AC (2010) A review of the environmental fate and effects of hazardous substances released from electrical and electronic equipments during recycling: examples from China and India. Environ Impact Assess Rev 30:28–41

Sitaramaiah Y, Kumari MK (2014) Impact of electronic waste leading to environmental pollution. J Chem Pharm Sci 3:39–42

Song Q, Li J (2014) Environmental effects of heavy metals derived from the e-waste recycling activities in China: a systematic review. Waste Manag 34:2587–2594

Song B, Shuai S, Hou W, Yang J, Hou Y (2012) CDNA, genomic sequence cloning and overexpression of ribosomal protein S9 gene (RPS9) from Ailuropoda. https://doi.org/10.1109/CSIP.2012.6308914

Step (2014) One global definition of e-waste. Solving the e-waste problem (step) white paper series. UnitedNationsUniversity/StepInitiative(2014):www.stepinitiative.org/files/step/_documents/StEP_WP_One%20Global%20Definition%20of %20E-waste_20140603_amended

Stewart PW, Reihman J, Lonky EI, Darvill TJ, Pagano J (2003) Cognitive development in preschool children prenatally exposed to PCBs and MeHg. Neurotoxicol Teratol 25:11–22

Sthiannopkao S, Wong MH (2013) Handling e-waste in developed and developing countries: initiatives, practices, and consequences. Sci Total Environ 463:1147–1153

Tue NM, Sudaryanto A, Minh TB, Isobe T, Takahashi S, Viet PH, Tanabe S (2010) Accumulation of polychlorinated biphenyls and brominated flame retardants in breast milk from women living in Vietnamese e-waste recycling sites. Sci Total Environ 408:2155–2162

Tue NM, Goto A, Takahashi S, Itai T, Asante KA, Kunisue T, Tanabe S (2016) Release of chlorinated, brominated and mixed halogenated dioxin-related compounds to soils from open burning of e-waste in Agbogbloshie (Accra, Ghana). J Hazard Mater 302:151–157

UNEP (2008) Bali declaration on waste management for human health and livelihood. Ninth meeting of the conference of the parties to the basel convention on the control of transboundary movements of hazardous wastes and their disposal, 23–27 June 2008, Bali, Indonesia. United Nations Environment Programme. Available http://www.basel.int/Portals/4/Basel%20Convention/docs/meetings/cop/cop9/bali-declaration/BaliDeclaration

Wang T, Fu J, Wang Y, Liao C, Tao Y, Jiang G (2009) Use of scalp hair as indicator of human exposure to heavy metals in an electronic waste recycling area. Environ Pollut 157(8–9):2445–2451. https://doi.org/10.1016/j.envpol.2009.03.010

Wang H, Zhang Y, Liu Q, Wang F, Nie J, Qian Y (2010) Examining the relationship between brominated flame retardants (BFR) exposure and changes of thyroid hormone levels around e-waste dismantling sites. Int J Hyg Environ Health 213:369–380

Wang Q, He AM, Gao B, Chen L, Yu QZ, Guo H, Shi BJ, Jiang P, Zhang ZY, Li PL, Sheng YG, Fu MJ, Wu CT, Chen MX, Yuan J (2011) Increased levels of lead in the blood and frequencies of lymphocytic micronucleated binucleated cells among workers from an electronic-waste recycling site. J Environ Sci Health A Tox Hazard Subst Environ Eng 46(6):669–676. https://doi.org/10.1080/10934529.2011.563176

Wang J, Liu L, Wang J, Pan B, Fu X, Zhang G, Zhang L, Lin K (2015) Distribution of metals and brominated flame retardants (BFRs) in sediments, soils and plants from an informal e-waste dismantling site, South China. Environ Sci Pollut Res 22:1020–1033

Wang Y, Hu J, Lin W et al (2016) Health risk assessment of migrant workers' exposure to polychlorinated biphenyls in air and dust in an e-waste recycling area in China: indication for a new wealth gap in environmental rights. Environ Int 87:33–41

WHO (2009) Busan pledge for action on children's health and environment. The 3rd WHO international conference of children's health and the environment, 7–10 June 2009, Busan, Republic of Korea. World Health Organization. Available http://www.who.int/ifcs/meetings/3ce

WHO (2013) E-waste and child health initiative. Geneva [Online]. Available http://www.who.int/ceh/risks/ewaste/en/. Accessed 01.09.18

Wong CSC, Duzgoren-Aydin NS, Aydin A, Wong MH (2007) Evidence of excessive releases of metals from primitive e-waste processing in Guiyu, China. Environ Pollut 148(1):62–72. https://doi.org/10.1016/j.envpol.2006.11.006

World Health Organization International Agency for Research on Cancer (IARC) (2010) Some non-heterocyclic polycyclic aromatic hydrocarbons and some related exposures. IARC Monogr Eval Carcinog Risks Hum 92:754–773

World Health Organization International Agency for Research on Cancer (IARC) (2011) Agents classified by the IARC monographs. IARC Monogr Eval Carcinog Risks Hum 1–106: 1–17

World Health Organization International Agency for Research on Cancer (IARC) (2012) A review of human carcinogens: chemical agents and related occupations. IARC Monogr Eval Carcinog Risks Hum 100:249–378

Wu K, Xu X, Liu J, Guo Y, Li Y, Huo X (2010) Polybrominated diphenyl ethers in umbilical cord blood and relevant factors in neonates from Guiyu, China. Environ Sci Technol 44:813–819

Wu K, Xu X, Liu J, Guo Y, Huo X (2011) In utero exposure to polychlorinated biphenyls and reduced neonatal physiological development from Guiyu, China. Ecotoxicol Environ Saf 74: 2141–2147

Wu K, Xu X, Peng LJ, Guo Y, Huo X (2012) Association between maternal exposure to perfluorooctanoic acid (PFOA) from electronic waste recycling and neonatal health outcomes. Environ Int 48:1–8

Wu Q, Leung JYS, Geng X, Chen S, Huang X, Li H, Huang Z, Zhu L, Chen J, Lu Y (2015) Heavy metal contamination of soil and water in the vicinity of an abandoned e waste recycling site: implications for dissemination of heavy metals. Sci Total Environ 506–507:217–225

Xing GH, Chan JKY, Leung AOW, Wu SC, Wong MH (2009) Environmental impact and human exposure to PCBs in Guiyu, an electronic waste recycling site in China. Environ Int 35(1):76–82. https://doi.org/10.1016/j.envint.2008.07.025

Xu X, Yang H, Chen A, Zhou Y, Wu K, Liu J, Zhang Y, Huo X (2012) Birth outcomes related to informal e-waste recycling in Guiyu, China. Reprod Toxicol 33:94–98

Xu L, Huo X, Zhang Y, Li W, Zhang J, Xu X (2015a) Polybrominated diphenyl ethers in human placenta associated with neonatal physiological development at a typical e-waste recycling area in China. Environ Pollut 196:414–422

Xu X, Zeng X, Boezen HM, Huo X (2015b) E-waste environmental contamination and harm to public health in China. Front Med 9(2):220–228. https://doi.org/10.1007/s11684-015-0391-1

Xu X, Chen Z, Zhang J, Guo P, Fu T, Dai Y, Lin SL, Huo X (2015c) Decreased blood hepatitis B surface antibody levels linked to e-waste lead exposure in preschool children. J Hazard Mater 298:122–128

Xu X, Liu J, Huang C, Lu F, Chiung YM, Huo X (2015d) Association of polycyclic aromatic hydrocarbons (PAHs) and lead co-exposure with child physical growth and development in an e-waste recycling town. Chemosphere 139:295–302

Xu X, Yekeen TA, Liu J, Zhuang B, Li W, Huo X (2015e) Chromium exposure among children from an electronic waste recycling town of China. Environ Sci Pollut Res 22:1778–1785

Yu Y, Lin B, Liang W, Li L, Hong Y, Chen X, Xu X, Xiang M, Huang S (2018) Associations between PBDEs exposure from house dust and human semen quality at an e waste areas in South China. A pilot study. Chemosphere 198:266–273

Yuan J, Chen L, Chen D, Guo H, Bi X, Ju Y, Jiang P, Shi J, Yu Z, Yang J, Li L, Jiang Q, Sheng G, Fu J, Wu T, Chen X (2008) Elevated serum polybrominated diphenyl ethers and thyroid-stimulating hormone associated with lymphocytic micronuclei in Chinese workers from an e-waste dismantling site. Environ Sci Technol 42(6):2195–2200. https://doi.org/10.1021/es702295f

Zeng X, Li J (2013) Implications for the carrying capacity of lithium reserve in China. Resour Conserv Recycl 80:58–63

Zeng X, Li J, Singh N (2014) Recycling of spent lithium-ion battery: a critical review. Crit Rev Environ Sci Technol 44:1129–1165

Zeng X, Xu X, Boezen HM, Huo X (2016a) Children with health impairments by heavy metals in an e-waste recycling area. Chemosphere 148:408–415

Zeng Y, Zhang R, Lim TJ (2016b) Wireless communications with unmanned aerial vehicles: opportunities and challenges. arXiv preprint arXiv:1602.03602

Zhang Y, Chen L, Ju Y, Chen Y, Jiang Q, Shi J (2007) Levels of cadmium, lead and copper in blood and urine samples from residents exposed to the e-waste recycling environment. J Environ Health 24(8):563–566

Zhang J, Jiang Y, Zhou J et al (2010) Elevated body burdens of PBDEs, dioxins, and PCBs on thyroid hormone homeostasis at an electronic waste recycling site in China. Environ Sci Technol 44:3956–3962

Zhang XL, Luo XJ, Liu HY, Yu LH, Chen SJ, Mai BX (2011) Bioaccumulation of several brominated flame retardants and dechlorane plus in waterbirds from an e-waste recycling region in South China: associated with trophic level and diet sources. Environ Sci Technol 45(2):400–405

Zhang A, Hu H, Sanchez BN et al (2012) Association between prenatal lead exposure and blood pressure in children. Environ Health Perspect 120:445–450

Zhao GF, Wang ZJ, Dong MH, Rao KF, Luo JP, Wang DH, Zha JM, Huang SB, Xu YP, Ma M (2008) PBBs, PBDEs, and PCBs levels in hair of residents around e-waste disassembly sites in Zhejiang Province, China, and their potential sources. Sci Total Environ 397(1–3):46–57. https://doi.org/10.1016/j.scitotenv.2008.03.010

Zhao G, Zhou H, Wang D, Zha J, Xu Y, Rao K, Ma M, Huang S, Wang Z (2009a) PBBs, PBDEs, and PCBs in foods collected from e-waste disassembly sites and daily intake by local residents. Sci Total Environ 407(8):2565–2575. https://doi.org/10.1016/j.scitotenv.2008.11.062

Zhao GF, Wang ZJ, Zhou HD, Zhao Q (2009b) Burdens of PBBs, PBDEs, and PCBs in tissues of the cancer patients in the e-waste disassembly sites in Zhejiang, China. Sci Total Environ 407(17):4831–4837. https://doi.org/10.1016/j.scitotenv.2009.05.031

Zhao G, Zhou H, Wang Z (2010) Concentrations of selected heavy metals in food from four e-waste disassembly localities and daily intake by local residents. J Environ Sci Health A Tox Hazard Subst Environ Eng 45(7):824–835. https://doi.org/10.1080/10934521003709016

Zhao W, Ding L, Gu X, Luo J, Liu Y (2015) Levels and ecological risk assessment of metals in soils from a typical e-waste recycling region in southeast China. Ecotoxicology 24(9):1947–1960. https://doi.org/10.1007/s10646-015-1532-7

Zheng L, Wu K, Li Y, Qi Z, Han D, Zhang B, Gu C, Chen G, Liu J, Chen S, Xu X, Huo X (2008) Blood lead and cadmium levels and relevant factors among children from an e-waste recycling town in China. Environ Res 108(1):15–20. https://doi.org/10.1016/j.envres.2008.04.002

Zheng J, Luo X-J, Yuan J-G, He L-Y, Zhou Y-H, Luo Y, Chen S-J, Mai B-X, Yang Z-Y (2011) Heavy metals in hair of residents in an e-waste recycling area, South China: contents and assessment of bodily state. Arch Environ Contam Toxicol 61(4):696–703. https://doi.org/10.1007/s00244-011-9650-6

Zheng G, Xu X, Li B, Wu K, Yekeen TA, Huo X (2013a) Association between lung function in school children and exposure to three transition metals from an e-waste recycling area. J Expo Sci Environ Epidemiol 23:67–72

Zheng J, Chen KH, Yan X, Chen SJ, Hu GC, Peng XW et al (2013b) Heavy metals in food, house dust, and water from an e-waste recycling area in South China and the potential risk to human health. Ecotoxicol Environ Saf 96:205–212

Zheng X, Xu X, Yekeen TA et al (2016) Ambient air heavy metals in PM2.5 and potential human health risk assessment in an informal electronic-waste recycling site of China. Aerosol Air Qual Res 16:388–397

Zheng J, He CT, Chen SJ, Yan X, Guo MN, Wang MH, Yu YJ, Yang ZY, Mai BX (2017) Disruption of thyroid hormone (TH) levels and TH-regulated gene expression by polybrominated diphenyl ethers (PBDEs), polychlorinated biphenyls (PCBs), and hydroxylated PCBs in e-waste recycling workers. Environ Int 102:138–144

Chapter 9
Health Effects of E-waste Pollution

Naseer Ali Shah, Yasir Rasheed, and Raja Muhammad Waqas Anjum

Abstract Human population health is at high risk with the increasing load of electronic waste (e-waste) on earth globally. E-waste has resulted in severe health issues such as endocrine disruption, reproduction anomalies, abnormalities of brain, and gene expression. Here we summarize, the different studies reporting the association of exposure of electronic waste and its harmful effects on human health.

Keywords Health effects · E-waste · Pollution

9.1 Introduction

Electrical and electronic waste (e-waste) is also known as a waste of electronic and electrical tools or equipment. It is also defined as any disposal or end life of such tools or equipment that relay on electric current or electromagnetic fields for their proper working. This category includes household appliances, electrical and electronic tools, toys, lighting equipment, telecommunications tools, information technology, and leisure and sports stuff, medical equipment and devices, controlling instruments, and dispensers. There are many other types of equipment and tools that are included in e-waste. Electrical and electronic tool's components and equipment like, plastic casings, batteries, cathode rays emitting tubes, circuit boards, lead capacitors, and activated breakers. Exposure routes to these health-threatening components can vary and it depends on the matter and the way it is recycled. Usually, exposure to these life-threatening harmful components of e-waste is probably through dermal contacts, or via inhalation or ingestion. People come across these e-waste equipment and pollutants through direct exposure with contaminated soil, air, dust, water, and food resources that include meat. High-risk groups include children, pregnant women, fetuses, old age community; disable people, those

N. A. Shah (✉) · Y. Rasheed · R. M. W. Anjum
Department of Biosciences, COMSATS University Islamabad, Islamabad, Pakistan
e-mail: drnaseer@comsats.edu.pk

M. Z. Hashmi, A. Varma (eds.), *Electronic Waste Pollution*, Soil Biology 57,
https://doi.org/10.1007/978-3-030-26615-8_9

workers who use informal techniques of recycling e-waste. Children, in particular, are susceptible because of supplementary routes of exposure and contact (e.g., placental, breastfeeding contacts), high-threat behaviors (e.g., hand-to-mouth behavior within adolescence and high risk activities in teenage), their altered functioning (e.g., ingestion of air, food, water, and toxin removal at minimum rates) (Grant et al. 2013).

9.2 E-waste Effect on Health

Strong connections have been reported by 16 out of 23 studies between e-waste exposure and physical health issues that include alterations to normal cell functioning, thyroid function, lung function, growth, and reproductive health concerns. Chemical constituents of e-waste analyzed are different in these studies, including polycyclic aromatic hydrocarbons, perfluoroalkyls, polychlorinated dibenzodioxins, polychlorinated dibenzofurans, polybrominated diphenyl ethers, polychlorinated biphenyls (dioxin-related polychlorinated biphenyls), and other chemical elements, for example, chromium, manganese, nickel, and lead (Grant et al. 2013). Some of the health effects have been described to evaluate the consequences of e-waste on human health (Figs. 9.1 and 9.2).

9.2.1 *Endocrine Function*

Many studies have reported the detrimental effects of e-waste on thyroid function. In these studies, TSH profile of inhabitants from towns having e-waste recycling areas is compared with unexposed towns (control). These studies showed that TSH profile of people living in the e-waste recycling areas was disrupted in comparison to normal values (Ju et al. 2008). Similar observations have also been reported for tetraiodothyronine concentrations (T4) (Grant et al. 2013).

9.2.2 *Reproduction*

Several studies have been conducted to analyze the outcome of e-waste on the pregnancy-related parameters. Although the exposure situations and toxicants assessed were different in these studies, but the effects were consistent with high rise in spontaneous abortion (Wu et al. 2012), stillbirths (Guo et al. 2012; Wu et al. 2012; Xu et al. 2012), premature births (Guo et al. 2012; Wu et al. 2012; Xu et al. 2012), low birth weights (Guo et al. 2012; Wu et al. 2012; Xu et al. 2012), and different birth lengths (Guo et al. 2012; Wu et al. 2012; Xu et al. 2012) in nearly all of the research studies.

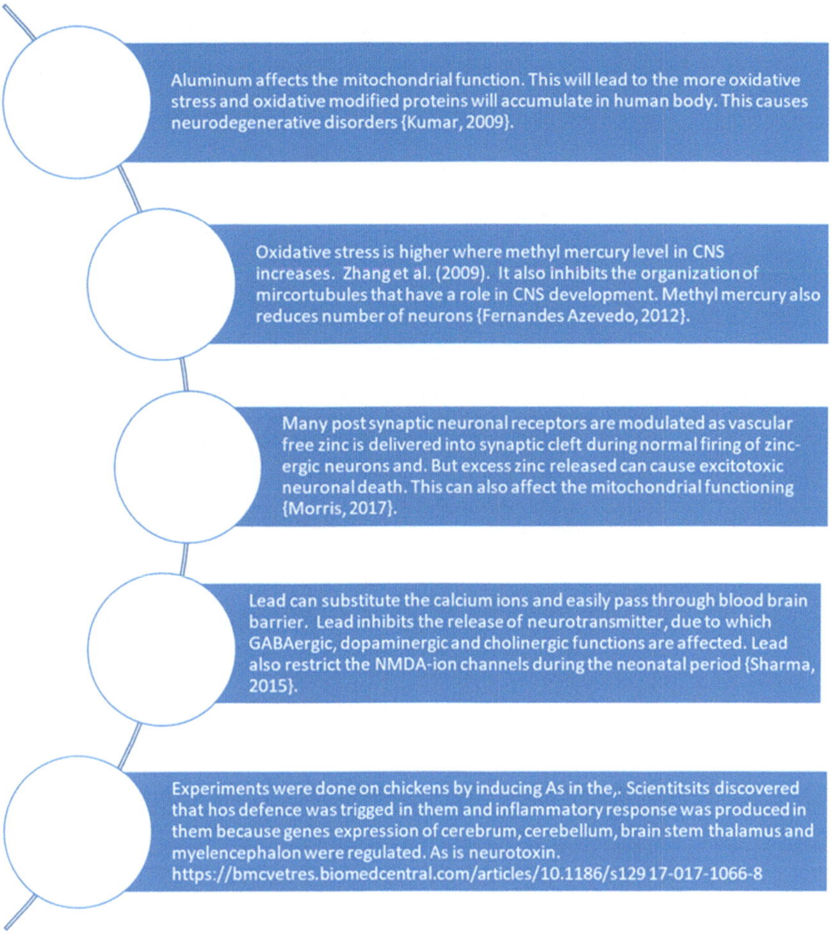

Fig. 9.1 Some important effect of heavy metals

Adverse and complicated birth defects have been related to increased level of exposures to polycyclic aromatic hydrocarbons (Guo et al. 2012) and organic pollutants that include polychlorinated biphenyls, polybrominated diphenyl ethers (Guo et al. 2012), and perfluoroalkyls (Wu et al. 2012).

Results of two ecological research studies (Guo et al. 2012; Lechner et al. 2004) confirmed that physical growth indicators, that is, body mass index, height, and weight have been notably lesser in exposed population of "Guiyu" as compared to population of control area "Liangying" (Grant et al. 2013).

Infertility in both men and women has been reported because of exposure to lead at high doses (Potti et al. 2006). Lead is positively associated with increased rates of hypospermia, asthenospermia, and malformed sperm, as well as decrease in motility,

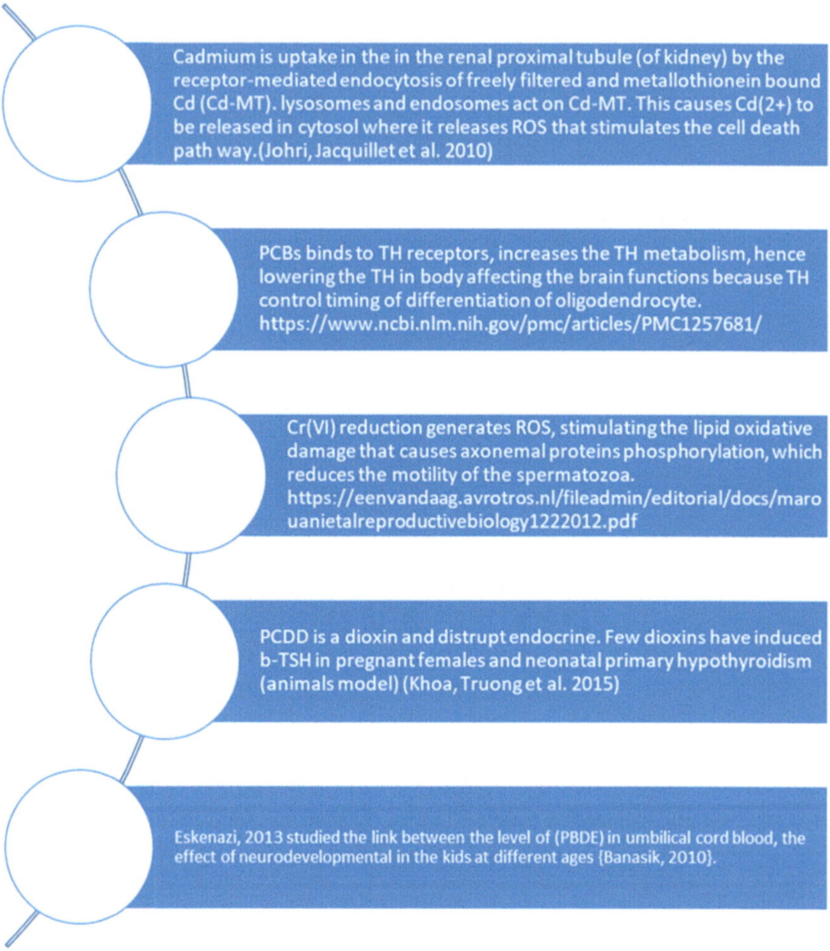

Fig. 9.2 Mechanism of heavy metal in human body

penetrability, and number of sperms (Meneses et al. 2013). A significant association has been observed among e-waste exposure and alteration in microRNA (small non-coding microRNAs that regulate gene translation) profiling of gene expression in spermatozoa. Out of 182 microRNAs that had been isolated from population affected by e-waste, 109 miRNAs have been downregulated and 72 were upregulated. These results indicate that there is a possible relationship between ecological exposure to e-waste and sperm counts and quality (Zhang et al. 2010).

9.2.3 Brain Damage

Lead has been confirmed the major chemical source that showed association with neural health problems linked as exposure to e-waste. A research evaluated potential in children and showed newborn babies behavioral neurological evaluation rates in others. About 45 children in Guiyu showed increased lead contents in their blood as compared to those living in regions with no e-waste recycling. It was also correlated with the living location of residence in Guiyu where a workshop on e-waste recycling had been organized (Xu et al. 2012). Newborn babies had an increased concentration of lead in blood meconium and cord that have been related to maternal and paternal association in e-waste recycling time-span spent in Guiyu before and during pregnancy and e-waste management. Newborn behavioral and neurological evaluation levels were altered significantly between the e-waste affected and control population. High concentration levels of lead have been related to lower newborn behavioral and neurological estimation levels and abnormalities in temperament levels (Grant et al. 2013). PCDD/Fs, PBDEs, PCBs, Fe, Hg, Pb, Se, and Zn are neurotoxic (Karaman and Frazzoli 2010).

9.2.4 DNA Damage

There is a difference of DNA damage in people having an exposure of e-waste during working or recycling waste. People with exposure to e-waste showed more DNA damage than those who are not exposed. The group exposed to e-waste showed increased damage and abnormalities of chromosome as compared to control group. Percentage of exposed and control group is (5.5% and 1.7%). The considerable differences in their DNA damage showed in the comet assay among the two populations were recorded (Robinson 2009). Effects of smoking cigarette had not been included in this study. DNA damage rate was higher in women than in men. Women had greater exposure to pollutants as they spent more time indoors. There are significant differences showed in lymphocytic DNA damage in newborn babies in Guiyu and Chaonan towns. Neonates from Guiyu showed more DNA damage. There was significantly higher rate of injuries and difference in tails length in the comet assay in Guiyu than neonates from Chaonan (Grant et al. 2013; Zheng et al. 2008).

9.2.5 Gene Expression

There is a significant association between DNA damage and blood chromium concentrations in neonates (Li et al. 2008). Changes in miRNA expression profiles are correlated with e-waste exposure in spermatozoa (Li et al. 2008). Differential expression profiling of 182 miRNAs is shown in this study by comparative analysis. They found a significant association between sperm quality and environmental exposure to e-waste (Wu et al. 2012). Newborn babies from Guiyu showed higher expression of placental metallothionein (biomarker of metals causing toxicity) as compared to those in Chaonan region (Li et al. 2008). Zhang and his fellows reported significant downregulation of "S100P" proteins and upregulation of metallothionein. They observed increased placental cadmium levels in mothers who were exposed to e-waste from Guiyu than those mothers from the control town region Shantou (Zhang et al. 2010). Pregnant women from Guiyu town showed decreased mRNA and "S100P" levels as compared to women from Shantou town. Metallothionein expression level had been increased in Guiyu as compared to Shantou town. S100P protein concentrations have been negatively associated with gestational age. Metallothionein expression was positively associated with gestational age and Apgar score (Grant et al. 2013).

Important relationships have been shown among the chemical components present in e-waste and physical health. These chemical pollutants of e-waste have significant effects on human health. International Agency for Research on Cancer has identified some metals as highly carcinogenic including cadmium, hexavalent chromium, "TCDD" (2,3,7,8-tetrachlorodibenzo-*p*-dioxin), and beryllium. Some other carcinogenic compound had also been identified included, polychlorinated biphenyls and polycyclic aromatic hydrocarbons and polybrominated diphenyl ethers, metallic nickel, and polycyclic aromatic hydrocarbons. Some other carcinogenic compounds had also been identified, including polychlorinated biphenyls, polycyclic aromatic hydrocarbons, polybrominated diphenyl ethers, metallic nickel and polycyclic aromatic hydrocarbons. Some compounds damage DNA in vitro such as polybrominated diphenyl ethers, polycyclic aromatic hydrocarbons, chromium, nickel and aluminum which are confirmed genotoxins. These chemical compounds also damage DNA of human beings and animal in tests. Some metals are cytotoxins like iron, copper and aluminum (Lankoff et al. 2006; Seth et al. 2004). Some other dioxins like cadmium, polychlorinated biphenyls, per fluoroalkyls and lead can cause obesity, type 2 diabetes, hypertension and cardiovascular diseases (Everett et al. 2011; Faulk et al. 2014; Halldorsson et al. 2012). Lung complications including lung cancer is also caused by those compounds, which are common in e-waste (Abakay et al. 2012; Hirano et al. 2013; Martin et al. 2011). There are adverse perinatal and neonatal results like loss in birth weight and birth height, low head circumference, stillbirth, and decreased childhood growth in mothers who are exposed to polybrominated diphenyl ethers, polychlorinated biphenyls, dioxins, perfluoroalkyls, polycyclic aromatic hydrocarbons, and cadmium (Al-Saleh et al. 2013; Bradman et al. 2012; Everett et al. 2011; Govarts et al. 2012; Kippler et al. 2012). Reproductive development and fertility are affected by exposure to hazardous e-waste compounds. Delayed puberty in girls is linked to lead while decreased sperm quality is linked to polychlorinated biphenyls, TCDD, and perfluoroalkyls (McAuliffe et al. 2012; Mocarelli et al. 2011) and female fertility is decreased by exposure to polybrominated diphenyl. Many constant organic pollutants disrupt endocrines, especially during early life exposure (Grant et al. 2013).

9.2.6 E-waste and Neurodevelopment

The harmful chemical compounds can create many neural complications, especially in children. Decreased intelligence and impaired cognitive functioning are associated with polybrominated diphenyl ethers, polychlorinated biphenyls, polycyclic aromatic hydrocarbons, lead, cadmium, and mercury (Axelrad et al. 2007; Li et al. 2012). While neurodevelopmental abnormalities are associated with polybrominated diphenyl ethers, polychlorinated biphenyls, mercury, and cadmium. Mental health, including behavioral disturbances, attention deficits, hyperactivity, and conduct issues are associated with exposures to polychlorinated biphenyls, lead, mercury, and aluminum. Some scientists reported a strong relationship between lead and polychlorinated biphenyl exposure that may lead to an elevated danger of aggressive behavior and high rates of brutal crime internationally (Carpenter and Nevin 2010). Data also suggest that schizophrenia can be triggered with an exposure to lead combined with genetic predisposition (Guilarte et al. 2012; Opler et al. 2008). Development of the Parkinson's disease and neurodegenerative Alzheimer's disease are associated to heavy metals exposure (Drago et al. 2008; Grant et al. 2013; Weisskopf et al. 2010).

Mothers living in Guiyu have higher level of PFOA exposure than other towns. Prenatal exposure to PFOA involved in decreased neonatal physical development and adverse birth outcomes (Wu et al. 2012).

9.2.7 Carcinogenicity

The study on adverse effects of carcinogenic agents of e-wastes was conducted in the Edo State, Metropolitan City of Benin, Nigeria, and south Nigeria. Metropolitan City of Benin is the present capital of Edo State, whose estimated population in 2006 was 1,147,188.

Comparative analysis has been done between occupationally unexposed and exposed groups. Chronic occupational exposure to e-waste chemicals has been associated with oxidative stress arising from increased lipid peroxidation and lowered antioxidant defenses observed in the studied population. This could be a key reason to chemical carcinogenesis in Nigerian e-waste employees and may be predictive of cancer in these individuals (Igharo et al. 2015).

Uric acid levels and elevated lipid peroxidation activities in the e-waste were compared between exposed and unexposed groups and they were significantly different at $P < 0.05$.

Increased lipid peroxidation and oxidative damage, a mechanism that contributes to carcinogenesis has been directly related to the significantly raised MDA concentration in the e-waste exposed groups.

Lipid peroxidation and reactive oxygen species (ROS) can easily attack proteins directly or indirectly therefore, this can have several effects in the introduction of carcinogenesis by mediating carcinogen establishment. It also causes DNA damage and interfere with the repair of the DNA damage (Igharo et al. 2015).

Several heavy metals, such as lead, mercury, cadmium, arsenic, and aluminum cause developmental neurotoxicity. Neurodevelopmental problems among children living in e-waste recycling region have increased fear of public around the globe. Children living and playing in e-waste recycling environment may have been exposed to harmful e-waste and byproducts. These byproducts cause major health risks to children from fetal development to the end of puberty (Chen et al. 2011; Kim et al. 2013; Wang et al. 2014). Additionally, there were significant associations among e-waste recycling and Blood lead levels (BLLs) temperament variations in Guiyu (Liu et al. 2014). Notable higher levels of meconium lead and umbilical cord blood lead was recorded in Guiyu and decreased newborn babies' behavioral neurological estimation than those in other towns. In addition, newborn behavioral neurological estimation levels were negatively associated with BLLs in newborn babies of Guiyu town. However, non-significant association has been observed among umbilical cord blood lead and NBNA scores. A negative association has been found between meconium lead, activity tone, behavioral score levels, and total NMNA (Li et al. 2008). Children blood levels of Cd, Pb, and Mn in Guiyu have been associated with behavioral defects, like antisocial behavior, conduct problems, and serum S100β have been linked with behavioral defects and blood heavy metals (Liu et al. 2014).

Hazardous materials exposure to mothers influences the unborn child, moreover, unpleasant health effects in generations are caused by e-waste. Mean BLLs of 15 μg/dL is observed in children 1–6 years of age, which is 50% higher than the reference site (10 μg/dL) (Huo et al. 2007; Zheng et al. 2008). Preschool child BLLs predicted neurologic abnormalities in young adults having profile higher than 10 μg/d (Hornung et al. 2009). Pb exposure in the early age of childhood was also been related to increased danger of hyperactivity disorder (Braun et al. 2006). The US Centers for Disease Control and Prevention (CDC) suggested that 5 μg/dL as a reference level can be used to activate interventions. They recognized that no level of lead was found to be secure (Betts 2012). Current Cd exposure concentration in the USA is 0.2 μg/L. A birth cohort study in China shows increased Cd exposure in cord blood that is greater than 0.6 μg/L. This has been associated with a four point's full-scale IQ abnormality at childhood level (Tian et al. 2009). In China, the average blood Cd level is 1.6 μg/L in children is increased than the control reference site 1.0 μg/L, reduced S100β, a Ca^{2+}-binding protein, and potential biological indicator of heavy metal environmental pollution in specific neural lesions was observed in children having high blood Cd (Wong et al. 2005). The main reason for high level Cr exposure in fetuses is e-waste recycling. The Mean cord blood Cr concentration of 99 μg/L is significantly higher than the mean cord blood Cr of the control reference site 32 μg/L (Li et al. 2008) (serum 0.5 μg/L) and Italian Cr people whole blood 6.9 μg/L. PCDD/Fs, PCBs, PAHs, Cr (VI), etc., were found cancer-causing agents (Frazzoli et al. 2010).

9.2.8 Cardiovascular System

From November to December 2016, 590 children 3–7 years of age from Guiyu e-waste exposed group ($n = 337$) and Haojiang control group ($n = 253$) that is 31.6 km away from Guiyu have been included. Both groups showed similar cultural and ethnicity background.

Generation of ROS is due to the exposure to the lead (Guo et al. 2014). Reactive oxygen species stimulate endothelial injury, smooth muscle cell loss, and elastin fragmentation. This decreases the peripheral vascular tension (Chistiakov et al. 2015; Ellinsworth 2015; Yu et al. 2015). The decrease of peripheral vascular resistance in Guiyu children is due to abnormal PP and SBP (Buckley and Ramji 2015).

Atherosclerosis is triggered by the accumulation of Pb in smooth muscle cells, infuriating lipid peroxidation, and inflammation (Di et al. 2016). The prevalence of Pb exposure in Guiyu children is higher. The minor physiological change could increase cardiovascular disease risk in the future (Lu et al. 2018; Frazzoli et al. 2010).

9.2.9 Growth

Healthy pregnant women have been recruited from hospitals in Guiyu and Chaoan. Chaoan is a reference city where no electronic recycling activities were found. They interviewed 154 volunteers and selected 101 people from Guiyu and 53 from Chaoan at the hospital clinics (Wu et al. 2009). Personal data was obtained by distributing questionnaires and clinical data was obtained at delivery indication for caesarian section, neonatal sex, weight, length, Apgar score, etc. Ethical approval was obtained by the Human Ethics Committee of Shantou, University Medical College.

This study includes demographic characteristics of subjects and has been analyzed. There has been no difference between the average mother's age, birth weight, and gestational ages of selected two groups. Mothers did not eat fried or smoked food during pregnancy. The infant Apgar scores and birth weights from Guiyu have been decreased as compared to Chaoan. Significant difference has been observed only between female newborns (Xu et al. 2013). Pb and Hg cause anemia while Se causes hair loss and nail brittleness. Al is toxic to fetus (Frazzoli et al. 2010).

9.2.10 Hearing

A cross-sectional study has been performed for a hearing test in 2014 including a total of 234 preschool children. Children between 3 and 7 years of age were willingly included to give a hearing test. From Guiyu town, 146 children were selected "an e-waste recycling region" and 88 children from Haojing region were selected. They measured urinary Cd and blood Pb levels in e-waste region and find the association of Cd and Pb exposure with children hearing abilities in this cross-sectional study. Results showed that hearing ability was reduced in children living in the e-waste area as compared to the controlled reference area and Pb causes a significant threat for child hearing defect. Recent advancement in research has shown that toxic chemicals increased the danger to pathologic changes in peripheral and central parts of the auditory pathway that result in hearing defects and loss (Godfrey et al. 2017; Sliwinska-kowalska 2015). This study indicates that children suffer more hearing impairment in the e-waste polluted area (Liu et al. 2018).

9.2.11 Immune Response

In December 2011, researchers recruited 294 preschool children 3–7 years of age from Guiyu ($n = 153$) and Haojiang ($n = 141$). Blood Cd and Pb levels have been calculated by graphite furnace atomic absorption spectrophotometer by using detec tion methods (He et al. 2013).

Studies suggest that Cd and Pb interaction can obstruct with the function of monocytes and lymphocytes (Fortier et al. 2008; Mishra 2009).

Cd and Pb increase inflammation as a result of ROS production, cell damage and apoptosis, and lead to increases in endogenous waste that requires exclusion by neutrophils and monocytes (Chen et al. 2006). Hence, it is proved that increased levels of monocytes are due to exposure of Pb (Zhang et al. 2017). The immune system is affected by PCDD/Fs (Frazzoli et al. 2010).

References

Abakay A, Gokalp O, Abakay O, Evliyaoglu O, Sezgi C, Palanci Y, Ekici F, Karakus A, Tanrikulu AC, Ayhan M (2012) Relationships between respiratory function disorders and serum copper levels in copper mineworkers. Biol Trace Elem Res 145:151–157

Al-Saleh I, Alsabbahen A, Shinwari N, Billedo G, Mashhour A, Al-Sarraj Y, Mohamed GED, Rabbah A (2013) Polycyclic aromatic hydrocarbons (PAHs) as determinants of various anthropometric measures of birth outcome. Sci Total Environ 444:565–578

Axelrad DA, Bellinger DC, Ryan LM, Woodruff TJ (2007) Dose–response relationship of prenatal mercury exposure and IQ: an integrative analysis of epidemiologic data. Environ Health Perspect 115:609

Betts KS (2012) CDC updates guidelines for children's lead exposure. Environ Health Perspect 120:a268

Bradman A, Castorina R, Sjödin A, Fenster L, Jones RS, Harley KG, Chevrier J, Holland NT, Eskenazi B (2012) Factors associated with serum polybrominated diphenyl ether (PBDE) levels among school-age children in the CHAMACOS cohort. Environ Sci Technol 46:7373–7381

Braun JM, Kahn RS, Froehlich T, Auinger P, Lanphear BP (2006) Exposures to environmental toxicants and attention deficit hyperactivity disorder in US children. Environ Health Perspect 114:1904

Buckley ML, Ramji DP (2015) The influence of dysfunctional signaling and lipid homeostasis in mediating the inflammatory responses during atherosclerosis. Biochim Biophys Acta (BBA)-Mol Basis Dis 1852:1498–1510

Carpenter DO, Nevin R (2010) Environmental causes of violence. Physiol Behav 99:260–268

Chen M, Lam BK, Kanaoka Y, Nigrovic PA, Audoly LP, Austen KF, Lee DM (2006) Neutrophil-derived leukotriene B4 is required for inflammatory arthritis. J Exp Med 203:837–842

Chen X, Zhu G, Jin T, Lei L, Liang Y (2011) Bone mineral density is related with previous renal dysfunction caused by cadmium exposure. Environ Toxicol Pharmacol 32:46–53

Chistiakov DA, Orekhov AN, Bobryshev YV (2015) Vascular smooth muscle cell in atherosclerosis. Acta Physiol 214:33–50

Di A, Mehta D, Malik AB (2016) ROS-activated calcium signaling mechanisms regulating endothelial barrier function. Cell Calcium 60:163–171

Drago D, Cavaliere A, Mascetra N, Ciavardelli D, Di Ilio C, Zatta P, Sensi SL (2008) Aluminum modulates effects of β amyloid1–42 on neuronal calcium homeostasis and mitochondria functioning and is altered in a triple transgenic mouse model of Alzheimer's disease. Rejuvenation Res 11:861–871

Ellinsworth DC (2015) Arsenic, reactive oxygen, and endothelial dysfunction. J Pharmacol Exp Ther 353:458–464

Everett CJ, Frithsen I, Player M (2011) Relationship of polychlorinated biphenyls with type 2 diabetes and hypertension. J Environ Monit 13:241–251

Faulk C, Barks A, Sánchez BN, Zhang Z, Anderson OS, Peterson KE, Dolinoy DC (2014) Perinatal lead (Pb) exposure results in sex-specific effects on food intake, fat, weight, and insulin response across the murine life-course. PLoS One 9:e104273

Fortier M, Omara F, Bernier J, Brousseau P, Fournier M (2008) Effects of physiological concentrations of heavy metals both individually and in mixtures on the viability and function of peripheral blood human leukocytes in vitro. J Toxic Environ Health A 71:1327–1337

Frazzoli C, Orisakwe OE, Dragone R, Mantovani A (2010) Diagnostic health risk assessment of electronic waste on the general population in developing countries' scenarios. Environ Impact Assess Rev 30:388–399

Godfrey DA, Chen K, O'Toole TR, Mustapha AIAA (2017) Amino acid and acetylcholine chemistry in the central auditory system of young, middle-aged and old rats. Hear Res 350: 173–188

Govarts E, Nieuwenhuijsen M, Schoeters G, Ballester F, Bloemen K, De Boer M, Chevrier C, Eggesbø M, Guxens M, Krämer U (2012) Birth weight and prenatal exposure to polychlorinated biphenyls (PCBs) and dichlorodiphenyldichloroethylene (DDE): a meta-analysis within 12 European Birth Cohorts. Environ Health Perspect 120:162

Grant K, Goldizen FC, Sly PD, Brune M-N, Neira M, van den Berg M, Norman RE (2013) Health consequences of exposure to e-waste: a systematic review. Lancet Glob Health 1:e350–e361

Guilarte TR, Opler M, Pletnikov M (2012) Is lead exposure in early life an environmental risk factor for Schizophrenia? Neurobiological connections and testable hypotheses. Neurotoxicology 33: 560–574

Guo Y, Huo X, Wu K, Liu J, Zhang Y, Xu X (2012) Carcinogenic polycyclic aromatic hydrocarbons in umbilical cord blood of human neonates from Guiyu, China. Sci Total Environ 427: 35–40

Guo S, Zhou J, Chen X, Yu Y, Ren M, Hu G, Liu Y, Zou F (2014) Bystander effects of PC12 cells treated with Pb2+ depend on ROS-mitochondria-dependent apoptotic signaling via gap-junctional intercellular communication. Toxicol Lett 229:150–157

Halldorsson TI, Rytter D, Haug LS, Bech BH, Danielsen I, Becher G, Henriksen TB, Olsen SF (2012) Prenatal exposure to perfluorooctanoate and risk of overweight at 20 years of age: a prospective cohort study. Environ Health Perspect 120:668

He H-Z, Leung K-H, Yang H, Chan DS-H, Leung C-H, Zhou J, Bourdoncle A, Mergny J-L, Ma D-L (2013) Label-free detection of sub-nanomolar lead (II) ions in aqueous solution using a metal-based luminescent switch-on probe. Biosens Bioelectron 41:871–874

Hirano M, Tanaka S, Asami O (2013) Classification of polycyclic aromatic hydrocarbons based on mutagenicity in lung tissue through DNA microarray. Environ Toxicol 28:652–659

Hornung RW, Lanphear BP, Dietrich KN (2009) Age of greatest susceptibility to childhood lead exposure: a new statistical approach. Environ Health Perspect 117:1309

Huo X, Peng L, Xu X, Zheng L, Qiu B, Qi Z, Zhang B, Han D, Piao Z (2007) Elevated blood lead levels of children in Guiyu, an electronic waste recycling town in China. Environ Health Perspect 115:1113

Igharo G, Anetor J, Osibanjo O, Osadolor H, David M, Agu K (2015) Oxidative stress and antioxidant status in Nigerian E-waste workers: a cancer risk predictive study. J Adv Med Med Res 13:1–11

Ju Y, Xu G, Chen L, Shi J (2008) Effects of the electronic waste pollution on the levels of thyroid hormone and its receptor expression in the parturient woman and the newborn. Wei sheng yan jiu= J Hyg Res 37:536–539

Karaman S, Frazzoli E (2010) Optimal kinodynamic motion planning using incremental sampling-based methods. In: 49th IEEE conference on Decision and Control (CDC), 2010. IEEE, pp 7681–7687

Kim S, Arora M, Fernandez C, Landero J, Caruso J, Chen A (2013) Lead, mercury, and cadmium exposure and attention deficit hyperactivity disorder in children. Environ Res 126:105–110

Kippler M, Tofail F, Gardner R, Rahman A, Hamadani JD, Bottai M, Vahter M (2012) Maternal cadmium exposure during pregnancy and size at birth: a prospective cohort study. Environ Health Perspect 120:284

Lankoff A, Banasik A, Duma A, Ochniak E, Lisowska H, Kuszewski T, Góźdź S, Wojcik A (2006) A comet assay study reveals that aluminium induces DNA damage and inhibits the repair of radiation-induced lesions in human peripheral blood lymphocytes. Toxicol Lett 161:27–36

Lechner A, Yang Y-G, Blacken RA, Wang L, Nolan AL, Habener JF (2004) No evidence for significant transdifferentiation of bone marrow into pancreatic β-cells in vivo. Diabetes 53: 616–623

Li Y, Xu X, Liu J, Wu K, Gu C, Shao G, Chen S, Chen G, Huo X (2008) The hazard of chromium exposure to neonates in Guiyu of China. Sci Total Environ 403:99–104

Li Y, Li M, Liu Y, Song G, Liu N (2012) A microarray for microRNA profiling in spermatozoa from adult men living in an environmentally polluted site. Bull Environ Contam Toxicol 89: 1111–1114

Liu W, Huo X, Liu D, Zeng X, Zhang Y, Xu X (2014) S100β in heavy metal-related child attention-deficit hyperactivity disorder in an informal e-waste recycling area. Neurotoxicology 45: 185–191

Liu Y, Huo X, Xu L, Wei X, Wu W, Wu X, Xu X (2018) Hearing loss in children with e-waste lead and cadmium exposure. Sci Total Environ 624:621–627

Lu X, Xu X, Zhang Y, Zhang Y, Wang C, Huo X (2018) Elevated inflammatory Lp-PLA2 and IL-6 link e-waste Pb toxicity to cardiovascular risk factors in preschool children. Environ Pollut 234:601–609

Martin AK, Mack DG, Falta MT, Mroz MM, Newman LS, Maier LA, Fontenot AP (2011) Beryllium-specific CD4+ T cells in blood as a biomarker of disease progression. J Allergy Clin Immunol 128:1100–1106. e5

McAuliffe ME, Williams PL, Korrick SA, Altshul LM, Perry MJ (2012) Environmental exposure to polychlorinated biphenyls and p, p-DDE and sperm sex-chromosome disomy. Environ Health Perspect 120:535

Meneses M, Fernández A, González V, Areco D, Montes de Oca O (2013) Tratamiento con heparina Na en un paciente con tromboembolismo pulmonar y trombo en tránsito a través de foramen oval permeable. Revista Uruguaya de Cardiología 28:382–386

Mishra K (2009) Lead exposure and its impact on immune system: a review. Toxicol In Vitro 23:969–972

Mocarelli P, Gerthoux PM, Needham LL, Patterson DG Jr, Limonta G, Falbo R, Signorini S, Bertona M, Crespi C, Sarto C (2011) Perinatal exposure to low doses of dioxin can permanently impair human semen quality. Environ Health Perspect 119:713

Opler MG, Buka SL, Groeger J, McKeague I, Wei C, Factor-Litvak P, Bresnahan M, Graziano J, Goldstein JM, Seidman LJ (2008) Prenatal exposure to lead, δ-aminolevulinic acid, and schizophrenia: further evidence. Environ Health Perspect 116:1586

Potti A, Dressman HK, Bild A, Riedel RF, Chan G, Sayer R, Cragun J, Cottrill H, Kelley MJ, Petersen R (2006) Genomic signatures to guide the use of chemotherapeutics. Nat Med 12:1294

Robinson BH (2009) E-waste: an assessment of global production and environmental impacts. Sci Total Environ 408:183–191

Seth R, Yang S, Choi S, Sabean M, Roberts E (2004) In vitro assessment of copper-induced toxicity in the human hepatoma line, Hep G2. Toxicol In Vitro 18:501–509

Sliwinska-kowalska M (2015) Hearing. Handb Clin Neurol 131:341–363

Tian L-L, Zhao Y-C, Wang X-C, Gu J-L, Sun Z-J, Zhang Y-L, Wang J-X (2009) Effects of gestational cadmium exposure on pregnancy outcome and development in the offspring at age 4.5 years. Biol Trace Elem Res 132:51–59

Wang Y, Chen A, Dietrich KN, Radcliffe J, Caldwell KL, Rogan WJ (2014) Postnatal exposure to methyl mercury and neuropsychological development in 7-year-old urban inner-city children exposed to lead in the United States. Child Neuropsychol 20:527–538

Weisskopf MG, Weuve J, Nie H, Saint-Hilaire M-H, Sudarsky L, Simon DK, Hersh B, Schwartz J, Wright RO, Hu H (2010) Association of cumulative lead exposure with Parkinson's disease. Environ Health Perspect 118:1609

Wong R-H, Kuo C-Y, Hsu M-L, Wang T-Y, Chang P-I, Wu T-H, Huang S (2005) Increased levels of 8-hydroxy-2′-deoxyguanosine attributable to carcinogenic metal exposure among schoolchildren. Environ Health Perspect 113:1386

Wu K, Xu X, Liu J, Guo Y, Li Y, Huo X (2009) Polybrominated diphenyl ethers in umbilical cord blood and relevant factors in neonates from Guiyu. China Environ Sci Technol 44:813–819

Wu K, Xu X, Peng L, Liu J, Guo Y, Huo X (2012) Association between maternal exposure to perfluorooctanoic acid (PFOA) from electronic waste recycling and neonatal health outcomes. Environ Int 48:1–8

Xu X, Yang H, Chen A, Zhou Y, Wu K, Liu J, Zhang Y, Huo X (2012) Birth outcomes related to informal e-waste recycling in Guiyu, China. Reprod Toxicol 33:94–98

Xu X, Yekeen TA, Xiao Q, Wang Y, Lu F, Huo X (2013) Placental IGF-1 and IGFBP-3 expression correlate with umbilical cord blood PAH and PBDE levels from prenatal exposure to electronic waste. Environ Pollut 182:63–69

Yu X-H, Zheng X-L, Tang C-K (2015) Peroxisome proliferator-activated receptor α in lipid metabolism and atherosclerosis. Adv Clin Chem 71:171–203

Zhang J, Jiang Y, Zhou J, Wu B, Liang Y, Peng Z, Fang D, Liu B, Huang H, He C (2010) Elevated body burdens of PBDEs, dioxins, and PCBs on thyroid hormone homeostasis at an electronic waste recycling site in China. Environ Sci Technol 44:3956–3962

Zhang Y, Xu X, Sun D, Cao J, Zhang Y, Huo X (2017) Alteration of the number and percentage of innate immune cells in preschool children from an e-waste recycling area. Ecotoxicol Environ Saf 145:615–622

Zheng L, Wu K, Li Y, Qi Z, Han D, Zhang B, Gu C, Chen G, Liu J, Chen S (2008) Blood lead and cadmium levels and relevant factors among children from an e-waste recycling town in China. Environ Res 108:15–20

Chapter 10
Effects of E-waste on Immune System of Preschool Children

Naseer Ali Shah, Jadoon Khan, Azmat Ullah Jan, and Ijaz Ali

Abstract Electronic waste (e-waste) is one of the growing issues globally. Various components of e-waste including lead, nickel, cadmium, aluminum, and various hazardous gases releasing from the burning of these materials are responsible for various health issues, especially for the impairment of immune system in preschool children. In this chapter, we report that various e-waste materials are involved in the impairment of the immune system, decrease level of immunoglobulins against viruses, defected level of CD4 and CD8 immune cells, hypersensitivity reactions, lower level of various interleukins which includes IL-6, IL-1b, and IL-27, and lower level of NK cells formation and CR1 level in preschool children. This chapter shows that children of preschool age are more susceptible to various e-waste materials and having more chances of immune impairment functions.

Keywords Effects · E-waste · Immune system · Preschool children

10.1 Introduction

Electronic waste (e-waste) is the fastest growing waste globally (Lundgren 2012) and about 50 million tons/year of e-waste are generated worldwide (Wang et al. 2013). In 2014, approximately 41.8 million tons of e-waste was produced which will expectedly increase to 49.8 million tons in 2018 (Balde et al. 2014). Various hazardous materials are present in e-waste whose proper disposal is necessary because they become a growing pollution issue. The collection and recycling of e-waste are very necessary to prevent serious health issues, environmental damage and loss of valuable resources.

N. A. Shah (✉) · A. U. Jan · I. Ali
Department of Biosciences, COMSATS University Islamabad, Islamabad, Pakistan
e-mail: drnaseer@comsats.edu.pk

J. Khan
Department of Microbiology, Quaid I Azam University Islamabad, Islamabad, Pakistan

M. Z. Hashmi, A. Varma (eds.), *Electronic Waste Pollution*, Soil Biology 57,
https://doi.org/10.1007/978-3-030-26615-8_10

The recycling of e-waste is different among various countries. In 2008, 14% of 3 million tons of e-waste was recycled in the United States of America (USA), as reported in a study, while 60% and 17% were recycled properly in some European countries and the United Kingdom (UK), respectively (Caravanos et al. 2011). According to an estimate, about 20–25% of e-waste was recycled properly in the USA and Europe (Cobbing 2008), while the rest of e-waste exported, landfilled, or stockpiled (Ongondo et al. 2011; Bhutta et al. 2011).

Besides China, Nigeria, and India, Pakistan is one of the world leading e-waste endpoint (Sthiannopkao and Wong 2013). Pakistan is a country where the earning of 60% of the population is less than US$ 2/day due to the low income; many of the people are engaged in the recycling of e-waste for the purpose of income (Bauer et al. 2008).

E-waste contains approximately 60 chemical elements, including aluminum, iron, zinc, arsenic, copper, mercury, nickel, manganese, chromium, cadmium, and lead, the majority of which are potentially hazardous. Due to various reasons, the exposure of children to heavy metals are frequent as compared to adults, including many exposure routes such as an object to mouth, hand to mouth, placental exposure, and breastfeeding. The basic metabolic rate of children are more than adults; they have decreased elimination rate of toxin and a higher rate of food uptakes. The inhalation rate of various heavy metallic substances is also higher in children in comparison to adults. Through their dermal surface, children can load high amount of toxicant substances; therefore, various tissues and organs of these are more sensitive. Similarly, the physical activities of children are higher as compared to adults; to avoid the exposure of children from various e-waste and ETS in the home is very difficult.

Heavy metals found in e-wastes cause many health problems in children. They have a bad effect on various organs and systems, with acute to chronic effects on the respiratory, reproductive, nervous, urinary, cardiovascular systems, and cause immune-impairment disorders (Xu et al. 2015b).

10.2 Effect of E-waste on Immune System

The exposure to each heavy metal in various doses causes various health problems in human especially on the health of preschool children. Exposure to heavy metals triggers the formation of autoantibodies against own neural proteins such as glial fibrillary acidic protein and basic myelin proteins which prove that lead is involved in the chronic neurological problems due to increase in immunogenicity of various proteins of the nervous system (Waterman et al. 1994).

Various effects of heavy metals from e-waste on immune functions in children and adults include immune cells expression, functions of erythrocytes, hypersensitivity, oral tolerance, and autoimmunity, which are given in Table 10.1. Aluminum, cadmium, lead, and some other heavy metals of e-waste cause the alteration of the immune system which is the challenging issue for public health (Waterman et al. 1994). According to the report of Fischbein et al., the increase in the level of lead

Table 10.1 Effect of e-waste on immune system

S. no	E-waste	Effect on immune system	References
1	Metals in e-waste	Activation of the inflammasome, induction of cellular death, metal-induced hypersensitivity, and autoimmunity	McKee and Fontenot (2016)
2	Elevated blood Pb level	Increased IL-6 and vascular inflammation	Lu et al. (2018)
3	Increase in the level of Pb in blood	Alteration of hemoglobin formation, morphology and expression of CR1, and threat development of erythrocyte immunity	Dai et al. (2017)
4	Heavy metal nickel	Rushes formation and allergic reactions in children	ATSDR (2005)
5	Elevated lead levels	Lower percentages of NK cells, but also alter the levels of platelets, IL-1b and IL-27, which might be unconducive to the activity and function of NK cells	Zhang et al. (2016)
6	Hexachlorobenzene (HCB) is formed because of recycling of e-waste	Damage immune system	Van Birgelen (1998)
7	Exposure to heavy metal	Formation of autoantibodies	Waterman et al. (1994)
8	Exposure to lead founds in e-waste	Alteration of immune functions	Xu et al. (2015a)
9	E-waste element antimony toxic compound	Alteration in the functions of immune system	Kim et al. (1999)
10	Lead	The reduction in the specific percentage and numbers CD4þ and CD3þ cells	Fischbein et al. (1993)
11	Exposed to lead	Decrease number of CD4þ T lymphocytes	Li et al. (2005)
12	Heavy metals of e-waste including aluminum, cadmium, and lead	Alteration of immune system, functions of erythrocytes and hypersensitivity, oral tolerance and autoimmunity	Waterman et al. (1994)

(>25 mg/dL) in the blood leads to the reduction in the specific percentage and numbers of CD4þ and CD3þ cells (Fischbein 1993). Similar to Fischbein et al.'s 1993 report, children who are exposed to lead will have a low number of CD4þ T lymphocytes (Zhuang et al. 2009). One of the important e-waste element antimony, a toxic compound, has been reported to be involved in the alteration of the immune system in children (Chen et al. 2011).

According to the study of Xu et al. in China, a negative association has been observed between the presence of lead and the level of anti-hepatitis B surface antigen (Hbs-Ag) antibody in the blood. However, a positive association was found with children age from an area of e-waste recycling (Xu et al. 2015a). Henceforth, it is concluded from the above study that exposure to lead found in e-waste is involved in the alteration of functions of the immune system in young children.

Hexachlorobenzene (HCB) is formed because of recycling of e-waste which belongs to 2B carcinogens group and damage organs and systems of the body, such as immune system, thyroid, liver, kidney, and nervous system, that is, CNS, in children living near e-waste recycling area (Van Birgelen 1998).

The children and infants living near e-waste recycling area are exposed to hazardous chemicals which can cause various health issues. Nickel is one of the heavy metals, which is involved in rushes formation and allergic reactions in children (ATSDR 2005). A study was carried on the effect of lead on immune system consist of 484 preschool children at Guiyu and Haojiang near an e-waste recycling area. They reported that the children exposed to lead shows a higher level of erythrocyte lead level (EPb) and blood level of lead (BPb); however, the CR1 level of erythrocytes and mean corpuscular hemoglobin concentration (MCHC) were decreased significantly as compared to reference children. The children exposed to lead had significantly high EPb and BPb which was associated with the level of CR1 of erythrocytes among all children. This study shows that increase in the level of lead in blood results in the alteration of hemoglobin formation, morphology, and expression of CR1 which is a serious threat for the development of erythrocyte immunity in the indigenous preschool children (Dai et al. 2017).

Elevated lead level was associated with higher interleukin (IL)-6 and vascular inflammation in preschool children (Lu et al. 2018). Taken together, the elevated level levels result in the lower percentages of NK cells, but also alter the levels of platelets, IL-1b and IL-27, which might be unconducive to the activity and function of NK cells (Zhang et al. 2016). According to another study, the preschool children group exposed to e-waste showed significant changes in their antibody titer, such as 40% and 45% of decrease were noted in the antibodies against measles and mumps, respectively. A significant decrease was also noted against rubella antibodies in preschool children (Lin et al. 2016).

Metals activate the innate immune system through direct binding to pathogen recognition receptors, activation of the inflammasome, or the induction of cellular death and release of alarmins. Certain metals can serve as adjuvants, promoting dendritic cell activation and migration as well as antigen presentation to metal-specific T cells. These T cells can recognize metals as haptens or as altered MHC-peptide complexes. The ability of metals to create these neoantigens emphasizes the similarity between metal-induced hypersensitivity and autoimmunity (McKee and Fontenot 2016).

10.3 Conclusion

It is concluded from the current study that the e-waste-containing various hazardous elements including lead, mercury, aluminum, nickel, and cadmium are responsible for various health issues, especially in preschool children. One of the most important and dangerous issues is the effects of e-waste on immune system in preschool children. The increase in the blood lead level leads to immune impairments including a reduction in Ig level and hypersensitivity. Nickel and antimony has a great impact on the blood CD4 and CD8 profile. Overall, we conclude that the e-waste-containing elements has a great impact on immune function impairment in preschool children.

References

Agency for Toxic Substances and Disease Registry (ATSDR) (2005) Toxicological profile for Nickel. U.S. Department of Health and Human Services, Public Health Service, Atlanta, GA

Balde CP, Wang F, Kuehr R, Huisman J (2014) E-waste monitor. https://doi.org/10.1007/s00705-012-1479-4

Bauer A, Hasan R, Magsombol R, Wan G (2008) The World Bank's new poverty data: implications for the Asian Development Bank. ADB, Manila

Bhutta MKS, Omar A, Yang X (2011) Electronic waste: a growing concern in today's environment. Econ Res Int 2011:1–8. https://doi.org/10.1155/2011/474230

Caravanos J, Clark E, Fuller R, Lambertson C (2011) Assessing worker and environmental chemical exposure risks at an e-waste recycling and disposal site in Accra, Ghana. J Health Pollut 1(1):16–25. https://doi.org/10.1016/j.jallcom.2016.03.280

Chen A, Dietrich KN, Huo X, Ho SM (2011) Developmental neurotoxicants in e-waste: an emerging health concern. Environ Health Perspect 119(4):431–438. https://doi.org/10.1289/ehp.1002452

Cobbing M (2008) Toxic tech: not in our backyard. Greenpeace International, (February). Retrieved from http://scholar.google.com/scholar?hl=en&btnG=Search&q=intitle:Toxic+Tech+:+Not+in+Our+Backyard#0

Dai Y, Huo X, Zhang Y, Yang T, Li M, Xu X (2017) Elevated lead levels and changes in blood morphology and erythrocyte CR1 in preschool children from an e-waste area. Sci Total Environ 592:51–59

Fischbein E (1993) The theory of figural concets. Educ Stud Math 24:139–162

Fischbein A, Tsang P, Luo JCJ, Roboz JP, Jiang JD, Bekesi JG (1993) Phenotypic aberrations of CD3+ and CD4+ cells and functional impairments of lymphocytes at low-level occupational exposure to lead. Clin Immunol Immunopathol 66(2):163–168

Kim HA, Heo Y, Oh SY, Lee KJ, Lawrence DA (1999) Altered serum cytokine and immunoglobulin levels in the workers exposed to antimony. Hum Exp Toxicol 18(10):607–613

Li, S., Zhengyan, Z., Rong, L. I., & Hanyun, C. (2005). Decrease of CD4+ T-lymphocytes in children exposed to environmental lead. Biological trace element research, 105(1-3), 19-25

Lin Y, Xu X, Dai Y, Zhang Y, Li W, Huo X (2016) Considerable decrease of antibody titers against measles, mumps, and rubella in preschool children from an e-waste recycling area. Sci Total Environ 573:760–766

Lu X, Xu X, Zhang Y, Zhang Y, Wang C, Huo X (2018) Elevated inflammatory Lp-PLA2 and IL-6 link e-waste Pb toxicity to cardiovascular risk factors in preschool children. Environ Pollut 234:601–609

Lundgren K (2012) The global impact of e-waste addressing the challenge. ILO, Geneva

McKee AS, Fontenot AP (2016) Interplay of innate and adaptive immunity in metal-induced hypersensitivity. Curr Opin Immunol 42:25–30

Ongondo FO, Williams ID, Cherrett TJ (2011) How are WEEE doing? A global review of the management of electrical and electronic wastes. Waste Manag 31(4):714–730. https://doi.org/10.1016/j.wasman.2010.10.023

Sthiannopkao S, Wong MH (2013) Handling e-waste in developed and developing countries: initiatives, practices, and consequences. Sci Total Environ 463–464:1147–1153. https://doi.org/10.1016/j.scitotenv.2012.06.088

Van Birgelen AP (1998) Hexachlorobenzene as a possible major contributor to the dioxin activity of human milk. Environ Health Perspect 106(11):683–688

Wang F, Huisman J, Stevels A, Baldé CP (2013) Enhancing e-waste estimates: improving data quality by multivariate input-output analysis. Waste Manag 33(11):2397–2407. https://doi.org/10.1016/j.wasman.2013.07.005

Waterman SJ, El-Fawal HA, Snyder CA (1994) Lead alters the immunogenicity of two neural proteins: a potential mechanism for the progression of lead-induced neurotoxicity. Environ Health Perspect 102(12):1052

Xu X, Shang Q, Chen X, Nie W, Zou Z, Huang A et al (2015a) Reversal of B-cell hyperactivation and functional impairment is associated with HBsAg seroconversion in chronic hepatitis B patients. Cell Mol Immunol 12(3):309

Xu X, Zeng X, Boezen HM, Huo X (2015b) E-waste environmental contamination and harm to public health in China. Front Med 9(2):220–228. https://doi.org/10.1007/s11684-015-0391-1

Zhang Y, Huo X, Cao J, Yang T, Xu L, Xu X (2016) Elevated lead levels and adverse effects on natural killer cells in children from an electronic waste recycling area. Environ Pollut 213:143–150

Zhuang P, McBride MB, Xia H, Li N, Li Z (2009) Health risk from heavy metals via consumption of food crops in the vicinity of Dabaoshan mine, South China. Sci Total Environ 407 (5):1551–1561. https://doi.org/10.1016/j.scitotenv.2008.10.061

Chapter 11
Metagenomics Approaches to Study Microbes in the E-waste Polluted Environment

Naseer Ali Shah, Imdad Kaleem, and Yasir Rasheed

Abstract Metagenomics approaches have tremendous application in encompassing ecological sustainability along with biotic and abiotic factors. The significance of metagenomics is well seen in establishing the role of microbes in sustainable environmental management tools such as bioremediation. This chapter highlights the recent innovative metagenomic techniques in determining the substantial role of microbes in environmental systems contaminated by e-waste dumping. We also describe modern metagenomic analysis for a variety of microbial communities and their key functions in e-waste soil. Moreover, culture-based and culture-independent integrated metagenomic analyses are discussed to authenticate microbial community taxonomic profiling and characterization of sustainable ecological development.

Keywords Metagenomics · Microbes · E-waste · Pollution · Environment

11.1 Introduction

Microorganisms have great chemical diversity that can monitor complicated life processes via a simple approach. Although more than 1.7 million extant classes have been reported, there are still many classes to be identified. The diversity and uniqueness of microbial fauna causes many limitations in their isolation and identification because bacteriological assessment is limited by existing distinguishing characteristics, which makes accurate identification difficult (Balkwill and Boone 2018).

Microbial products include an extensive range of significant medicinal complexes such as xanthan, anti-helminthics (albendazole), immune modulators, immunosuppressants, enzymes, antitumor agents, enzyme inhibitors, insecticides, and vitamins. During the past 50 years, these mediators have been shown to be only a small part of those likely existing in nature (Mérillon and Riviere 2018).

N. A. Shah (✉) · I. Kaleem · Y. Rasheed
Department of Biosciences, COMSATS University Islamabad, Islamabad, Pakistan
e-mail: drnaseer@comsats.edu.pk

M. Z. Hashmi, A. Varma (eds.), *Electronic Waste Pollution*, Soil Biology 57,
https://doi.org/10.1007/978-3-030-26615-8_11

In regard to plant nourishment and soil vigor, the variety of microbes has an important role in biogeochemical processes (Mohanty and Swain 2018; Wood and Bradford 2018), even in agrarian and risky surroundings (Rangel et al. 2018; Taketani et al. 2017). Soil microbes symbolize a substantial segment of the active biomass on Earth (with superficial soils containing 10^3–10^4 kg infectious biomass per hectare) (Brady and Weil 2002; Horwath 2017), notwithstanding this profusion and the significance of soil microbes for main ecological functions (Bardgett and van der Putten 2014; Bender et al. 2016; He et al. 2007; Schloter et al. 2018).

The advent of molecular procedures allows us now to review microbial variety, counting the massive communities of microorganisms (Grattepanche et al. 2018; Lynch et al. 2004; Theron and Cloete 2000; Wilcox et al. 2018). However, outmoded fungal tactics do not address extant boundaries, as only a small proportion of soil bacteria can be cultured using typical means (Meert et al. 1998). Consequently, in the past two eras, numerous molecular methods have been proposed (De Vries et al. 2006; Field et al. 2008; Kirk et al. 2004). Recently, for the examination of whole genomes existing in a soil mockup, metagenomics introduced a new method for comprehensive taxonomies (Finniss et al. 2010; Jansson and Hofmockel 2018; Loureiro et al. 2018). Many investigators have demonstrated possible ways to discover the "black box" of bacterial populations in mud (Kardol et al. 2015; Matt et al. 2017).

Associations with automated e-waste are developing difficulties in the twenty-first century (Heacock et al. 2016; Tansel 2017). Almost 50–80% of e-waste from the manufacturing countries is transferred to reprocessing centers in emerging nations such as Pakistan, India, Philippines, China, and Vietnam because costs are lower in the lower-income countries; also, there are fewer ecological guidelines in these countries (UNEP 2005). Currently, the e-waste occupation has also been rising in African states. Reprocessing and discarding of e-waste in developing countries are causing cumulative anxiety because of the effects on the surroundings as well as the related threats to human health.

Bi, Hg, Ag, Zn, In, Cd, Sb, Sn, Cu, and Pb were assessed in mud from e-waste reprocessing spots associated with visible locations. For Hg, Cu, Pb, Hg, and Sb in certain loams from e-waste spots, the results surpassed transmission standards anticipated by the US Environmental Protection Agency (EPA). Assessments of Bi, Cu, Pb, Mn, Tl, Co, Sb, Sn, In, and Cr from e-waste in the atmosphere following reprocessing were developed, compared to the stages in the city of Chennai. Great increases of Pb, Cd, Cu, Mo, Ag, Tl, Sb, and In were detected in the hairs of male laborers at e-waste reprocessing centers (Ha et al. 2009). Outcomes recommend that e-waste reprocessing and its exclusion may progress to ecological and human impurities through certain TEs.

11.2 Culture-Independent Techniques to Assess Microbial Diversity

Most microbes have only been identified with culture-independent methods. As microorganism cultivation is a significant task, investigators have dedicated themselves to this effort in numerous past eras. Several chief groups of Archaea and Bacteria are now recognized only by arrangements directly reported from ecological models, and these creatures are preserved until identification can be refined. Resources to determine their part in the atmosphere are simply culture-self-governing descriptions related to the purpose of in situ metabolic movements (Liu et al. 2018). Although most microbes are unproductive, inappropriately, the development of most of them was not known until now, owing not only to organizational boundaries but also the absence of taxonomic information (Kirk et al. 2004). The measurement of microbial and molecular ecosystems is clearly explained as the purpose of molecular equipment, classically based on examination of nucleic acids, to recognize detailed microorganisms in a specific situation, to disperse purposeful parts to these microorganisms, and to measure their importance or involvement in ecological developments sovereign from husbandry (Zoetendal et al. 2004).

Based on the idea of specific prevailing molecular methods, the exact changing aspects and occurrences of microbial populations can be scrutinized. Microbial naturalists at this time have bypassed the harmonizing culture-self-governing approaches that are capable of recycling to examine the microbial purpose in situ (Oguntoyinbo et al. 2011). Usually, molecular methods for measuring bacterial variety are based on the examination of total DNA isolated from ecological models (metagenomics DNA). Examination by polymerase chain reaction (PCR) is improved by 16S rDNA from metagenomics DNA using general bacterial readers duplicated into a path monitored by restriction fragment length polymorphism (RFLP) and sequencing. Most importantly, it is a foundation for phylogenetic studies of bacteria populations (Fernández-Arrojo et al. 2010) (Fig. 11.1).

Metagenomic analyses can provide widespread evidence on the structure, arrangement, and expected gene purposes of various ecological microbial collections. Every atmosphere presents its individual exclusive trials to metagenomic examination and needs an exact intended method to set exclusive biotic and physicochemical issues, so this can overcome methodical obstacles as well as prejudice the metagenomics technique. Specifically, mud can carry an astonishing assortment of prokaryotes; these are mainly unlabeled as to ribotype as well as possibly being a huge reserve for accepted development detection. The fruitful purpose of a loam meta-genomic method can be contingent on choosing suitable deoxyribose nucleic acid removal, cleansing, and, if essential, duplicating approaches for the planned downstream technique. Significant scientific considerations in a metagenomic study include gaining an adequate influx of elevated purified RNA expression. The embattled microorganisms are surrounded by an ecological model and (but essential) erecting a metagenomic collection in an appropriate path as well as a swarm (Parsley et al. 2010).

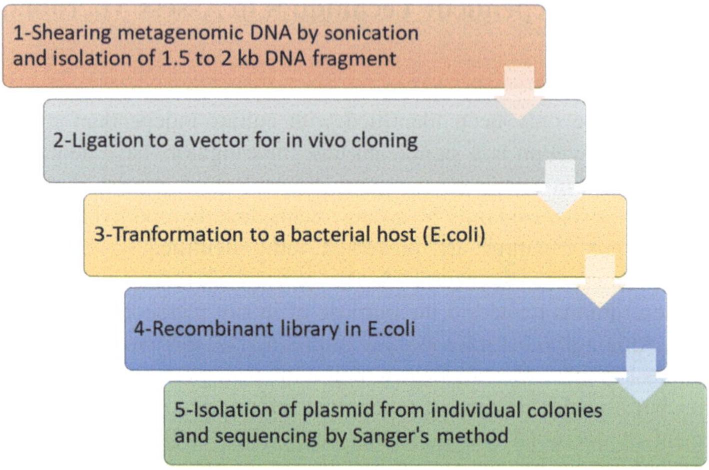

Fig. 11.1 General steps involved in a culture-independent study

11.3 PCR-Based Techniques

Polymerase chain reaction (PCR) is the elementary method in high molecular research laboratories. Through exact briefings, complete amplification of almost all nucleotide arrangements is conceivable. Useful genes from dissimilar microorganisms are able to augment, other than for microbes, the molecular documentation with the gene rRNA 16S that is typically recycled. Since the 16S rRNA gene has preserved sections, with a similar arrangement within countless microorganisms as well as mutable sections, which arrangements can be detailed, amplification and therefore recognition at diverse taxonomic rankings can be attained. The utilization of preserved rRNA 16S arrangements as well permits recognition of formerly unidentified as well as undefined microorganisms.

When associated with dot-blot hybridization, concurrent PCR constitutes greater agreement, more suitable, and also fewer classes for the quantification of designated bacterial inhabitants (Zhang et al. 2018). Other quantifiable PCR methods are reasonable (RT-PCR), and nearly all feasible figures of PCR (most probable number, MPN). Through inexpensive PCR, exact normal recognized absorption is possible in addition to the diverse quantity of the samples monitored by PCR amplification. On an agarose gel, change in the amount between the aims as well as the normal permits judgment and successive quantification.

11.4 Temperature Gradient Gel Electrophoresis

Denaturing gradient gel electrophoresis (DGGE), as well as the currently less utilized temperature gradient gel electrophoresis (TGGE), is based upon the weak actions of double-stranded deoxyribose nucleic acid fragments. Their saturated actions, typically labeled as the melting temperature, are reliant upon the nucleotide arrangement of a fragment. Through accumulation, a GC-rich tail (GC-clamp) is used for amplification of all the basic coverage; a fragment is produced that will partly melt in denaturing gradient polyacrylamide gel when it is electrophoresed. Generally, a formamide and urea drop is realistic in DGGE and a high-temperature incline in TGGE.

In accumulation, internal standards in molecular analysis of diversity (ISMAD) have a particularly valuable influence in relative ecotoxicological and environmental DGGE experiments, wherever behavior alterations of the identical unique principals are deliberated as well as some variations toward management are required (Camus et al. 2018). The utilization of ISMAD throughout PCR-DGGE variations allows analyzing changes in bacterial amount and variety owing to changes inside the unique models, or owing to presumptions from investigational changeability among mockups familiarized throughout deoxyribose nucleic acid removal, DGGE, and PCR (Petersen and Dahllöf 2005).

11.5 Techniques Based on Single-Strand Conformation Polymorphism

An additional method used for recognition of changes in arrangements of DNA by means of departure via electrophoresis is single-strand conformation polymorphism (SSCP). In a polyacrylamide gel, the single-stranded deoxyribose nucleic acid is detached on the basis of modifications during flexibility produced via their subordinate structure. Through the usage of a changeable element of the rRNA 16S gene (in support, for instance, the V3 region) from an ecological model, every SSCP at the highest point resembles a dissimilar microbial arrangement, representing occurrence of the microbial strain or genus recovered from the mockup (Peters et al. 2000).

11.6 Restriction Analysis-Based Approaches

A large number of molecular methods utilize exact nucleic acid-adapting enzymes, originally purified as well as categorized by microorganisms. Heat-constant Taq deoxyribose nucleic acid polymerase, a castoff from PCR, is however more resistant to enzymes and is extensively recycled. Constraint enzymes identify exact DNA arrangements and expurgate these in a replicable method. Unsolicited or main deoxyribose nucleic acid patterns can be augmented in an original bout of PCR, products manufactured by means of constraint enzymes, causing an assimilated pattern no longer individually accessible for PCR amplification (Inbar et al. 2005).

In amplified ribosomal DNA restriction analysis (ARDRA), the rRNA gene is augmented through PCR as well as assimilated in exact remnants via limit enzymes (frequently by means of 4-bp spots). Behind the development through limited enzymes, fragments are detached through elevated determination gel electrophoresis, subsequent to precise designs by dissimilar arrangements. ARDRA could be recycled for a quick comparison of rRNA genes (Knowles and Cole 2008). The characteristic examination of constraint resumes for separates or replicas is accomplished on agarose gels, although for communal examination the hypothetically huge quantity of fragments can be determined by means of polyacrylamide gels to create a community-precise design (Martinez-Murcia et al. 1995), excluding the original large determination mediums that are also obtainable.

Terminal restriction fragment length polymorphism (T-RFLP) is an additional consequent fingerprinting method as well as formulating usage of constraint enzymes; however, only terminal restriction fragments (T-RF) are noted and selected for quantitative and qualitative examination (Nanda et al. 2018).

11.7 Cloning and Sequencing Approaches

The duplicating as well as ordering of rRNA 16S genes augmented by surroundings from side-to-side culture-independent methods established that microbial variety is more widespread than we ever expected from culture based studies (Handelsman 2004). For an appropriate phylogenetic description of microorganisms, the arrangement of the indicator gene remnant should be resolved. Qualified ordering of the rRNA 16S gene has developed via the majority usually used as the extent of ecological variety (Handelsman 2004). Preferably, the entire genome should be ordered, and actually this is completed more often now. Still, the metagenome of an entire community could be exposed to duplicating as well as ordering (Tringe and Rubin 2005; Tringe et al. 2005). Pyrosequencing has been utilized for assessment of the entire variety in mud as well as that caused within the recognition of the added 50,000 operational taxonomic units (OTUs) (Roesch et al. 2007). rRNA arrangements can be gained via PCR amplification moreover by programming genes or deviations from rRNA. Within the amplification by rRNA, the opposite transcription-polymerase chain reaction (RT-PCR) has to be engaged (Bordukalo-Nikšić 2007).

11.8 Techniques Based on Fluorescence In Situ Hybridization

Phylogenetically based oligonucleotide reviews could complete precise quantification through fluorescence in situ hybridization (FISH) and dot-blot hybridization. This final whole-cell hybridization along rRNA-targeted lines exploits an appropriate instrument for formative, phylogenetic, and ecological learning in microbiology (Amann et al. 1990). These days it is possible to characterize seven microbial collections instantaneously, thus called rainbow-FISH (Sunamura and Maruyama 2006). Under this inspiration, this compact replicated in situ hybridization method comprises compartments in RNA protection as well as convenience toward enquiries; type and quality of these probes; the competence of probe–target hybrid creation; the permanence of hybrids molded in situ throughout post-hybridization management; the technique of recognition of hybrids; and the related noise masking the hybridization signal (Pernthaler and Amann 2004). The availability of rRNA has been methodically examined, and now it has been shown that a balanced probe enterprise approach, connecting ΔG^0 overall, hybridization kinetics, and fluorophore extinguishing, stemmed in no totally unreachable object areas in the 16S rRNA of *Escherichia coli* (Yilmaz et al. 2006). The cells of dissimilar types having diverse ribosome substances reach about 10^3 to 10^5 ribosomes per cell; for one strain, cellular rRNA could still vary meaningfully although they are connected by rapid growth. Nevertheless, the comparative rRNA profusion should signify a sensible dimension of the comparative functional movement of the particular populations, as it is the result of the quantity of reported cells and usual rRNA combined (Yilmaz et al. 2006).

11.9 Techniques Based on Dot-Blot Hybridization

The drawback of PCR-based methods is that they do not continuously deliver impartial measurable facts because of the amplification procedures. Quantification of 16S rRNA order compared to the whole 16S rRNA substance of certain models can be attained by dot-blot hybridization of inaccessible nucleic acids with general and precise oligonucleotide probes (Albuquerque et al. 2017). For this effort, the entire RNA is accessed from the model, immobilized on a membrane, and hybridized with selected oligonucleotide probes. The comparative quantities can be read as a proportion of the quantity of the exact probe assigned to a particular sample compared to the quantity of hybridized general probe (Albuquerque et al. 2017). Although quantification is more precise, these results of comparative rRNA amounts cannot be exactly interpreted as cell numbers; cells of diverse types contain dissimilar ribosome contents, extending unevenly at about 10^3 to 10^5 ribosomes per cell (Amann et al. 1995).

11.10 Phylogenetic Microarrays

DNA microarrays (deoxyribose nucleic acid chips or biochips) classically contain thousands of restrained deoxyribose nucleic acid fragments (PCR derived, oligonucleotides, or other DNA fragments) that exist on the exterior, for example, on glass coverslips or membranes (Venter et al. 2001). It can be chosen as an opposite outdated dot blot; subsequently, the individual dotted probes are recognized and the sample is characterized. Labeled nucleic acids will spot the precise locations on the collection where hybridization occurred. Microarray research outcomes contain a list of hybridization results, representing the occurrence or the comparative profusion of the precise deoxyribose nucleic acid or ribose nucleic acid arrangements present in the sample. Microarrays were previously extensively utilized for recognition of transcriptional outlines (expression assortment) or the resemblances and changes of genetic content among dissimilar microorganisms, and they can be utilized to separate (by fingerprinting comparative to location strain) bacterial types as well as for the documentation of the original diagnostic genetic markers (Mastriani and Giraldez 2018). Use of microarrays for formative studies provides numerous compensations compared to conservative hybridization setups. Thousands of dissimilar oligonucleotides can be restrained on a particular assortment, permitting synchronized recognition of a countless diversity of dissimilar microorganisms in a single sample. Thus, simple sample preparation, mechanization, and high-throughput examination along microarrays can be readily understood (Fig. 11.2).

11.11 Conclusion

Metagenomics approaches are being used for in-depth study of the presence of microbial colonies on e-waste-contaminated soil. These techniques allow the optimization of the process to target groups of microbes, particularly in the bioremediation process. Moreover, these approaches open a gateway for unbiased examination of the microbes. Metagenomics has the ability for suitable application of these methods to report the unknown species of microbes present at e-waste dumping sites. Moreover, there is an immediate need to study the advances involved in these protocols. The great discoveries in the process of sequencing give a huge amount of data that may aid in the discovery of various uncultured microbes. There is a need to merge the metagenomics techniques with culture-based protocols to understand microbe populations at a high level.

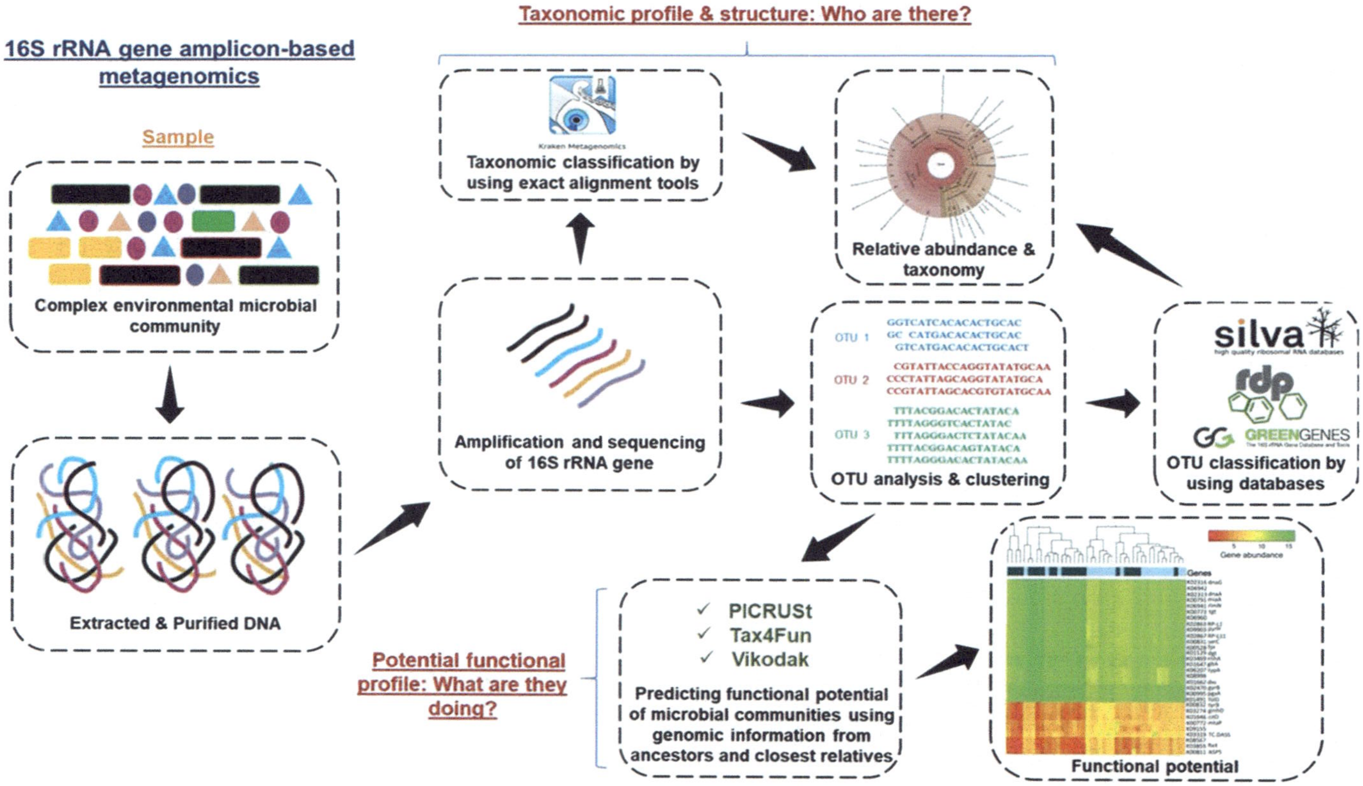

Fig. 11.2 Overall pipeline and bioinformatics connections measured for the examination of metagenomic 16S rRNA record sets to acquire taxonomic and useful outlines from any multipart ecological microbial community. (Picture courtesy of Morgan and Huttenhower 2012)

References

Albuquerque P, Ribeiro N, Almeida A, Panschin I, Porfirio A, Vales M, Tavares F (2017) Application of a dot blot hybridization platform to assess Streptococcus uberis population structure in dairy herds. Front Microbiol 8:54

Amann RI, Binder BJ, Olson RJ, Chisholm SW, Devereux R, Stahl DA (1990) Combination of 16S rRNA-targeted oligonucleotide probes with flow cytometry for analyzing mixed microbial populations. Appl Environ Microbiol 56(6):1919–1925

Amann RI, Ludwig W, Schleifer K-H (1995) Phylogenetic identification and in situ detection of individual microbial cells without cultivation. Microbiol Rev 59(1):143–169

Balkwill DL, Boone DR (2018) Identity and diversity of microorganisms cultured from subsurface environments. In: Microbiology of the terrestrial deep subsurface. CRC, Boca Raton, pp 105–118

Bardgett RD, van der Putten WH (2014) Belowground biodiversity and ecosystem functioning. Nature 515(7528):505

Bender SF, Wagg C, van der Heijden MG (2016) An underground revolution: biodiversity and soil ecological engineering for agricultural sustainability. Trends Ecol Evol 31(6):440–452

Bordukalo-Nikšić, T. (2007) Reverse transcription-polymerase chain reaction (RT-PCR). Metode u molekularnoj biologiji, Institut Ruđer Bošković

Brady NC, Weil RR (2002) The nature and properties of soils, 13th edn. Agrofor Syst 54(3):249

Camus C, Faugeron S, Buschmann AH (2018) Assessment of genetic and phenotypic diversity of the giant kelp, *Macrocystis pyrifera*, to support breeding programs. Algal Res 30:101–112

De Vries FT, Hoffland E, van Eekeren N, Brussaard L, Bloem J (2006) Fungal/bacterial ratios in grasslands with contrasting nitrogen management. Soil Biol Biochem 38(8):2092–2103

Fernández-Arrojo L, Guazzaroni M-E, López-Cortés N, Beloqui A, Ferrer M (2010) Metagenomic era for biocatalyst identification. Curr Opin Biotechnol 21(6):725–733

Field D, Garrity G, Gray T, Morrison N, Selengut J, Sterk P, Angiuoli SV (2008) The minimum information about a genome sequence (MIGS) specification. Nat Biotechnol 26(5):541

Finniss DG, Kaptchuk TJ, Miller F, Benedetti F (2010) Biological, clinical, and ethical advances of placebo effects. Lancet 375(9715):686–695

Grattepanche JD, Walker LM, Ott BM, Paim Pinto DL, Delwiche CF, Lane CE, Katz LA (2018) Microbial diversity in the eukaryotic SAR clade: illuminating the darkness between morphology and molecular data. Bioessays 40(4):1700198

Ha NN, Agusa T, Ramu K, Tu NPC, Murata S, Bulbule KA, Tanabe S (2009) Contamination by trace elements at e-waste recycling sites in Bangalore, India. Chemosphere 76(1):9–15

Handelsman J (2004) Metagenomics: application of genomics to uncultured microorganisms. Microbiol Mol Biol Rev 68(4):669–685

He JZ, Shen JP, Zhang LM, Zhu YG, Zheng YM, Xu MG, Di H (2007) Quantitative analyses of the abundance and composition of ammonia-oxidizing bacteria and ammonia-oxidizing archaea of a Chinese upland red soil under long-term fertilization practices. Environ Microbiol 9(9): 2364–2374

Heacock M, Kelly CB, Suk WA (2016) E-waste: the growing global problem and next steps. Rev Environ Health 31(1):131–135

Horwath WR (2017) The role of the soil microbial biomass in cycling nutrients. In: Microbial biomass: a paradigm shift in terrestrial biogeochemistry. World Scientific, London, pp 41–66

Inbar E, Green SJ, Hadar Y, Minz D (2005) Competing factors of compost concentration and proximity to root affect the distribution of streptomycetes. Microb Ecol 50(1):73–81

Jansson JK, Hofmockel KS (2018) The soil microbiome—from metagenomics to metaphenomics. Curr Opin Microbiol 43:162–168

Kardol P, Veen G, Teste FP, Perring MP (2015) Peeking into the black box: a trait-based approach to predicting plant–soil feedback. New Phytol 206(1):1–4

Kirk JL, Beaudette LA, Hart M, Moutoglis P, Klironomos JN, Lee H, Trevors JT (2004) Methods of studying soil microbial diversity. J Microbiol Methods 58(2):169–188

Knowles JG, Cole AL (2008) Handbook of the arts in qualitative research: perspectives, methodologies, examples, and issues. Sage, Thousand Oaks

Liu Y-F, Galzerani DD, Mbadinga SM, Zaramela LS, Gu J-D, Mu B-Z et al (2018) Metabolic capability and in situ activity of microorganisms in an oil reservoir. Microbiome 6(1):5
Loureiro C, Medema MH, van der Oost J, Sipkema D (2018) Exploration and exploitation of the environment for novel specialized metabolites. Curr Opin Biotechnol 50:206–213
Lynch J, Benedetti A, Insam H, Nuti M, Smalla K, Torsvik V, Nannipieri P (2004) Microbial diversity in soil: ecological theories, the contribution of molecular techniques and the impact of transgenic plants and transgenic microorganisms. Biol Fertil Soils 40(6):363–385
Martinez-Murcia A, Acinas S, Rodriguez-Valera F (1995) Evaluation of prokaryotic diversity by restrictase digestion of 16S rDNA directly amplified from hypersaline environments. FEMS Microbiol Ecol 17(4):247–255
Mastriani M, Giraldez A (2018) Microarrays denoising via smoothing of coefficients in wavelet domain. arXiv preprint arXiv:1807.11571
Matt M, Gaunand A, Joly PB, Colinet L (2017) Opening the black box of impact: ideal-type impact pathways in a public agricultural research organization. Res Policy 46(1):207–218
Meert JG, Torsvik TH, Eide EA, Dahlgren S (1998) Tectonic significance of the Fen Province, S. Norway: constraints from geochronology and paleomagnetism. J Geol 106(5):553–564
Mérillon J-M, Riviere C (2018) Natural antimicrobial agents, vol 19. Springer, Cham, Switzerland
Mohanty S, Swain CK (2018) Role of microbes in climate smart agriculture. In: Microorganisms for green revolution. Springer, Singapore, pp 129–140
Morgan XC, Huttenhower C (2012) Human microbiome analysis. PLoS Comput Biol 8(12): e1002808
Nanda DK, Chaudhary R, Kumar D (2018) Molecular approaches for identification of lactobacilli from traditional dairy products. In: Advances in animal biotechnology and its applications. Springer, Singapore, pp 181–196
Oguntoyinbo FA, Tourlomousis P, Gasson MJ, Narbad A (2011) Analysis of bacterial communities of traditional fermented West African cereal foods using culture independent methods. Int J Food Microbiol 145(1):205–210
Parsley LC, Consuegra EJ, Kakirde KS, Land AM, Harper WF, Liles MR (2010) Identification of diverse antimicrobial resistance determinants carried on bacterial, plasmid, or viral metagenomes from an activated sludge microbial assemblage. Appl Environ Microbiol 76(11): 3753–3757
Pernthaler A, Amann R (2004) Simultaneous fluorescence in situ hybridization of mRNA and rRNA in environmental bacteria. Appl Environ Microbiol 70(9):5426–5433
Peters S, Koschinsky S, Schwieger F, Tebbe CC (2000) Succession of microbial communities during hot composting as detected by PCR–single-strand-conformation polymorphism-based genetic profiles of small-subunit rRNA genes. Appl Environ Microbiol 66(3):930–936
Petersen DG, Dahllöf I (2005) Improvements for comparative analysis of changes in diversity of microbial communities using internal standards in PCR-DGGE. FEMS Microbiol Ecol 53(3): 339–348
Rangel DE, Finlay RD, Hallsworth JE, Dadachova E, Gadd GM (2018) Fungal strategies for dealing with environment- and agriculture-induced stresses. Fungal Biol 122(6):602–612
Roesch LF, Fulthorpe RR, Riva A, Casella G, Hadwin AK, Kent AD et al (2007) Pyrosequencing enumerates and contrasts soil microbial diversity. ISME J 1(4):283
Schloter M, Nannipieri P, Sørensen SJ, van Elsas JD (2018) Microbial indicators for soil quality. Biol Fertil Soils 54(1):1–10
Sunamura M, Maruyama A (2006) A digital imaging procedure for seven-probe-labeling FISH (rainbow-FISH) and its application to estuarine microbial communities. FEMS Microbiol Ecol 55(1):159–166
Taketani RG, Kavamura VN, dos Santos SN (2017) Diversity and technological aspects of microorganisms from semiarid environments. In: Diversity and benefits of microorganisms from the tropics. Springer, Cham, pp 3–19
Tansel B (2017) From electronic consumer products to e-wastes: global outlook, waste quantities, recycling challenges. Environ Int 98:35–45

Theron J, Cloete T (2000) Molecular techniques for determining microbial diversity and community structure in natural environments. Crit Rev Microbiol 26(1):37–57
Tringe SG, Rubin EM (2005) Metagenomics: DNA sequencing of environmental samples. Nat Rev Genet 6(11):805
Tringe SG, Von Mering C, Kobayashi A, Salamov AA, Chen K, Chang HW et al (2005) Comparative metagenomics of microbial communities. Science 308(5721):554–557
UNEP (2005) http://wedocs.unep.org/handle/20.500.11822/183
Venter JC, Adams MD, Myers EW, Li PW, Mural RJ, Sutton GG et al (2001) The sequence of the human genome. Science 291(5507):1304–1351
Wilcox TM, Schwartz MK, Lowe WH (2018) Evolutionary community ecology: time to think outside the (taxonomic) box. Trends Ecol Evol 33(4):240–250
Wood SA, Bradford MA (2018) Leveraging a new understanding of how belowground food webs stabilize soil organic matter to promote ecological intensification of agriculture. In: Soil carbon storage. Elsevier, Amsterdam, pp 117–136
Yilmaz A, Javed O, Shah M (2006) Object tracking: a survey. ACM Comput Surv (CSUR) 38(4):13
Zhang L, Dai Y, Chen J, Hong L, Liu Y, Ke Q, Chen Z (2018) Comparison of the performance in detection of HPV infections between the high-risk HPV genotyping real time PCR and the PCR-reverse dot blot assays. J Med Virol 90(1):177–183
Zoetendal EG, Cheng B, Koike S, Mackie RI (2004) Molecular microbial ecology of the gastrointestinal tract: from phylogeny to function. Curr Issues Intest Microbiol 5(2):31–48

Chapter 12
Recycling Processes and Plastic in Electronic Waste Is an Emerging Problem for India: Implications for Future Prospect

K Ronnie Rex, Revathy Rajakumaran, Omkar Gaonkar, Elvis Dsouza, and Paromita Chakraborty

Abstract A steady rise in demand of electronic equipment is prevalent in both developed and developing nations. The high consumption rate of electronic goods has given rise to a new stream of waste known as electronic waste or e-waste. The informal sector is dominant and critically active in developing countries like India to recover precious metals from e-waste. The illegal trade of electronic waste products also adds up to a good proportion of e-waste. Recent studies from India reported that the crude processes involved in the informal e-waste recycling sector is a major cause for the emission of persistent toxicants into the environment. Some of those studies mentioned that the open burning of e-waste in dumpsites and landfills have

K. Ronnie Rex
Department of Civil Engineering, SRM Institute of Science and Technology, Kattankulathur, Kancheepuram District, Tamil Nadu, India

SRM Research Institute, SRM Institute of Science and Technology, Kattankulathur, Kancheepuram District, Tamil Nadu, India

R. Rajakumaran
Environmental and Water Resource Engineering Division, Department of Civil Engineering, IIT Madras, Chennai, Tamil Nadu, India

O. Gaonkar
Nuevo Chakra, Mumbai, Maharashtra, India

SRM Research Institute, SRM Institute of Science and Technology, Kattankulathur, Kancheepuram District, Tamil Nadu, India

E. Dsouza
EDPC Polymer Industries, Vasai, Maharashtra, India

P. Chakraborty (✉)
Department of Civil Engineering, SRM Institute of Science and Technology, Kattankulathur, Kancheepuram District, Tamil Nadu, India

Nuevo Chakra, Mumbai, Maharashtra, India

SRM Research Institute, SRM Institute of Science and Technology, Kattankulathur, Kancheepuram District, Tamil Nadu, India
e-mail: paromitc@srmist.edu.in

M. Z. Hashmi, A. Varma (eds.), *Electronic Waste Pollution*, Soil Biology 57,
https://doi.org/10.1007/978-3-030-26615-8_12

further contributed to the release of persistent organic pollutants in the environement. Burning of plastics present in e-waste (e-plastics) is another serious problem. This chapter presents a review of both formal and informal e-waste recycling and how informal sector is proving to be of a major environmental concern in India. Further, we have discussed the sustainable recycling methods for plastic and metal recovery and the future prospects of e-waste recycling sector.

Keywords E-waste · Persistent organic pollutants · Informal recycling sector · Metal recovery

12.1 Introduction

The revolutionary impact of information and technology due to globalization has commercialized a wide range of electronic products in the economy, thereby drastically changing the lifestyle and perspective of people all over the world (Hischier et al. 2005; Pradhan and Kumar 2014; Robinson 2009; Røpke 2001; Wagner 2009). The increase in population density and desire to improve the standard of life has increased the demand for electrical and electronic equipment (EEE), eventually leading to an enhanced production of EEE to meet the availability.

Electronic waste or e-waste are various forms of electronic gadgets at their end of serviceable life and are disowned by their users. E-waste is generically referred to all forms of electrical and electronic waste which has ceased the value period and is of minimal use to the consumer (Mohan and Bhamawat 2008; Raghupathy et al. 2010). Among the different types of wastes defined in the Basel Convention, e-waste is an emerging concern due to the incremental rates of production and disposal in the waste stream. Due to inexpensive manpower in developing countries, e-waste is imported for recycling purposes mostly by the informal sector and therefore possess serious public health concern (Chakraborty et al. 2016a, b; Robinson 2009; Terazono et al. 2006).

In India, the electronic industry stepped up into action in the year 1965 with the preliminary focus on the technological advancement in space, defence and communication divisions. Thus, with the increasing growth of information and technology industry, the production rate of such products started to increase keeping pace with the economic growth of the country (Ganguly 2016; GOI 2011).

E-waste in the year 2010 constituted nearly 2% of the municipal solid waste generated, that is, about 50 million tonnes (MT) in developed countries (Ganguly 2016; GOI 2011). In the year 2017, it was estimated to increase to 65 MT and is expected to perpetually increase in the forthcoming years (Balde et al. 2017; Heacock et al. 2016). Among the top five e-waste-generating countries, the United States ranks first, with 3 MT in the year 2007 which increased to 11.7 MT in the year 2016 and still accounting to perish the concurrent limit (GOI 2011; Kumar et al. 2017). China is in the second position with a production rate of 2.3 MT in 2007 with a considerable increase of 6.1 MT in the year 2016 (Kumar et al. 2017). Recent studies have shown that in the present decade, contribution of e-waste in the global expanse is parentally by the United States and European Union (EU) which is roughly estimated to be 30% and 28%, respectively (GOI 2011; Kumar et al. 2017). According to the study conducted in EU in 2007, it has been found that 14–15 kg per capita e-waste is generated and contributed to an annual production of roughly 7 MT and is expected to propagate at a cumulative increase of 3–5% per year (Kumar et al. 2017; Zoeteman et al. 2010). Plastic in e-waste (e-plastic) is the end of life plastic component covering or integrated in any discarded and thrown away electronic equipment, such as refrigerators, air conditioners, computers, televisions, VCRs, stereos, copiers and fax machines (Yang et al. 2013). Annually 50 million tons of e-waste is generated, resulting in 10 million tons of plastics in e-waste (Jiang et al. 2012).

Given the fact that there is e-waste transported from developed countries with a reasonably good quantity of waste generated in India, we are presenting a review of scientific literature and articles from the government and nongovernment organizations with the aim to report the current problems associated with e-waste recycling in India. This chapter is based on the following objectives in Indian context to elucidate the: (1) generation and production of e-waste, (2) recycling of e-waste by the formal and informal sectors, (3) plastic waste in e-waste as an emerging problem and (4) prospect for sustainable recycling of e-waste for recovery of metals and plastics.

12.2 Generation and Production of E-waste in India

India ranks fifth in e-waste generation. Central Pollution Control Board (CPCB) estimated it to be 0.573 MT per day in the year 2005 and 1.8 MT in 2016 (Khattar et al. 2007; Ganguly 2016). India is estimated to generate below 1 kg per capita e-waste. United Nations (UN) reports estimated that e-waste such as old computers will increase up to 500% in terms of quantity from 2007 to 2020 (Ganguly 2016). Mobile phones, being the most popular and consumable product, are estimated to increase by 18 times from 2007 to 2020 because of the rise in unusable and discarded phones. During the year 2007, 382,979 tonnes of waste were generated in India with major contribution from discarded televisions, computers and mobile phones. It is noteworthy that about 50,000 tonnes were found to be imported illegally from developed nations where the United States and the EU were the major benefactors with 80% and 20% of contribution, respectively (Ganguly 2016; Khattar et al. 2007). There has been an increasing trend in illegal export of almost 50–80% of e-waste from these countries to developing countries, where the environmental regulations are not so stringent and labour wages are low (Agoramoorthy 2006; Chan et al. 2007; Pradhan and Kumar 2014). Predominant hot spots include China, India, Pakistan, Vietnam, and the Philippines where the informal recycling of e-waste is conducted on a larger scale (ToxicsLink 2016).

In India, there are ten states which majorly contribute to over 70% of total e-waste produced in the country. Among the Indian states, Maharashtra tops the list, Tamil Nadu is in the second position, followed by Andhra Pradesh, Uttar Pradesh, West Bengal, Delhi, Karnataka, Gujarat, Madhya Pradesh and Punjab (Begum 2013).

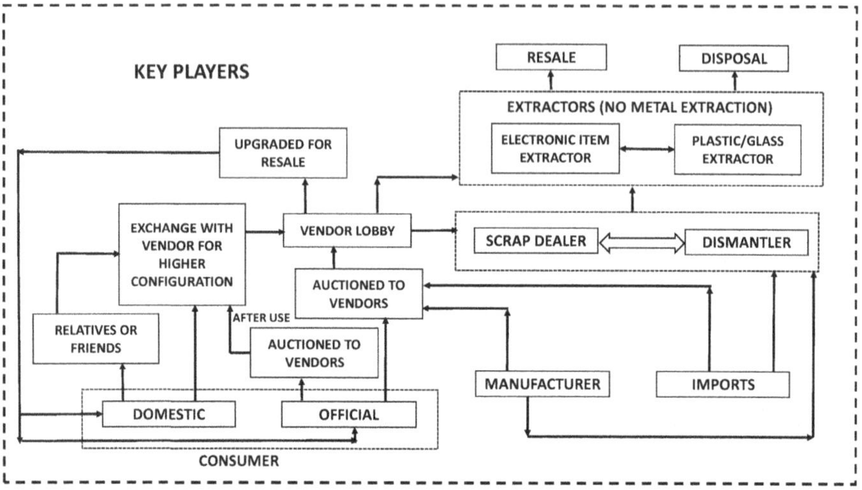

Fig. 12.1 E-waste recycling in India as explained by Wath et al. (2010)

Mumbai tops the list of 65 cities contributing nearly 60% of e-waste generated in India followed by New Delhi, Bengaluru, Chennai and Kolkata (Ganguly 2016; Jog 2008). Only 5% of e-waste generated in India is recycled by the formal sector (Chakraborty et al. 2016b; Khattar et al. 2007). But the high import rates and predominant e-waste generation is being extensively supported by the informal e-waste recycling sectors in India, where unrefined primitive methods are used to recover precious elements such as expensive valuable metals primarily gold, silver and copper (Chakraborty et al. 2018, 2019; GOI 2011; Ha et al. 2009; ToxicLink 2014). The extraction process involves dismantling of discarded e-waste materials which are shredded and treated with acid for metal recovery (Fig. 12.1) (Wath et al. 2010). Metal recovery processes are carried out by the unorganized sector and poses environmental and health threats due to release of persistent toxic substances (Chakraborty et al. 2016b).

12.3 Recycling of E-waste by the Formal Sector and Regulatory Framework in India

Out of 41.8 MT e-waste generated in India, informal sector receives 35.3 MT of e-waste (Baldé et al. 2015; Chakraborty et al. 2016b). Formal sectors are licensed organizations regulating recycling activities according to strong technical and environmental standards (Wang et al. 2013). Formal sector is expensive in setting up and have better control on health and environmental protection aspects. They usually deal with refurbishment, reselling, dismantling and shredding electronics to plastic and metals which are further recycled separately through sound environmental processes (Ceballos and Dong 2016).

The growing environmental concern in India has initiated the development of environmentally sound methods by organized sectors engaged in the reduction of the accumulated e-waste in a cost-effective manner by recycling and retrieving raw materials beneficial to the society. Initiatives are taken by the organized sector regarding proper training on recycling and reusing strategies with a major focus to develop such practices in the unorganized sector. This will help to reduce environmental pollution and in turn lead to conservation of natural resources and even employment generation. Training regarding proper recycling and reusing is initially being incorporated into unorganized sector with emphasis on reducing environmental pollution and natural resource conservation leading to employment and revenue generation.

India faces a major problem in the management of e-waste not only because of the import of e-waste but also due to the huge amount of domestically produced e-waste. The lack of appropriate cost-effective technology and uncontrolled disposal in the waste stream has affected the growth of formal sectors in India. Due to higher profit margin, the unorganized sector plays an important role in hampering the collection of e-waste by the formal sector (Ganguly 2016).

The three rules governing e-waste management in India are: Hazardous Wastes Management, Handling and Transboundary Movement Rules of 2008 and The Batteries Management and Handling Rules. The e-waste Management and Handling Rules 2010 is the most recent attempt to regulate e-waste in India. However, a final outcome of this rule is yet to come and is much awaited (Ganguly 2016). The Ministry of Environment, Forest and Climate Change (MoEFCC) updated the e-waste management rules in 2016 which requires producers to have Producer Responsibility Organisation (PRO) and can therefore ensure sustainable collection and disposal of e-wastes (Chakraborty et al. 2018). The implementation of these rules by the formal e-waste recycling sector becomes difficult as the responsibilities of certain activities fall within the state jurisdiction while other activities are governed by the central government. Hence, proper monitoring becomes increasingly difficult, sometimes leading to increasing role of informal sector. The rules put forward by MoEFCC in 2016 has created a powerful awareness in the government level regarding e-waste generation (Chakraborty et al. 2018).

12.4 Recycling of E-waste by the Informal Sector in India

The informal sector recycles about 95% of e-waste in India (Chakraborty et al. 2018). Urban slums and suburbs are the major locations where unorganised recycling sector endeavours in and around major metropolitan cities of India. Such localities have unskilled labour and reduces labour investment. Hence, maximum profit can be extracted from this business sector (Fig. 12.2) (Williams et al. 2013).

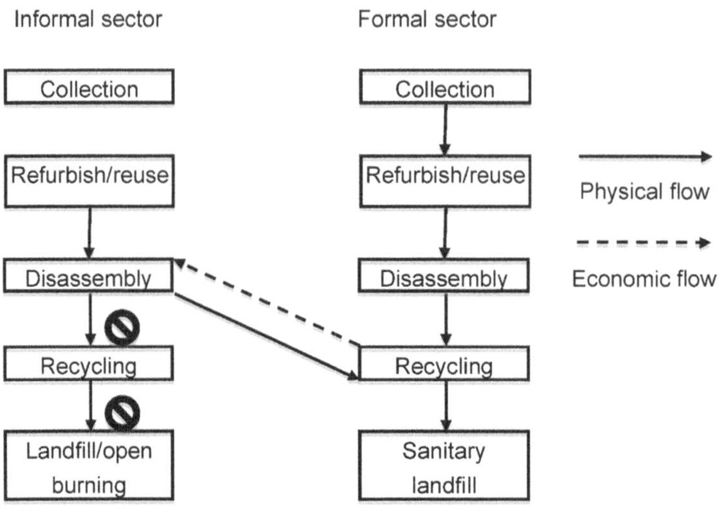

Fig. 12.2 Formal versus informal recycling of e-waste as described by Williams et al. (2013)

The unorganized nature and the revenue generation from e-waste explains why this sector adopts rudimentary methods to extract precious metals. Less complicated methods involve three major stages—dissembling of e-waste, followed by shredding and finally the metal recovery process involving the extraction of precious metals from e-waste (Chakraborty et al. 2016b). Remnants are discarded directly into the environment and majority of it is plastic waste. Retrieval of iron and steel, present in 48% of electronic equipment, has prime significance. The plastic component in e-waste contributes about 15% of non-flame-retardant plastics and 5% flame-retardant plastics with an average gauging of 25–35% of its weight making it the second largest component of EEE (ToxicsLink 2016).

The drop in price of electronic products has not only benefitted mankind but also increased the production of electronic goods. This has significantly affected the product life span overshooting the quantitated proportion of electronic detritus. All the reusable components are refurbished and sent to the market again. Hence, the waste generation rate do not subside and the quantitative flow is further increased by the inflow of illegal e-waste from developed countries. The extraction of precious metals from e-waste is a major profit-earning component which in turn leads to high import rates of e-waste in India. This exploitation has contributed to the dangerous health hazards associated with each activity during informal e-waste recycling process. The most affected class include women and children who are somewhat forcefully drifted into this industry due to poor employment options and sustenance. Environment is impacted heavily due to the release of toxicants from each step carried out during the crude recycling process and related waste disposal.

The plastic components in e-waste, especially the ones having potential of forming smaller fragments and are recyclable in nature, are removed by means of

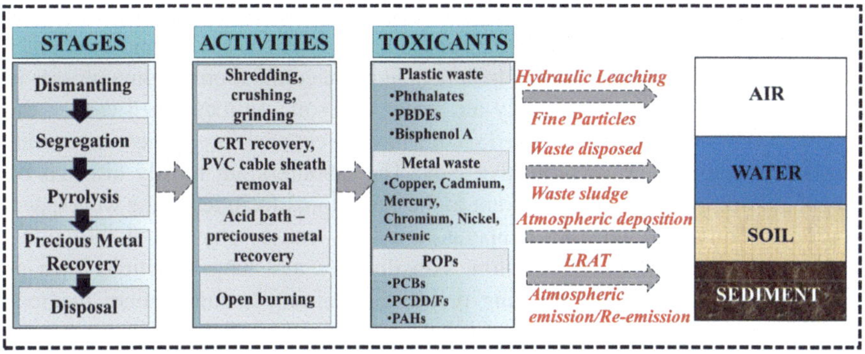

Fig. 12.3 Stages, activities and toxicants released during informal e-waste recycling in India as described by Chakraborty et al. (2016b)

mechanical shredding (GreenPeace 2005). A major proportion of e-waste plastics, rich in both secondary and primary toxicants, are disposed essentially using crude methods by the informal recycling sector, thus extensively promoting emission levels of several toxicants to a precarious limit.

Acid recovery is the most toxic step of all the processes where the emission of toxicants are expected to be predominant (Chakraborty et al. 2016b). Once the recycling materials are removed in the preliminary stage after collection, segregated e-waste is processed through shredding and then transferred to acid leaching sites where precious metals including gold, silver and copper are recuperated (GreenPeace 2005). Since e-waste mainly constitute of metals and plastics, this procedure of acidic leaching is performed by dissolving the preprocessed waste in strong acidic solutions. Cathode ray tubes (CRTs) are mostly disassembled and recycled so that valuable materials can be recovered. Printed circuit boards are major contributors of e-waste category since they are the principal components in most electrical equipments (Begum 2013). After recyclable components are extracted from the printed circuit boards via dismantling or heat treatment, the rest is subjected to acid treatment for metal extraction (Chakraborty et al. 2016a, b). The remaining parts are openly burnt (GreenPeace 2005). The burning of e-waste is also practiced in the recycling workshops to remove the plastic or polyvinyl chloride (PVC) coatings and recover copper (Chakraborty et al. 2018, 2019; Chakraborty et al. 2016b; GreenPeace 2005). Rudimentary methods used during informal e-waste recycling activities in India subsequently release toxic pollutants like polychlorinated biphenyls (PCBs) (Chakraborty et al. 2013), polycyclic aromatic hydrocarbons (PAHs) (Chakraborty et al. 2019), polybrominated diphenyl ethers (PBDEs) (Chakraborty et al. 2017) and polychlorinated dibenzo-*p*-dioxins and furans (PCDDs/Fs) (Chakraborty et al. 2016a, b; 2018) (Fig. 12.3). These semi-volatile persistent organic pollutants (POPs) are capable of long range atmospheric transport (LRAT).

The most severe concern regarding the informal e-waste recycling sector in India (Chakraborty et al. 2018, 2019; Chakraborty et al. 2016b; Ha et al. 2009) and China

(Ha et al. 2009) is the metal extraction, involving various crude methods leading to the release of toxic substances. The disposed wastes such as sludge and discharges from industries due to its improper treatment led to accumulation of contaminants in environmental matrices dominantly air, water and soil. As mentioned earlier, the people working in these unorganized sectors have direct exposure to these hazardous substances due to lack of protective aids. The labourers are mostly affected by irregular working schedule concurrent to the operational facilities in the workspace. Improper ventilation and high temperature usually causes physical abnormalities including injuries, vision obstruction, skin diseases and lung afflictions such as asthma. Thus, e-waste is a growing problem pertaining to rudimentary methods practiced during informal recycling for revenue generation.

12.5 E-plastic: An Emerging Issue

Plastics can be efficiently processed into light, durable and affordable goods having very less thermal and electrical conductivity (Brebu et al. 2004). Hence plastics constitute an integral component of EEE and majority of applications include insulation purpose, reduce noise, sealing, housing, interior structural, functional and interior electronic parts. The usage of plastics in electronic sector over the last decade has helped in lowering raw material consumption and minimizing the total manufacturing cost, which resulted in reduction of the total weight of the equipment (Fisher et al. 2004).

The waste e-plastics constitute complex mixture of wide variety of substances including a large range of polymers which are mostly incompatible. E-plastics consist of more than fifteen non-identical polymers which makes it difficult to recycle. To characterize plastic composition, a total of 3417 appliances including large cooling equipment (refrigerators, freezers), small waste electrical and electronic equipment (WEEE), central processing units (CPU), coping equipment, printers, cathode ray tube (CRT) monitors and CRT televisions were characterized by Martinho et al. (2012). The percentage of plastic in large cooling appliances, small WEEE, printers, coping equipment, CPU, CRT monitors and CRT televisions were reported to be 10.4%, 49.1%, 23.6%, 15.7%, 3.5%, 17.5% and 16.5%, respectively. The types of polymers analysed in appliances are given in Fig. 12.4 (Martinho et al. 2012). The most predominantly found polymers were polystyrene (PS), acrylonitrile-butadiene-styrene (ABS), polycarbonate blends (PC/ABS), high-impact polystyrene (HIPS) and polypropylene (PP) (Martinho et al. 2012).

Flame retardants (FR) are used in e-plastics to increase the fire-resisting property of the electronic products which in turn reduces the threat of fire. More than 175 types of FR plastics are present, which include about 27% halogenated organic compounds and the rest are non-halogenated FRs. Around 30% FR plastics used in electronic equipment comprises of halogenated or non-halogenated plastics (Vehlow et al. 2002). The FR plastics accounts for 4,50,000 tonnes containing about 41% of halogenated FRs as shown in Fig. 12.5 (Vehlow et al. 2002).

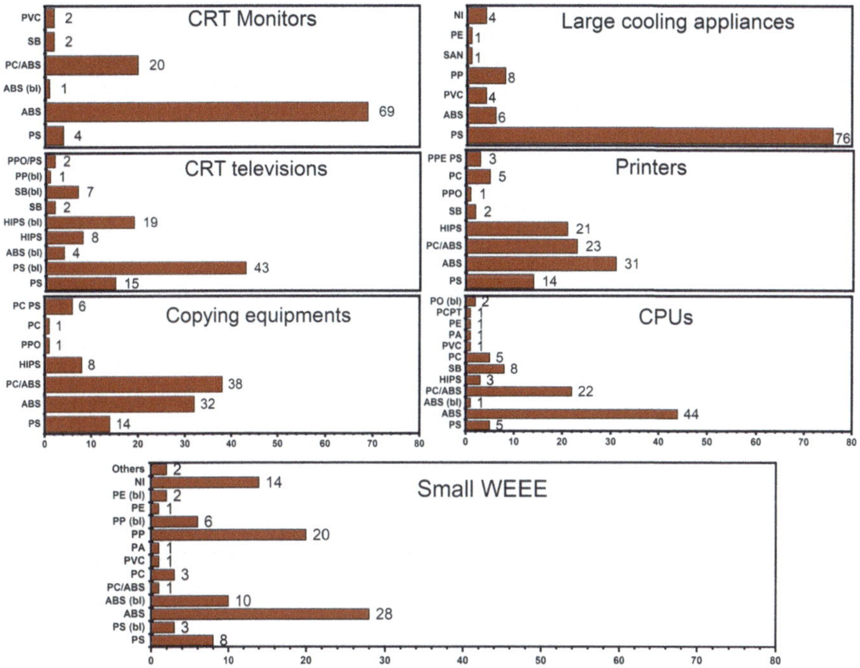

Fig. 12.4 Characterization of different polymer types from waste electrical and electronic equipment (WEEE) (values are in percentage) (Martinho et al. 2012)

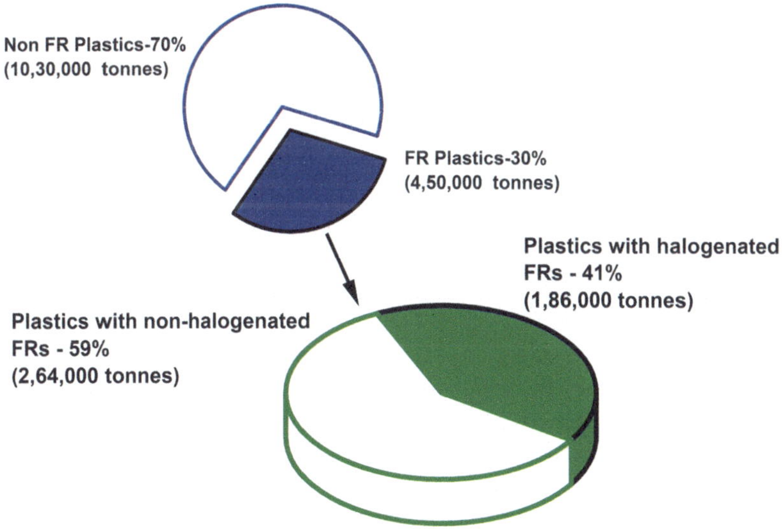

Fig. 12.5 Usage of flame retardants plastics in EEE (Vehlow et al. 2002) (FR—flame retardants, E&E—electrical and electronic)

Table 12.1 Health risks of some BFRs

BFRs	Health risks	References
PBDE	Human milk associated with cryptorchidism and increased serum LH	Main et al. (2007)
PBDE	Elevated levels in the human milk correlated with lower birth weight, length, lower head and chest circumference	Chao et al. (2010)
PBB	Diabetes	Lee et al. (2010)
TBBPA	Thyroid hormone interference and neurotoxicity	Van der Ven et al. (2008)
TBBPA	Binds to oestrogen hormone receptors at high concentrations	Gosavi et al. (2013)
HBCD	Evidence of biomagnification in aquatic food webs	Morris et al. (2004)
HBCD	TH-mediated brain development is disruption by HBCD	Ibhazehiebo et al. (2011)

Brominated flame retardants (BFRs) are the most frequently used FRs in e-plastics, containing about 50–90 wt% of bromine. BFRs are applied to 2.5 million tonnes of polymers annually (Tohka and Zevenhoven 2002). Among the 75 categories of BFRs available, only around 30–40 are extensively used. Tetrabromobisphenol A (TBBPA), decabromodiphenyl oxide (DBDPO), 1,2,5,6,9,10-hexabromocyclodecane (HBCD) and decabromodiphenyl ethane (DBDPE) (health risks as given in Table 12.1) are the highly used BFRs (Freegard et al. 2006; Tohka and Zevenhoven 2002).

Generally, these BFRs are persistent in nature and have the prime susceptibility to bioaccumulate in the ecosystem and human beings. Further, they are presumed to be major carcinogens, neurotoxins and endocrine disruptors (Brown et al. 2004; De Wit 2002). Liver tumours, neurodevelopmental problems and thyroid dysfunctions are the major expected disorders caused by BFRs (Siddiqi et al. 2003). All the major matrixes including air, water, sewage sludge, sediments and biota have been quantified for BFRs. The water solubility of BFR is exceptionally small. When plastic wastes with FR composition is reprocessed, BFRs predominantly PBDEs have the tendency to form brominated dibenzo dioxins and furans (PBDD/Fs) (Siddiqi et al. 2003). It is evident that PBDE with less brominated fractions have high potential to bioaccumulate and biomagnify in human, fish and other animal adipose tissues. PBDEs have been detected in human blood, serum, adipose tissue, breastmilk, placental tissue and the brain (Bjermo et al. 2017; Meironyté et al. 1999; Sellström et al. 1993; Siddiqi et al. 2003; Sjödin et al. 2001). Studies conducted in Sweden quantified the concentration of PBDE for a time span of thirty years in human milk samples reported that the concentration doubled approximately after every 5 years (Meironyté et al. 1999).

Regardless of the existence of Basal Convention which regulates and controls the transboundary movement of hazardous wastes from the United States, Canada, Australia, EU, Japan and South Korea to Asian countries such as China and India, the rate of inflow of toxic waste is still increasing in developing countries

Table 12.2 Management options of e-plastics

Options	Definitions	Comments
Landfilling	Landfilling is the least preferred option, e-plastics is dumped and interlayer with municipal solid wastes	Landfilling is a risky method. Leaking of landfills is a problem because pollutants will infiltrate and contaminate surrounding soil and aquifers
Material recycling	Plastics with FRs can be reprocessed mechanically to form new consumer plastic products or can be brought back to the system with close properties of the principle product	It is essential to segregate heterogeneous e-plastics before material recycling. In addition, the reprocessing leads to formation of toxic PBDD/Fs when heat treated (Schlummer et al. 2007), special measures are adopted during recycling process
Energy recycling	Objective of energy recycling is to produce thermal energy by incineration of e-plastics	Due to the emission of toxic substances like polycyclic aromatic hydrocarbons (PAHs), polychlorinated biphenyls (PCBs) and polychlorinated dioxins (PCDs) incineration is considered to be very dangerous (Wang and Xu 2014)
Feedstock recycling	Method of converting plastics to their original chemical constituents. These are applicable for mixed polymers in e-plastics, for example, pyrolysis and gasification	Pyrolysis is one of the feedstock recycling methods; 10% of energy is produced and the rest is used to transform material scrap into useable and costly hydrocarbon products (Brebu et al. 2004)

(Schlummer et al. 2007). However, in most of the developing countries, e-plastic is presently managed based on unsafe disposal practices majorly burning or burying. This inadequate waste management approach is no more secure on international level. Rising knowledge about the environmental issues in population has resulted in most developing countries adopting sound waste management practices. The risk associated with open burning of e-plastics and emission of various toxins from it is oblivious to majority of population in the developing countries and corresponding regulating authorities. Landfilling, material recycling, energy recycling and feedstock recycling are the current options available for managing e- plastics (Table 12.2).

The pyrolysis technology is proven to be consuming lowest energy and the oil generated in the process can be a raw material substitute in petrochemical industry. Nevertheless, BFRs present in plastics make this process somewhat problematic. However, the dehalogenation measures are mandatory to avoid the impact on reuse of pyrolysis oil. The debromination efficiency of the process can be improved by adding catalyst or incorporating supercritical fluid extraction (SCF) technology. The two processes can be combined to avoid the release of brominated dioxins and furans (Onwudili and Williams 2009).

12.6 Prospect for Sustainable Recycling of E-waste

Due to high market value, precious metals in e-waste are recycled to a credible extent but e-plastic waste is not significantly recycled to that level. Although recycling of e-plastics may seem to be impractical today because of high energy required and significant financial costs involved but, upcoming technologies and systems are making it a viable option for the long term. China is an ideal example where large recycling options for e-plastics are available. In the United States, over 99% of their e-plastics are collected and shipped to China. This is because of the high labour costs associated with dismantling and dissociation of e-waste in the United States. A recycling facility in Qing dao, northern China, managed by e-plastic consolidators ParcCorporation, USA, uses recycled e-plastics for devices like water coolers. A large US-based reclaimer Fortune Metals and Plastics has set up a Chinese joint venture to sort and upgrade e-waste metals and plastics 10 years ago in Nanjing. Other companies in this sector which have developed new techniques for e-waste and e-plastic recycling include MBA polymers, USA; Plas-Sep Ltd., Canada; Electrocycling GmbH in Goslar, Germany; Plastic Herverwerking Brabant in Waalwijk, The Netherlands (Schut 2007).

Axion Recycling Ltd., UK, is a recycling consultant engaged in testing several methods to separate brominated FRs from e-plastics and developed its own Centrevap method. Separation and recycling of e-plastic scrap generated by WEEE regulations can process 28 million lb/year of e-plastics via size reduction, wet and dry cleaning, automatic polymer identification and separation and repalletizing of e-scrap plastics. The polystyrene recycled from refrigerators is supplied by Axion Polymers, UK, to Brookhaven Instruments Corporation (BIC) for manufacturing pencils that is being marketed under its 'Ecosolutions' product lines. About 640 pencils could be typically made from one refrigerator (Recycling Today Staff 2011). Thus, various players in the recycling business and technological innovations can be the right way forward in handling e-plastic waste.

In India, there are some well-known companies involved in e-waste recycling. An e-waste management company named 'Attero', based in Roorkee, handles about 500 tonnes of e-waste per month. Another company based in Bangalore 'E-Parisaraa Pvt Ltd' is authorized by the government and involved in extraction of precious metals, such as gold and silver, from e-waste since 2005 (Ecoideaz 2016).

E-plastic recycling usually takes place along with the normal plastic recycling by the informal sector in India. The e-plastic recycling line should be separated because of the toxic nature of e-plastics and more environmentally friendly recycling techniques with minimal health impacts needs to be adopted. EDPC Polymer Industries located in Dadra and Nagar Haveli in western India is a pencil manufacturing company using recycled plastic for manufacturing pencils under the brand Ecocils. Their unique material and special extrusion technology supports production of more pencils and can be compared with the conventional pencil manufacturing methods. Presently, the company is exploring the potential use of e-plastics for the production of pencils in addition to other recycled products.

12.7 Conclusion

Being a developing nation, India stands reasonably high in e-waste generation at local level as well as import of e-waste from developed nations. The increasing trend of recycling e-waste by informal sector has also weighed in the illegal import of e-waste. Due to the lack of properly practised environmental regulations in India, the disposal is diverted towards landfills and dumpsites. The rules and regulations are stringent and the growth of formal sectors considering the legal constraints are really poor.

Being economically viable and environmentally sound, recycling is considered as the major significant policy for end-of-life electronic products. However, there are challenges that exist both on technological aspects and social issues associated with collection and processing of recyclable e-plastics. The increase in the use of plastics in electronic products is forcing companies to look for alternatives apart from landfilling.

Absence of stringent rules adds up to the concern regarding e-waste recycling in India. The informal recycling trend has affected the environmental matrices making it susceptible to organic and inorganic contaminants both directly and indirectly. Proper channel for collection and segregation of waste need to be practised with sustainable cost-effective recycling techniques. Separate recycling of plastic components into useful products or useable raw materials should be encouraged by providing subsidies from government.

Acknowledgement This work was supported by the grant from the Environmental Health Cell, Ministry of Environment, Forest and Climate Change, Government of India (File No.Q-14011/43/2013-CPW(EHC).

References

Agoramoorthy G (2006) Computer 'recycling' builds garbage dumps overseas. Nature 441:25

Baldé C, Wang F, Kuehr R, Huisman J (2015) The global e-waste monitor 2014, Institute for the Advanced Study of Sustainability (IAS) and Sustainable Cycles (SCYCLE). Technical report, United Nations University and Institute for Advanced Study of Sustainability

Balde CP, Forti V, Gray V, Kuehr R, Stegmann P (2017) The global e-waste monitor 2017: quantities, flows and resources. United Nations University, International Telecommunication Union, and International Solid Waste Association, Bonn

Begum DKJA (2013) Electronic waste (e-waste) management in India: a review. IOSR J Human Soc Sci (IOSR-JHSS) 10:46–57

Bjermo H, Aune M, Cantillana T, Glynn A, Lind PM, Ridefelt P, Darnerud PO (2017) Serum levels of brominated flame retardants (BFRs: PBDE, HBCD) and influence of dietary factors in a population-based study on Swedish adults. Chemosphere 167:485–491

Brebu M, Bhaskar T, Murai K, Muto A, Sakata Y, Uddin MA (2004) The individual and cumulative effect of brominated flame retardant and polyvinylchloride (PVC) on thermal degradation of acrylonitrile-butadiene-styrene (ABS) copolymer. Chemosphere 56:433–440

Brown DJ, Van Overmeire I, Goeyens L, Denison MS, De Vito MJ, Clark GC (2004) Analysis of Ah receptor pathway activation by brominated flame retardants. Chemosphere 55:1509–1518
Ceballos DM, Dong Z (2016) The formal electronic recycling industry: challenges and opportunities in occupational and environmental health research. Environ Int 95:157–166
Chakraborty P et al (2013) Atmospheric polychlorinated biphenyls in Indian cities: levels, emission sources and toxicity equivalents. Environ Pollut 182:283–290
Chakraborty P, Prithiviraj B, Selvaraj S, Kumar B (2016a) Polychlorinated biphenyls in settled dust from informal electronic waste recycling workshops and nearby highways in urban centers and suburban industrial roadsides of Chennai city, India: levels, congener profiles and exposure assessment. Sci Total Environ 573:1413–1421
Chakraborty P, Selvaraj S, Nakamura M, Prithiviraj B, Ko S, Loganathan BG (2016b) E-waste and associated environmental contamination in the Asia/Pacific region (Part 1): an overview. In: Persistent organic chemicals in the environment: status and trends in the pacific basin countries. I: Contamination status. ACS Symposium Series, vol 1243. American Chemical Society, pp 127–138. doi: https://doi.org/10.1021/bk-2016-1243.ch006
Chakraborty P, Zhang G, Cheng H, Balasubramanian P, Li J, Jones KC (2017) Passive air sampling of polybrominated diphenyl ethers in New Delhi, Kolkata, Mumbai and Chennai: levels, homologous profiling and source apportionment. Environ Pollut 231:1181–1187
Chakraborty P, Selvaraj S, Nakamura M, Prithiviraj B, Cincinelli A, Bang JJ (2018) PCBs and PCDD/Fs in soil from informal e-waste recycling sites and open dumpsites in India: levels, congener profiles and health risk assessment. Sci Total Environ 621:930–938
Chakraborty P, Sampath S, Mukhopadhyay M, Selvaraj S, Bharat GK, Nizzetto L (2019) Baseline investigation on plasticizers, bisphenol A, polycyclic aromatic hydrocarbons and heavy metals in the surface soil of the informal electronic waste recycling workshops and nearby open dumpsites in Indian metropolitan cities. Environ Pollut 248:1036–45
Chan JK et al (2007) Body loadings and health risk assessment of polychlorinated dibenzo-p-dioxins and dibenzofurans at an intensive electronic waste recycling site in China. Environ Sci Technol 41:7668–7674
Chao HA et al (2010) Concentrations of polybrominated diphenyl ethers in breast milk correlated to maternal age, education level, and occupational exposure. J Hazard Mater 175:492–500
De Wit CA (2002) An overview of brominated flame retardants in the environment. Chemosphere 46:583–624
Ecoideaz (2016) E-waste recycling in india – managing a growing crisis. http://www.ecoideaz.com/expert-corner/managing-the-e-waste-crisis-in-india. Accessed 21 Jan 2019
Fisher M, Kingsbury T, Headley L (2004) Sustainable electrical and electronic plastics recycling. In: Electronics and the Environment, 2004. Conference record, IEEE International Symposium on, 2004. IEEE, pp 292–297
Freegard K, Tan G, Morton R (2006) Develop a process to separate brominated flame retardants from WEEE polymers. Final report. Waste & Resources Action Programme (WRAP), UK
Ganguly R (2016) E-waste management in India–an overview. Int J Earth Sci Eng 9(2):574
GOI (2011). http://rajyasabha.nic.in/rsnew/publication_electronic/E-Waste_in_india.pdf. Acessed 21 Jan 2019
Gosavi RA, Knudsen GA, Birnbaum LS, Pedersen LC (2013) Mimicking of estradiol binding by flame retardants and their metabolites: a crystallographic analysis. Environ Health Perspect 121:1194
GreenPeace (2005). http://www.greenpeace.org/international/en/publications/reports/recyclingelectronicwasteindiachinafull/. Acessed 21 Jan 2019
Ha NN et al (2009) Contamination by trace elements at e-waste recycling sites in Bangalore, India. Chemosphere 76:9–15
Heacock M et al (2016) E-waste and harm to vulnerable populations: a growing global problem. Environ Health Perspect 124:550
Hischier R, Wäger P, Gauglhofer J (2005) Does WEEE recycling make sense from an environmental perspective? The environmental impacts of the Swiss take-back and recycling systems

for waste electrical and electronic equipment (WEEE) environmental impact. Assess Rev 25:525–539
Ibhazehiebo K, Iwasaki T, Xu M, Shimokawa N, Koibuchi N (2011) Brain-derived neurotrophic factor (BDNF) ameliorates the suppression of thyroid hormone-induced granule cell neurite extension by hexabromocyclododecane (HBCD). Neurosci Lett 493:1–7
Jiang P et al (2012) Improving the end-of-life for electronic materials via sustainable recycling methods. Procedia Environ Sci 16:485–490
Jog S (2008) Ten states contribute 70% of e-waste generated in India. The financial express, 13 March 2008
Khattar V, Kaur J, Chaturvedi A, Arora R (2007) E-waste assessment in India: specific focus on Delhi. http://www.weeerecycle.in/publications/reports/GTZ_MAIT_E-waste_Assessment_Report.pdf
Kumar A, Holuszko M, Espinosa DCR (2017) E-waste: an overview on generation, collection, legislation and recycling practices. Resour Conserv Recy 122:32–42
Lee D-H, Steffes MW, Sjödin A, Jones RS, Needham LL, Jacobs DR Jr (2010) Low dose of some persistent organic pollutants predicts type 2 diabetes: a nested case–control study. Environ Health Perspect 118:1235
Main KM et al (2007) Flame retardants in placenta and breast milk and cryptorchidism in newborn boys. Environ Health Perspect 115:1519
Martinho G, Pires A, Saraiva L, Ribeiro R (2012) Composition of plastics from waste electrical and electronic equipment (WEEE) by direct sampling. Waste Manag 32:1213–1217
Meironyté D, Norén K, Bergman A (1999) Analysis of polybrominated diphenyl ethers in Swedish human milk. A time-related trend study, 1972–1997. J Toxicol Environ Health Part A 58:329–341
Mohan D, Bhamawat P (2008) E-waste management-global scenerio: a review. J Environ Res Dev 2(4):817–823
Morris S et al (2004) Distribution and fate of HBCD and TBBPA brominated flame retardants in North Sea estuaries and aquatic food webs. Environ Sci Technol 38:5497–5504
Onwudili JA, Williams PT (2009) Degradation of brominated flame-retarded plastics (Br-ABS and Br-HIPS) in supercritical water. J Supercrit Fluids 49:356–368
Pradhan JK, Kumar S (2014) Informal e-waste recycling: environmental risk assessment of heavy metal contamination in Mandoli industrial area, Delhi, India. Environ Sci Pollut Res 21:7913–7928
Raghupathy L, Krüger C, Chaturvedi A, Arora R, Henzler MP (2010) E-waste recycling in India: bridging the gap between the informal and formal sector. In: International Solid Waste Association, World Congress, Hamburg, Germany
Recycling Today Staff (2011) BIC makes pencil with recycled plastic supplied by axion polymers - recycling today. In: Recycling today. https://www.recyclingtoday.com/article/bic-axion-partner-on-pencil/. Accessed 21 Jan 2019
Robinson BH (2009) E-waste: an assessment of global production and environmental impacts. Sci Total Environ 408:183–191
Røpke I (2001) New technology in everyday life–social processes and environmental impact. Ecol Econ 38:403–422
Schlummer M, Gruber L, Mäurer A, Wolz G, Van Eldik R (2007) Characterisation of polymer fractions from waste electrical and electronic equipment (WEEE) and implications for waste management. Chemosphere 67:1866–1876
Schut JH (2007) Recycling E-plastics: new material stram brings its own set of problems. Plastics Technol 53:48–53
Sellström U, Jansson B, Kierkegaard A, de Wit C, Odsjö T, Olsson M (1993) Polybrominated diphenyl ethers (PBDE) in biological samples from the Swedish environment. Chemosphere 26:1703–1718
Siddiqi MA, Laessig RH, Reed KD (2003) Polybrominated diphenyl ethers (PBDEs): new pollutants–old diseases. Clin Med Res 1:281–290

Sjödin A, Patterson DG, Bergman Å (2001) Brominated flame retardants in serum from US blood donors. Environ Sci Technol 35:3830–3833

Terazono A et al (2006) Current status and research on E-waste issues in Asia. J Mater Cycles aste Manag 8:1–12

Tohka A, Zevenhoven R (2002) Brominated flame retardants–a nuisance in thermal waste processing. In: TMS extraction and processing division meeting on recycling and waste treatment in mineral and metal processing: technical and economic aspects, Lulea, Sweden.

ToxicLink (2014). http://toxicslink.org/docs/Impact-of-E-waste-recycling-on-Soil-and-Water.pdf. Accessed 21 Jan 2019

ToxicsLink (2016). http://toxicslink.org/docs/Report-WEEE-plastics.pdf. Accessed 21 Jan 2019

Van der Ven LT et al (2008) Endocrine effects of tetrabromobisphenol-A (TBBPA) in Wistar rats as tested in a one-generation reproduction study and a subacute toxicity study. Toxicology 245:76–89

Vehlow J et al (2002) Recycling of bromine from plastics containing brominated flame retardants in state-of-the-art combustion facilities. APME Technical Report No. 8040, APME, Brussels

Wagner TP (2009) Shared responsibility for managing electronic waste: a case study of maine, USA. Waste Manag 29:3014–3021

Wang R, Xu Z (2014) Recycling of non-metallic fractions from waste electrical and electronic equipment (WEEE): a review. Waste Manag 34:1455–1469

Wang F, Huisman J, Stevels A, Baldé CP (2013) Enhancing e-waste estimates: improving data quality by multivariate input-output analysis. Waste Manag 33:2397–2407

Wath SB, Vaidya AN, Dutt P, Chakrabarti T (2010) A roadmap for development of sustainable E-waste management system in India. Sci Total Environ 409:19–32

Williams E, Kahhat R, Bengtsson M, Hayashi S, Hotta Y, Totoki Y (2013) Linking informal and formal electronics recycling via an interface organization. Challenges 4:136–153

Yang X, Sun L, Xiang J, Hu S, Su S (2013) Pyrolysis and dehalogenation of plastics from waste electrical and electronic equipment (WEEE): a review. Waste Manag 33:462–473

Zoeteman BC, Krikke HR, Venselaar J (2010) Handling WEEE waste flows: on the effectiveness of producer responsibility in a globalizing world. Int J Adv Manuf Technol 47:415–436

Chapter 13
Techniques Used for Recycling E-waste Worldwide

Naseer Ali Shah, Raja Muhammad Waqas Anjum, and Yasir Rasheed

Abstract There is a revolution in the field of material science and information technology. The whole world has become a global village due to the distribution of technology all around the globe. Electronic products have been welcoming to everyone as cheap and updated versions are being introduced day by day. People are enjoying the perks of being modernized due to technology, but at the same time, it is creating drastic effects as disposal of electronic waste has become a major challenge these days. All countries are trying to find new ways to dispose of e-waste because burning would create incurable environmental pollution. In this chapter, we will study the techniques which are being used by developing and developed countries to dispose of e-waste.

Keywords Techniques · Recycling · E-waste · Worldwide

13.1 Introduction

In the recent years, many significant efforts are being made for the improvement of technology. This has made recycling of e-waste essential. The improvement in telecommunication and information technology has given rise to International markets, which in turn has also enlarged globalization. In different countries, there exist different rules and regulations regarding management of e-waste (Amankwah-Amoah 2016; Goldsmid and Douglas 1954; Singh et al. 2016). For example, in the USA, certain apparatuses which are used on daily basis such as Toasters, fridges, microwaves, coffee makers are not considered under e-waste whereas, Swiss law on the return, getting back and dumping of Electrical and electronic equipment (Swiss ORDEE) and European union squander of Electrical and electronic gear (EU WEEE) considers them as e-waste.

N. A. Shah (✉) · R. M. W. Anjum · Y. Rasheed
Department of Biosciences, COMSATS University Islamabad, Islamabad, Pakistan
e-mail: drnaseer@comsats.edu.pk

M. Z. Hashmi, A. Varma (eds.), *Electronic Waste Pollution*, Soil Biology 57,
https://doi.org/10.1007/978-3-030-26615-8_13

In addition to these internationally implemented rules and regulations, commonly certain apparatuses which are unproductive such as old appliances are included in e-waste. These can be computers, mobile phones, etc. or large equipments such as refrigerators and other things (Garlapati 2016; Haynes et al. 2014; Julander et al. 2014). In addition, improvement of unused advances to buyers has largely increased, but management of e-waste is still a very slow process. Technology revolution has caused many issues, as the management of a large amount of e-waste produced is becoming a serious issue. Storage of such e-waste and handling and processing them have posed a real danger. Moreover, certain metals having high market values are being processed by unregulated little level operations. This situation of handling Electronic waste is becoming a threat for secure handling of materials and work conditions (Bousquet et al. 2008; Das et al. 2011; Haynes et al. 2014; Robinson 2009; Sepúlveda et al. 2010).

13.1.1 Growth of E-waste

The rapid growth of e-waste produced in the developing countries continued development of a persistent and economical informal processing sector, well organized and efficient but inherently hazardous.

In spite of plans of local controls and unsafe disposal laws, e-waste is considered as common deny, or roughly handled, frequently by burning or corrosive showers, with recuperation of as it were some materials of esteem. As furans, dioxins, and heavy metals are discharged, damage to the environment, laborers, and zone inhabitants is unavoidable. The more rapid growth of e-waste in the developing countries than in the developed countries indicates that the improvement of a certain and easygoing planning division, proficient in its case way, however naturally danger ridden.

E-waste issues related to exchange in waste and casual reusing within the advanced nations address natural, social, financial impacts. Besides, given on numerous viewpoint contemplations, it is found that right now reusing crumbling exchange presents. This paper to begin with surveys the dynamic forces of universal exchange in waste in reusing mob. Within the preface that situations and economic or social gains can be traded among nations, we offer administrative situations universal participation arrangement that increments e-waste treatment participation and fragmentation and contribute to compelling e-waste administration. The essential problems associated with e-waste issue (disposal and supply of electrical and electronic waste, WEEE) emerge by unequal amount of dangerous as well as poisonous constituents. Furthermore, reusing a tremendous amount of electronic transfers comes about in natural treachery and transcontinental contaminations as significant amounts of wastes are not reused locally, however, maybe dispatched to creating districts, in which unseemly and casual reusing causes extreme harms to situations and human wellbeing (Pellow 2010; Tu 2005; Widmer et al. 2005). Separated through financial perspective, present contamination impact on e-waste is worldwide and not nearby which requests for agreeable tries on overseeing e-waste emergency. Hence, distinctive approaches and activities show up to be planned and actualized at individually domestic and worldwide stages which are comprising: WEEE in Europe, EPR policy, the Basel Convention, RoHS in Europe, StEP initiative, and Basel Ban. In any case, prove appears that existing arrangements headings will relieve but not resolve issues of lawful as well as illicit trans edge development of wastes, casual reusing in industrializing nations, and a worldwide point of view on e-waste administration and economical advancement (Bousquet et al. 2008; Widmer et al. 2005).

Exceeding requirements for crude materials (particularly for uncommon soil and small components) and expanding amounts of unrestrained e-waste reusing tasks donate to the developing disquiets for e-waste manageement. The cross boundary transport of e-waste with low labor expenditures at the worldwide to ranges makes human health hazard as well as natural issues at these areas.

The main issues associated with e-waste problem arise not from both hazardous and toxic materials and quantities. They are also associated with recycling of massive quantity of electronic waste involved in transnational pollutions and environmental inequality. Considerable quantities of e-wastes are shipped to developing countries but they are not recycled domestically. That is the reason why improper recycling causes unfavorable effects to human health and rigorous damages to environment (Widmer et al. 2005. E-waste damages economy and effects current pollution. These adverse effects are global that need joint activities to get rid of e-waste crisis. Unique strategies are designed and applied at local and worldwide stages. It includes RoHS in Europe, EPR policy, WEEE in Europe, the Basel Convention, Basel Ban, and StEP plan. Existing data and other evidences confirmed that already presented strategies will not be able to solve problems of transboundary movement of waste products informal recycling at industrial level in different countries (Widmer et al. 2005; Williams et al. 2008).

Global quantities of e-waste products are growing. In the USA (2005), the total amount of discarded e-waste has reached up to 1.36 million metric tons. Waste products were discarded mostly in landfills. The major problems related to management of e-waste management system in the USA are discussed in this paper. Globally, electronic recycling systems have already been legislated. E-waste management systems have been explored and established in Japan, South Korea, Taiwan, and the European Union. Specific conditions that are discussed in this research article are expected to influence the implementation and acceptability in the USA. For market-driven solutions that enable competition, cultural imperative is a key consideration in the USA. For recycling and reuse services, establishment of a competitive market is an option to solve this issue. E-Market for Returned Deposit is a solution for reuse and recycle service establishment that begins with a deposit paid by consumers to sellers during purchasing the product. Device is placed on the product that is electronically registered and detected through radio frequency detection. The solution, termed e-market for returned deposit, begins with a deposit paid by consumers to sellers at the time of purchase, electronically registered and tracked via a radio-frequency identification device (RFID) placed on the product. At end-of-life, consumers consult an Internet-enabled market in which firms compete to receive the deposit by offering consumers variable degrees of return on the deposit. After collection of the computer by the selected firm, the cyberinfrastructure utilizes the RFID to transfer the deposit to the winning firm when recycled. If the firm chooses to refurbish or resell the computer in lieu of recycling, the transfer is deferred until true end-of-life processing (Kahhat et al. 2008).

13.1.2 Ultimate Destination of E-waste (Developing Countries)

An important stage for proper e-waste administration is to gauge its flow. A few studies pointed to gauge this. The flow of 5 electronic items from their production as waste to their last goals (the destiny): reuse, storage, recycling, and landfill were being measured by Peralata and Fontanos in 2006. Adams et al. (2016) focused on the expectation of e-waste production as well as the rate transportation of e-waste in India. Andarani and Goto (2013) applied the material flow analysis to Indonesia and showed that a large trend of reuse is currently happening.

One way in which the flow of e-waste to recycling and disposal is being decreased is via reuse because reuse reduces the amount of e-waste that is potentially being generated. Other studies by Ibrahim et al. (2013), Jacob et al. (2014), Lau et al. (2014), and Liu et al. (2006) reported similar findings.

The following are the problems concerning e-waste in developing countries:

- The exportation of e-waste from developed nations to developing nations for reusing leads to a situation in which EEE products become e-waste in developing nations, which compounds e-waste management in those nations (ESDO 2011).
- Some elements of e-waste are frequently blended with Municipal Solid Waste (MSW) as well, and both may be treated inefficiently.
- The major concern is about total amount of e-waste production. This is because of lack of infrastucture to appropriately manage e-waste. Moreover, the significance of second hand EEE is pressing issue because its functionality may not necessarily be verified and its content may be hazardous.
- In many developing nations, stock appraisal of e-waste is run-down or does not exist.
- The absence of data with respect to the unsafe nature of e-waste.
- Lacking information about the impact to human welfare as well as in the environment where e-waste management practices followed.
- In developing nations, Law to control the import as well as transfer of the produced e-waste doesn't occur. Such laws, if does exist, can decrease the insecure nature of e-waste management in those countries (Ikhlayel 2018).

13.1.3 E-waste Recycling Procedures

Traditional methods of recycling WEEE are inceineration, hydraulic shaking bed separation, and acid leaching process. For incineration, WEEE is directly burned within the blast furnace. There is 70–85 wt. % black copper within the blast furnace product. The black copper is fed into the converter to be oxidized and then is reduced in the anode furnace. The anode copper may further purified in H_2SO_4 electrolyte with other components, like Zn, Ni, and Fe (Kang and Schoenung 2005). The hydraulic shaking bed separation major focus was in the recycling of metals from waste PCBs in the previous years. We can collect unrefined copper elements from this process. By this process there is excess generation of waste water and residues (Thiery et al. 2009). If these events are not controlled, it leads to secondary contamination. Moreover, the other metals and non-metals are difficult to recycle by this process. Conventional hydrometallurgy processes are acid leaching to recover metals. For the extraction of metals such as Cu, Zn, and Pb, HClO4/HCl/HNO_3 is utilized as leaching solvent from WEEE. Additionally, for the adsorption of base and valuable metals (Zn, Cu, Pb, Ag, etc.) from WEEE, nitric acid with hydrochloric acid were mixed in the form of aqua regia (Beelen et al. 2014; Sheng and Etsell 2007). Cyanide leaching used to valuable metal Au for more than one century due to the selectivity as well as stability of dicyanoaurate complex (Collard et al. 2008; Raman et al. 2003). In this method, cyanide used as leaching agent cause a major issue, its toxic quality, which will cause many issues. These old techniques are not only unproductive but also critically harmful to the environment. In recent years, on the basis of old machineries, some innovative recycling machinery is put forward which has increased significant progress.

13.2 Technologies to Recycle Metals from WEEE at Advanced Level

13.2.1 Pyrometallurgical Technology

Pyrometallurgical advancement has been proved very important in recycling innovations and treatment of toxic products. Pyrometallurgy can be very effective in industrialization, including different arrangement stages, such as dismantling, smelting in a plasma arc furnace, drossing, sintering, melting and reactions in a gas phase at high temperature. This methodology includes scraps having metals that are burned at high temperature heater. Metals are volatilized by a chemical response or by warm and impurities are changed over into slag. Mechanical physical pretreatment like, grinding forms for size lessening, disassembling primary parts, liberation of constituents etc are required to treat WEEE initially. Then rich metals are smelted in furnaces and coarse metal ingot can be obtained. Mostly, smelting procedure is most significant process for pyrometallurgy. Flash smelting and bath smelting are two primary and largely functional smelting forms. Oxygenated gas is used in flash smelting to progress autogenous conditions. Bath smelting depends upon the roasting and smelting steps (Santos et al. 2013). (Kucharski et al. 2014; Moskalyk and Alfantazi 2003; Zhang and Xu 2016) (Fig. 13.1).

13.2.2 Electrochemical Technology

Electrochemical approach is a very promising advancement. It offers a kind of strategy to regain valuable metals from WEEE. It is done by tall natural affinity,

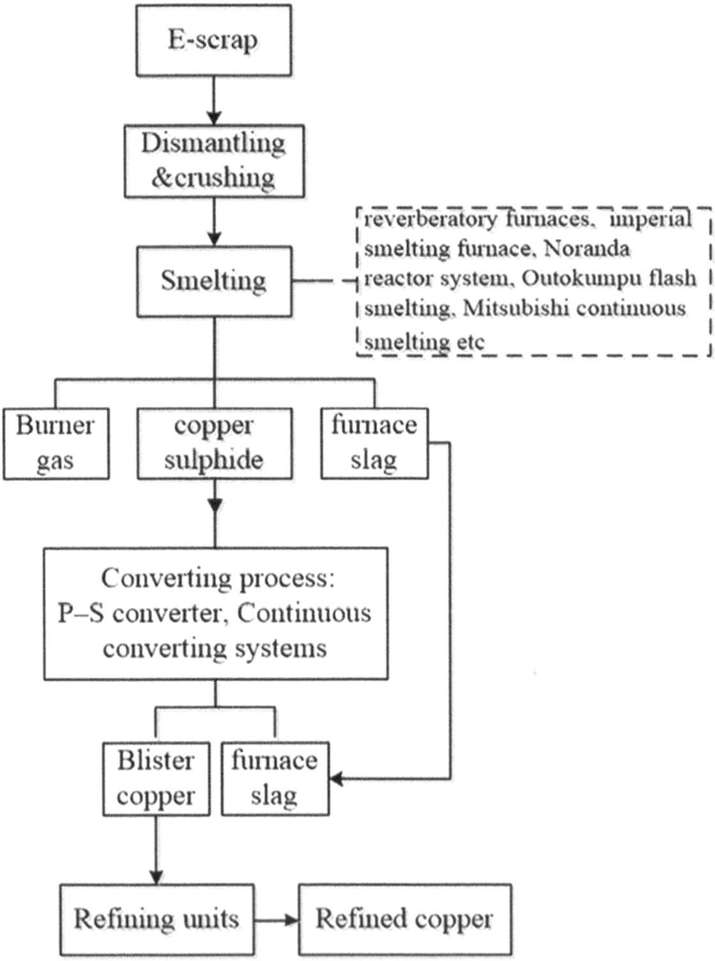

Fig. 13.1 A flow sheet shows recycling copper in WEEE by pyrometallurgy (Figure adopted from Zhang and Xu 2016)

tall vitality proficiency, and negligible chemical utilization. A few analysts detailed this advancement to recover metals from WEEE.

13.3 Systems of Recycling and Management of E-waste Around the World

13.3.1 U.S. Recycling Systems

In 2006, 32.5% of produced municipal solid waste has been recycled in the USA (US EPA 2008). Nevertheless, communities have different recycling rates. It depends upon their waste collection options, recycling materials targeted, regulatory approaches, and other different factors. An example would be the 11 different states (California, Connecticut, Delaware, Hawaii, Iowa, Maine, Massachusetts, Michigan, New York, Oregon, and Vermont) which apply the beverage container deposit refund system. Beverage container recycling strategy includes "Bottle Bill". It is the illustration of a well-organized deposit refund structure. The "Bottle Bill" is a skeleton that motivates customers to return refreshment bottles as they are given discount on the store. This framework is measured by several partners as an efficient reusing program that must be extended as a government approach (US GAO 2005). The eleven states employing this framework have accomplished higher reusing rates and have reused more refreshment holders than the other thirty-nine states (DOC 2007). The USA produces 15 million end-of-life vehicles each year. Presently, 95% of cars within the USA are sent to present reusing offices for destroying and destroying at the conclusion of their benefit lives (Daniels and Whitener 2003). The administration of ELVs includes disassembling, spoiling, as well as reusing of portions and constituents. The parts that have sensible esteem are expelled by the demolishers, and at that point renovated and recycled. The shredders bunch the remaining resources into nonferrous as well as ferrous metals, which are all sent to recyclers. In 2001, 6000–7000 dismantlers were evaluated within the USA. In expansion, in 2006, around 75% of the resources found in vehicles were beneficially reused by the destroying industry (Kahhat et al. 2008).

In addition, around 75% of the materials found in vehicles were profitably recycled by the reuse or shredding industry in 2006 (Jody et al. 2006).

13.3.2 Europe

Electronic waste-related community level guidelines have been adopted by 25 associated States of the European Union. They are aimed at improving the environment to protect and ensure human health and promoting best use of natural resources

(European Commission-WEEE Directive 2003). For example, the European Commission; WEEE Directive (2003) adopted following guidelines in January 2003.

1. EEE product plan and design
2. E-waste collection
3. E-waste recovery
4. E-waste treatment and treatment financing
5. EEE user awareness

The basic objective of the directive is to collect, recover or recycle and again use of e-waste. During product design, the main objective of procedure is to increase understanding and understanding of end of life factors. Deconstruct of parts and reuse of materials are two factors. To decrease disposal in municipal waste streams, well organized collection systems is required that have the tendency to isolate the desire (e-waste). These are the finest method and procedure for treatment or management, retrieval and recycling of e-waste. Minimum recovery rates are (70–80% by weight) and substances and materials recycling rates are (50–80% by weight) according to e-waste directive (European Commission-WEEE Directive 2003) for producers. Differences are there or made which are depends on the basis of the e-waste, if it is private or public household and old or new products (European Commission-WEEE Directive 2003). Execution and assignment have not been simple. Like the Netherlands and Greece, some other European Union nations are already executing the guideline. Other states without e-waste administration system, have been requested to observe the set laws (Magalini and Huisman 2007; Savage 2006). The following are the two main reasons and factors that postpone the transposition or displacement of the directive: (1) transfer of existing set of laws (e.g., Austria, Belgium, Denmark, Sweden, and Luxemburg) and (2) discussions with stakeholders about tasks in the procedure (e.g., France) (Magalini and Huisman 2007). Six compounds that are hazardous (mercury, lead, hexavalent chromium, cad-mium, polybrominated biphenyls (PBB), polybrominated diphenyl ethers (PBDE) and polybrominated biphenyls (PBB), are controlled by the European Union Restriction of harmful and Hazardous Sub-stances (RoHS) Directive (2002/95/EC).

Some European nations that are not the part of the European Union are also managing e-waste efficiently. Switzerland, for example, has two e-waste management systems, (1) the Swiss Association for Information, Communication and Organization Technology (SWICO) for office, dental, graphic, and telecommunication equipments and (2) the Swiss Foundation for Waste Management for household appliances (Hischier et al. 2005; Streicher-Porte et al. 2005). Recycling companies associated with both systems recycled about 11 kg of e-waste per individual in 2004. It amounts to 75,000 metric tons of e-waste (Hischier et al. 2005; Streicher-Porte et al. 2005; SWICO 2005). The European Commission-WEEE Directive (2003) recommends 4 kg/hab/year (Kahhat et al. 2008).

13.3.3 Japan

Japan's e-waste management has laws to enforce e-waste management. Manufacturers and importers need to take back end of life electronic products for their recycling and waste management. They are meant to make sure separation of e-waste from the MSW stream (Widmer et al. 2005). Particularly, the Japanese "Home Appliance Recycling Law" is responsible for recycling home appliances. Law enforces manufacturers and importers to recycle four kinds of household e-waste. It includes televisions, refrigerators, washing machines, and air conditioners. Customers pay an end of life charges that cover transportation and recycling expenses. Customers pay US$23 and US$46 (US$1 = JPY 107) to cover the recycling charges and an extra US$4 to US$19 (US$1 = JPY 107) to cover the transportation charges of the e-waste products to selected collection sites. The system also enforces retailers to collect and transport useless goods from customers.

Japan began compulsory recycling of business personal computers (PCs) in April 2001, and expanded the requirement to residential PCs in the summer of 2003 with the "Law for Promotion of Effective Resource Utilization." Local authorities have been managing this system initially but for personal computers sold after October 2003, manufactures grouped in the PC3R Promotion Center collects and recycle computers. Those computers which are under the PC recycling program have a PC recycling mark. No additional charges are required. It includes hidden and nonrefundable recycling fee in the sale price. On the other hand, customers need to give a collection and recycling charges with no mark that ranges from US$29 to US$40 (PC3R, before October 2003, Terazono et al. 2006; Yoshida et al. 2007; Kahhat et al. 2008).

13.3.4 South Korea

Since 2003, local manufacturers, distributors, and importers of shopping products are required to get official recycling targets or face financial punishment under the "Extended Producer Responsibility Law" in South Korea (Yoon and Jang 2006). These goods include air conditioners, TVs, and PCs. To deposit recycling funds, they have to make an account with the government which is refundable. Manufacturers and importers can either set up their recycling services to meet the EPR requirements. Retailers and suppliers are also required to collect and transport used equipment for free if the customer purchases a similar product.

Roughly 70% of e-waste has been collected by manufacturers in 2003 reported by Yoon and Jang (2006). Additionally, 12% of collected e-waste have been reused, 69% of collected e-waste have been recycled, and 19% have been discarded to landfills or incinerators in South Korea in that same year (Ministry of the Environment Republic of Korea 2003). The local government collects e-waste generally from retailers and recyclers. Fifty percent are collected by the local government (Ministry of the Environment Republic of Korea 2003).

The recycling fee paid by manufacturers ranges from US$4 to US$17 (US $1 = 1160 won). However, consumers are responsible for the collection fee and have two options: (1) paying no collection fee if the consumer decides to buy a new replacement product, in which case the retailers collect the e-waste, or (2) paying a collection fee to the local government collection system. It is important to note that in the case of computers, there are companies that offer free collection of the e-waste; the fee is around US$8 to US$10 (Kahhat et al. 2008).

13.3.5 Taiwan

In 1998, Taiwan enforced the Waste Disposal Act (AWDA). The amendments make sure the financial duty to the manufacturers and importers by enforcing the payment of recycling payment similar to the Korean system. An organization that collected the funds is known as the "Recycling Management Fund". Refrigerators, washing machines, air conditioners, PCs, and others are included in e-waste. Around US$8 and US$21 (US$1 = TWD 32) recycling fee has been imposed (Terazono et al. 2006; Kahhat et al. 2008).

References

Adams DM, Trenor CC, Hammill AM, Vinks AA, Patel MN, Chaudry G et al (2016) Efficacy and safety of sirolimus in the treatment of complicated vascular anomalies. *Pediatrics* 137 (2):2015–3257

Amankwah-Amoah J (2016) Global business and emerging economies: towards a new perspective on the effects of e-waste. Technol Forecast Soc Chang 105:20–26

Andarani P, Goto N (2013) Potential e-waste generated from households in Indonesia using material flow analysis. J Mater Cycles Waste Manage 16:306–320

Beelen R, Raaschou-Nielsen O, Stafoggia M, Andersen ZJ, Weinmayr G, Hoffmann B et al (2014) Effects of long-term exposure to air pollution on natural-cause mortality: an analysis of 22 - European cohorts within the multicentre ESCAPE project. Lancet 383(9919):785–795

Bousquet J, Khaltaev N, Cruz AA, Denburg J, Fokkens WJ, Togias A et al (2008) Allergic rhinitis and its impact on asthma (ARIA) 2008. Allergy 63:8–160

Collard JM, Bertrand S, Dierick K, Godard C, Wildemauwe C, Vermeersch K et al (2008) Drastic decrease of Salmonella enteritidis isolated from humans in Belgium in 2005, shift in phage types and influence on foodborne outbreaks. Epidemiol Infect 136(6):771–781

Daniels AC, Whitener B (2003) Bringing out the best in people. American Media International

Das MC, Xu H, Wang Z, Srinivas G, Zhou W, Yue Y-F et al (2011) A Zn 4 O-containing doubly interpenetrated porous metal–organic framework for photocatalytic decomposition of methyl orange. Chem Commun 47(42):11715–11717

ESDO (2011) Illegal import and trade off of e-waste in Bangladesh. Bangladesh

Garlapati VK (2016) E-waste in India and developed countries: management, recycling, business and biotechnological initiatives. Renew Sust Energ Rev 54:874–881

Goldsmid HJ, Douglas RW (1954) The use of semiconductors in thermoelectric refrigeration. Br J Appl Phys 5(11):386

Haynes R, Lewis D, Emberson J, Reith C, Agodoa L, Cass A et al (2014) Effects of lowering LDL cholesterol on progression of kidney disease. J Am Soc Nephrol 25(8):1825–1833

Hischier R, Wager P, Gauglhofer J (2005) Does WEEE recycling make sense from an environmental perspective? the environmental impacts of the Swiss take-back and recycling systems for waste electrical and electronic equipment (WEEE). Environ Impact Assess Rev 25(5):525–539

Ibrahim FB, Adie DB, Giwa A-R, Abdullahi SA, Okuofu CA (2013) Material flow analysis of electronic wastes (e-wastes) in Lagos, Nigeria. J Environ Prot 04:1011–1017

Ikhlayel M (2018) An integrated approach to establish e-waste management systems for developing countries. J Clean Prod 170:119–130

Jacob P, Kashyap P, Suparat T, Visvanathan C (2014) Dealing with emerging waste streams: used tyre assessment in Thailand using material flow analysis. Waste Manag Res 32:918–926

Jody J, Daniels EJ, Duranceau CM, Pomykala JA Jr, Spangenberger JS (2006) End-of-life vehicle recycling: the state of the art of resource recovery from shredder residue. energy systems division. Argonne National Laboratory. Google Scholar

Julander A, Lundgren L, Skare L, Grandér M, Palm B, Vahter M, Lidén C (2014) Formal recycling of e-waste leads to increased exposure to toxic metals: an occupational exposure study from Sweden. Environ Int 73:243–251

Kahhat R, Kim J, Xu M, Allenby B, Williams E, Zhang P (2008) Exploring e-waste management systems in the United States. Resour Conserv Recycl 52(7):955–964

Kang H-Y, Schoenung JM (2005) Electronic waste recycling: a review of US infrastructure and technology options. Resour Conserv Recycl 45(4):368–400

Kucharski TJ, Ferralis N, Kolpak AM, Zheng JO, Nocera DG, Grossman JC (2014) Templated assembly of photoswitches significantly increases the energy-storage capacity of solar thermal fuels. Nat Chem 6(5):441

Lau WKY, Liang P, Man YB, Chung SS, Wong MH (2014) Human health risk assessment based on trace metals in suspended air particulates, surface dust, and floor dust from e-waste recycling workshops in Hong Kong, China. Environ Sci Pollut Res 21:3813–3825

Liu X, Tanaka M, Matsui Y (2006) Generation amount prediction and material flow analysis of electronic waste: a case study in Beijing, China. Waste Manag Res 24:434–445

Magalini F, Huisman J (2007) Management of WEEE & cost models across the EU: could the EPR principle lead US to a better environmental policy? In: Proceedings of the 2007 IEEE international symposium on electronics and the environment. IEEE, pp 143–148

Moskalyk RR, Alfantazi AM (2003) Processing of vanadium: a review. Miner Eng 16(9):793–805

Pellow DN (2010) 18 The global waste trade and environmental justice struggles. Handbook on trade and the environment, 225

Raman N, Muthuraj V, Ravichandran S, Kulandaisamy A (2003) Synthesis, characterisation and electrochemical behaviour of Cu (II), Co (II), Ni (II) and Zn (II) complexes derived from acetylacetone and p-anisidine and their antimicrobial activity. J Chem Sci 115(3):161–167

Robinson BH (2009) E-waste: an assessment of global production and environmental impacts. Sci Total Environ 408(2):183–191

Santos LHMLM, Gros M, Rodriguez-Mozaz S, Delerue-Matos C, Pena A, Barceló D, Montenegro MCBSM (2013) Contribution of hospital effluents to the load of pharmaceuticals in urban wastewaters: identification of ecologically relevant pharmaceuticals. Sci Total Environ 461:302–316

Savage M (2006) Implementation of the waste electric and electronic equipment directive in the EU European Commission. Office for Official Publications of the European Communities, Luxembourg

Sepúlveda A, Schluep M, Renaud FG, Streicher M, Kuehr R, Hagelüken C, Gerecke AC (2010) A review of the environmental fate and effects of hazardous substances released from electrical and electronic equipments during recycling: examples from China and India. Environ Impact Assess Rev 30(1):28–41

Sheng PP, Etsell TH (2007) Recovery of gold from computer circuit board scrap using aqua regia. Waste Manag Res 25(4):380–383

Singh JA, Saag KG, Jr B, Louis S, Akl EA, Bannuru RR, Sullivan MC et al (2016) 2015 American College of Rheumatology guideline for the treatment of rheumatoid arthritis. Arthritis Rheumatol 68:1), 1–1),26

State of California Department of Conservation-Division of Recycling (DOC) (2007) Beverage container recycling market development and expansion grant program. Sacramento, California; February

Streicher-Porte M, Widmer R, Jain A, Bader HP, Scheidegger R, Kytzia S (2005) Key drivers of the e-waste recycling system: assessing and modeling e-waste processing in the informal sector in Delhi. Environ Impact Assess Rev 25(5):472–491

Terazono A, Murakami S, Abe N, Inanc B, Moriguchi Y, Sakai S-i et al (2006) Current status and research on e-waste issues in Asia. J Mater Cycles Waste Manag 8(1):1–12

Thiery JP, Acloque H, Huang RYJ, Nieto MA (2009) Epithelial-mesenchymal transitions in development and disease. Cell 139(5):871–890

Tu W-L (2005) Challenges of environmental governance in the face of IT industrial dominance: a study of Hsinchu Science-based industrial park in Taiwan. Int J Environ Sustain Dev 4 (3):290–309

United States Environmental Protection Agency (U.S. EPA) (2008) International Waste Activities. Import/Export Requirements. Collection of Materials on Import/Export Regulatory Requirements Chapter V: Basel Convention 2000. http://www.epa.gov/epaoswer/osw/internat/basel3.htm

United States Government Accountability Office (US GAO) (2005) Electronic waste: strengthening the role of the federal government in encouraging recycling and reuse, GOA-06-47. http://www.gao.gov/new.items/d0647.pdf

Widmer R, Oswald-Krapf H, Sinha-Khetriwal D, Schnellmann M, Böni H (2005) Global perspectives on e-waste. Environ Impact Assess Rev 25(5):436–458

Williams E, Kahhat R, Allenby B, Kavazajian E, Kim J, Xu M (2008) Environmental, social and economic implications of global reuse and recycling of personal computers. Environ Sci Technol 42:6446–6454

Yoon H, Jang Y (2006) The practice and challenges of electronic waste recycling in South Korea with emphasis on extended producer responsibility (EPR). In: Proceedings of the 2006 IEEE international symposium on electronics and the environment. IEEE, pp 326–330

Yoshida A, Tasaki T, Terazono A (2007) Material flow of used PCs in Japan. In: Proceedings of the 2007 IEEE international symposium on electronics and the environment. IEEE, pp 46–51

Zhang L, Xu Z (2016) A review of current progress of recycling technologies for metals from waste electrical and electronic equipment. J Clean Prod 127:19–36

Chapter 14
Ecotoxicological Risk Assessment of E-waste Pollution

Biljana Balabanova

Abstract Electronics have unique characteristics and can be a source of significant environmental degradation, making these devices problematic toxic waste. Over the past few decades, e-waste management has not been well organized, meaning that the human population and the environment have suffered the consequences of improper treatment of e-waste. For instance, much ends up in landfill, where it can cause leaching of hazardous materials, mercury vaporization, and fires, which lead to atmospheric pollution and toxic ash residues. This study reviews recent reports on human exposure to e-waste, with particular focus on exposure routes and toxicities of humans. Specific e-associated chemical elements and compounds exist in the form of components of the equipment, target. Pieces of evidence that associate e-waste exposure with human health effects are assessed. The role of toxic heavy metals (lead, cadmium, chromium, and mercury) and organic pollutants (polybrominated diphenyl ethers (PBDEs), polychlorinated biphenyls (PCBs), polycyclic aromatic hydrocarbons (PAHs), polybrominated biphenyls (PBBs), polyhalogenated aromatic hydrocarbons (PHAHs), and biphenyl A (BPA)) on human health is also briefly discussed.

Keywords Toxic substances · Electronic waste · Environmental pollution · Toxicological effects

14.1 Introduction: Terminology and Definitions

The constant advancement of society is conditioned by the rapid development of modern technologies leading to the increasing production and use of various electrical and electronic devices and equipment without which everyday life cannot be imagined. As electrical and electronic equipment (EE equipment) and appliances can

B. Balabanova (✉)
Faculty of Agriculture, University "Goce Delčev", Štip, North Macedonia
e-mail: biljana.balabanova@ugd.edu.mk

M. Z. Hashmi, A. Varma (eds.), *Electronic Waste Pollution*, Soil Biology 57,
https://doi.org/10.1007/978-3-030-26615-8_14

be defined all products that are completely dependent on electricity or electromagnetic fields for their proper operation. Furthermore, in EE equipment can be also included equipment for production, transmission, and measurement of current, or equipment for measuring the strength of electromagnetic fields. This equipment is intended for use at a voltage not higher than 1000 V for alternating current and 1500 V for direct current (Widmera et al. 2005; UN Environment Programme 2007; Robinson 2009). Electrical and electronic waste (EE waste) is waste electrical and electronic equipment including the assemblies and components that occur in the economy (industry, processing, etc.), electronic waste from households, that is, waste electrical and electronic equipment incurred in households or in production/service activities when by type and quantity is similar to EE household waste (UN Environment Programme 2007; Robinson 2009).

Electrical and electronic waste has the characteristics of hazardous waste, and, according to the Law on Waste Management, it cannot be mixed with other types of waste. Electrical and electronic waste contains toxic and very often carcinogenic substances, and therefore it is very dangerous for human health, the environment, and our planet. Exposure to e-waste might occur directly via recycling or indirectly via ecological exposure (Frazzoli et al. 2010). Huge amounts of electronic waste are transferred to less developed countries at certain dump sites or are subjected for the recycling processes (Frazzoli et al. 2010; Sthiannopkao and Wong 2013; Julander et al. 2014). Therefore, it is necessary for EE to handle the waste properly and that its recycling procedure is in accordance with the prescribed legal provisions. The goal of electrical and electronic waste management is to establish a system for separate collection of electrical and electronic waste. It needs to be classified on the basis of the categories of EE waste for its further use, care and protection of the environment and the health of the citizens.

14.2 Origin, Processes, and Circumstances of Environmental Risks Related to E-waste

The electronic waste category includes all types of computers and computer equipment (keyboards, mice), monitors, printers, cartridges, toners, scanners, mobile phones, cameras, TVs, video, and audio devices. For electrical waste are considered all larger devices, i.e., refrigerators, stoves, washing machines and dryers, boilers, air conditioners, and other household appliances. This category also includes assembles and device component parts that are accessible in the industry. Electrical and electronic equipment from which natural and legal persons want to be released because it is dysfunctional or obsolete is considered waste and should be handed over to an authorized recycler.

Due to the toxicity of the components in the electronic equipment, they should be treated very carefully. Therefore, it is extremely important how to deal with it. Most electronic and electrical devices contain dangerous substances such as mercury, freon, and arsenic that are harmful to the environment (water, soil, air) that directly affect human health and animal health. In order to reduce the risk of environmental pollution and eliminate the negative effects on our health, electronic and electrical devices should be properly recycled, i.e., not disposed of with municipal waste, but delivered to a company authorized to recycle EE waste (Frazzoli et al. 2010; Ramesh et al. 2007; Robinson 2009).

Households are the most significant users of the of electrical and electronic equipment, which means that this part of the society generates the largest percentage of electronic waste. Many households store non-functional waste electrical and electronic equipment, not knowing that it has a negative effect on their health. Most of that equipment contains mercury, freon, liquid porous crystals, mercury, etc. This type of waste is a toxic inheritance to our digital era and very quickly manages to pollute the water, soil, air, etc. (Ramesh et al. 2007). According to the Law on Electronic Waste Management, natural persons must not throw old electronic devices together with other municipal waste, but should leave them in appropriate locations marked for their waste. These electronic waste disposal bins are most commonly available in places where electrical and electronic equipment is sold. More complex or interdisciplinary approach should be applied for understanding the interactions between the chemical components contained in the electrical devices left as waste. Electronic waste on a global scale is cluttered with a rate of 40 million tons per year. E-waste is delayed by three times the usual rate waste. In one landfill, electronic waste is responsible for 70% of toxic chemicals like lead, cadmium, and mercury.

E-waste contains heavy metals, plastics, and organic compounds. Moreover, the electronics are made from valuable resources, such as precious metals, plastics, glass, and other materials—and all in need of energy to produce. When the equipment is discharged these resources cannot be recovered and caused additional pollution for the production of new products from unprocessed raw materials. Recycling these materials will help protect landfills and our environment (Sthiannopkao and Wong 2013). When e-waste is disposed of at the landfill, toxic substances can be released from the waste and distributed to the environment. These substances can infiltrate groundwater, contaminate the soil, and penetrate the food chain. In contact with these toxins, there is a risk in people to develop health problems, which includes respiratory, endocrine, and carcinogenic agent diseases, developmental abnormalities, and impairments of the reproductive, developmental, and nervous system.

14.2.1 Settings and Routes of Exposure

Electronics possesses certain unique properties that make their production and it use associated with high environmental risks. Therefore, these products as waste constitute a major environmental problem, which should be carefully managed.

The electronics industry is characterized, on the one hand, by rapid technological advancement, and on the other, there is an increasing phenomenon that manufacturers fail to design goods on some of the ways that take into account the effects that these products would have on the whole of their life cycle, including the period when they become waste. The consequence of this are that devices have an incredibly complex design and composition and include a palette breathing material, some of which are dangerous and it is not easy to handle them when they become waste. Also, their production requires a huge amount of energy, materials, and water, and the emission of harmful gases is high (Sthiannopkao and Wong 2013).

Supply of raw materials for the production process also has its negative impacts. Raw materials used in the production of electronic equipment are primarily supplied by mining, which requires a large area of land and a large amount of energy. Also, a large amount of harmful gases is generated, for example, 10,000 tons of carbon dioxide is released into the atmosphere to produce one ton of gold, palladium, or platinum. In addition, negative effects on society (areas in poor countries from which most of these materials originate, Africa, Asia) as a result of the extraction of materials used in the electronics industry are increasingly worrying. The production of electronic devices is a job that requires a lot of materials and energy. For example, small semiconductors (which are also known as microchips or integrated circuits) that are used in almost all devices and have introduced a revolution in electronics in terms of application and functionality require an exceptional amount of material and energy to make. A good example is a silicone (or some other material) used for making thin layers. The amount of chemical products used for the production of these key components is enormous—only one semiconductor can contain 500–1000 different chemical ingredients, and several analyses estimated that production of a 2 g of microchip can create up to 26 kg of waste, from which part is very toxic. Therefore, the e-waste dump sites can create a very high-risk environment, especially for the sensitive population group, such as children, pregnant women, elderly people, and people with disabilities (Sthiannopkao and Wong 2013).

The rapid development of technology had the trend of shortening the lifetime of products and this way it will produce more waste. Even in cases where efficiency is increased in the area of products or processes themselves production, these improvements have not done much to reduce the creation of e-waste, as we buy more and more electronics. Each country uses hundreds of thousands, if not hundreds of millions, of electrical and electronic devices every year. In the last decade, waste computers were collected from more than a million households. On a global level, there are about 4.6 billion mobile phones in use today. Many electrical devices are in use today (about a billion devices in use), this way a large amount of electronic waste will be created in the near future.

Some of the ways in which bad design and other practices make modern electronic devices obsolete fast include:

- Faults of physical parts—24% of laptops will break down in the first 3 years due to failures of physical parts.
- Improved software—new software can increase the amount of e-waste due to incompatibility with older computers. For example, many computers do not have large enough memory or fast processors which could support new operating programs like Windows Vista.
- Digital conversions—switching from analogue TV to digital has resulted in more eco-pads in the form of analogue TVs. The development of HD TV was also the cause of this.
- Batteries—many smaller electronics with rechargeable batteries have a limited cycle charging before it's time to replace the batteries. Consumers often choose to buy a new device instead of replacement of batteries in small devices.
- Upgrading of mobile phones—companies often offer cheap new models of mobile phones to consumers at regular intervals, thus causing consumers to engage in more frequent phone conversations.

When this quantity of goods is no longer in use, we will face a major challenge for the disposal of waste generated by it. While sales of computers and electronics products are growing steadily around the world, the same cannot be said for the infrastructure needed to safely manage electronic waste. A small number of manufacturers are willing to modify the manufacturing process in order to achieve a more efficient product. As a consequence, we are witnessing the creation of a "digital landfill" that adversely affects human health and leads to environmental pollution.

Over 1000 different types of materials are used to make our electronic devices and their components—semiconductor chips, printed circuits, disk drives, etc. Many of them are poisonous, including chlorine-based solvents, antimony based on bromine, PVC, heavy metals (such as lead, mercury, arsenic, cadmium, and hexavalent chromium), plastics, and gases. The listed substances are harmful to human health and life environment if they are not treated with care. Over 1450 tons of brominated carbon derivatives known as TBBPA was used for production of 991 million mobile phones sold in 2006. This chemical is classified in neurotoxin group of hazards (Grant et al. 2013).

The recycling process includes collecting, extracting, processing, and manufacturing new products. Waste cannot be reduced and avoided. It should be distributed to the place of origin by type of waste, since only separately collected waste can be used. Recycling is a collection of activities that ensure the recycling of waste materials. Regardless of the fact that millions of tons of precious metals lie on the offices' desk absolutely useless, recycling of electronic and electrical waste leads to problems of toxic and polluting substances that occur in the recycling process.

More than 38 different chemical elements can appear in the process of recycling EE waste, some of which have a high degree of environmental pollution and must be treated in special plants until their destruction, which is a skilled technological process. Today, the recycling activity of EE waste in all areas of the developed world is very widespread and rapidly developing. Electronic waste processing systems have matured in recent years, following an increase in regulatory, public, and commercial initiatives in proportion to the growth of transitory interests. Part of this evolution caused the development process of electronic and electrical waste from energy-intensive processes, unlike conventional recycling, where the equipment was based on forms of raw materials. This is achieved by diversion and regeneration of various chemical and physical processes. The following objectives can be achieved within recycling:

1. Conservation of raw materials (all materials originate from nature and are in limited quantities)
2. Energy saving (there is no consumption in primary processes, the transport mechanisms that drive these processes and the extra energy is generated by combustion of non-recyclable materials)
3. Environmental protection (waste materials degrade the ambient life, and with recycling we save the environment)
4. Opening new jobs (a process called material class involves investing knowledge and work, which opens the need for new jobs)

In terms of reuse options, the materials can be divided into four groups: (1) recyclable (can be used with a return to the production process), (2) non-critical (cannot be returned to the production process and used to obtain energy incineration or are settled in a safe environment), (3) hazardous (materials harmful to man and his surroundings), and (4) non-dangerous (materials that are not harmful to man and his surroundings).

In our environment, large amounts of recyclable waste are generated on a daily basis. We can generalize and summarize the life cycle of the electronic and electrical waste as is given in Fig. 14.1.

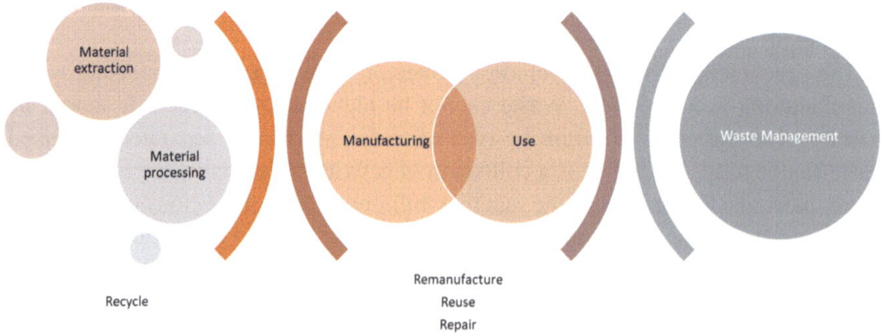

Fig. 14.1 Life cycle of electronics

14.3 Identification of Most Common Hazardous Chemicals Potentially Released

Disposal of waste to landfills is one of the most common ways to store e-waste, despite being dangerous. In all depots it was observed the release of matter into the environment to some extent, which means that it's heavy metals and other toxic substances from e-waste that can leak and pollute the land and water. About 40% of heavy metals that were found at landfills in the United States, including lead, mercury, and cadmium, come from e-waste.

Landfill can also pollute the environment by emitting harmful gases from the vapor and the resulting vapor's uncontrolled fire. Toxicity of substances used in the field electronics makes their safe recycling extremely difficult. Before recycling, lead must be removed from the cathode ray tube (CRT) monitor and the lamp with liquid from the screen based on liquid crystal, as well as PVC, antipyrene and other additives from plastic components. Electronic equipment is manufactured in such a way that, in the course of recycling, some materials remain that can cause pollution and emit gases harmful to humans and the entire planet (Van den Berg et al. 2006; Frazzoli et al. 2010).

There are toxic substances that are built into your TVs that can seriously pollute the environment if it is deposited in landfills. Old TVs with cathode ray tubes (CRT) contain between 2 and 4 kg of lead that can leak into the soil if it comes to the landfill damage and decomposition and thus pollute groundwater sources. New liquid crystal displays (LCDs) contain mercury lamps to illuminate the screen. Each LCD uses microgram mercury. This liquid metal is very toxic, if only 1 g of mercury is released per year the water in a lake with 1.61 ha surface can be polluted. If manufacturers would adopt safer design methods to eliminate toxic substances, risks arising from inadequate treatment of electronic products after cessation of production and use would be significantly reduced. In that case the fish cannot be included in human dietary. It is estimated that from the total annual production of mercury to around 22% worldwide is used in the production of electrical and electronic devices. In addition to containing many harmful and toxic substances, electronics also contains substances that are very valuable. Most of these valuable substances are found in printed circuit boards that connect electronic components. Printed boards in the home computer contain iron, aluminum, copper, lead, nickel, tin, gold, silver, platinum, and palladium. Metals and other valuable components exist in other electrical components, such as copper in wires and iron and aluminum in enclosures. Many materials that are key in the ICT industry are very rare and deficient. Supply with a type of minerals used in modern electronics has reached a critical rate, and it is expected that demand for them will triple in 2030. Some of them, such as tantalum, tungsten, and rare earth elements, come from other sources. When taking into account the risks associated with the supply of these materials and with the fact that some of them are so rare and deficient, it is a great pity that they go backward when old appliances are thrown into the landfill or when they are improperly recycled (Van den Berg et al. 2006; Frazzoli et al. 2010).

We need to require manufacturers to minimize the amount of these raw materials in production and to produce electronic devices so we can reuse and recycle. We also have to keep e-waste away from landfills, to recover as much raw materials as we can from our discarded electronic equipment.

E-waste contents can be recovered and brought back into the production cycle, such as heavy metals, polycyclic aromatic hydrocarbons (PAHs), polychlorinated biphenyls (PCBs), and brominated flame retardants and a number of plastic components (Van den Berg et al. 2006).

Potentially toxic heavy metals/metalloids can be also contained in e-waste (Frazzoli et al. 2010). Heavy metals present metallic elements that have a relatively high density compared to water. With the assumption that heaviness and toxicity are interrelated, heavy metals also include metalloids, such as arsenic, that are able to induce toxicity at low level of exposure. Although heavy metals are naturally occurring elements that are found throughout the earth's crust, most environmental contamination and human exposure result from anthropogenic activities. Environmental contamination can also occur through several pathways of releasing from e-waste materials. For example, lead (Pb) can be released from printed circuit boards, cathode ray tubes, light bulbs, televisions, solder, and batteries. Chromium (hexavalent chromium) usually is contained in anticorrosion coatings, data tapes, and floppy disks. Cadmium can be released from switches, springs, connectors, printed circuit boards, batteries, infrared detectors, semiconductor chips, ink or toner photocopying machines, cathode ray tubes, and mobile phones. Thermostats, sensors, monitors, cells, printed circuit boards, cold cathode fluorescent lamps, and liquid crystal display backlights are the most common sources for releasing mercury. Zinc is contained in the cathode ray tubes and metal coatings, while nickel and lithium can be released from batteries. Cathode ray tubes and fluorescent lamps are also sources of barium. Beryllium (light metal as lithium) in higher content can indicate toxicity. Beryllium can be released from power supply boxes, computers, x-ray machines, and ceramic components of electronics. Arsenic can also be released in small quantities in the form of gallium arsenide within light-emitting diodes. Even some of the rare earth elements, such as yttrium and europium, can be released from the fluorescent layer (CRT screen). Some trace quantities of radioactive americium can be released from e-waste. Main compartments, which contain trace quantities of americium, are medical electronic equipment, fire detectors, and active sensing element in smoke detectors. Other substances, such as toner dust, can be released from the toner cartridges for laser printers or copiers.

14.4 Evidence of Exposure

Typical contamination scenarios for creation of the e-waste are (a) dumping sites at/or near the riverbanks, (b) villages situated along rivers that receive e-waste, (c) manually disassembling and repairing, (d) open burning the e-waste to extract valuable metals, and (e) final disposal sites or landfills.

Due to the toxicity of the components in the electronic equipment, the use of the backward ones. Waste treatment methods pose a major risk to the health of workers—some of them even children—and the environment. Since e-waste contains a lot of valuable materials, there is an illegal export that is motivated by a desire for profit. E-waste safety treatment and preservation regulations in place are at very low level. In these cases, the possibility of getting out is maximal. The profit from e-waste is brought to its maximum, but all of this is done on the threat of the population and the planet. Furthermore, a relevant example is the export of toxic waste: informal recycling in China. Workers who are often children work with bare hands with large amounts of electronic devices and use hammers and chisels to disassemble monitors and equipment to extract valuable materials. The most dangerous parts of the "recycling" process in terms of environmental pollution are electronic boards, which create heated pools with molten lead and tin alloys. Daily exposure to tin vapors is extremely dangerous. Removing valuable materials is often done in an open flame. In the Guiyu region, an informal village is formed at a place where the wires are burned to separate the copper. Because of these processes toxic ash and residues covers the places where people live and children play. This informal recycling is a direct threat not only to many individuals who do this job but also throughout the community and the environment. For example, openly the incineration and disposal of wasted parts in water polluted the source of drinking water in Guiyu (Guo et al. 2009).

For example, although lead is extremely toxic to humans and the environment, it is still contained in CRT monitors and televisions, as well as luminescent and printed circuits in older computers and mobile phones. The disposal of lead to the landfill is prohibited in many countries. For example, cathode ray tubes and old cell phones, which should be handled in a special way for safely recycle, are dismantled manually at e-waste landfills in countries such as China. Even in the European Union, in which there is certainly a very developed system of regulations in the whole world about the treatment of electronic devices after their production stops, only one-third of e-waste is handled in accordance with directive for waste electrical and electronic equipment (Waste Electrical and Electronic Equipment (WEEE) Directive). This included illegal exports to developing countries where unregistered recycling facilities process waste in an unsafe way while risking their own health workers and polluting the environment. Under these circumstances, developing countries have a

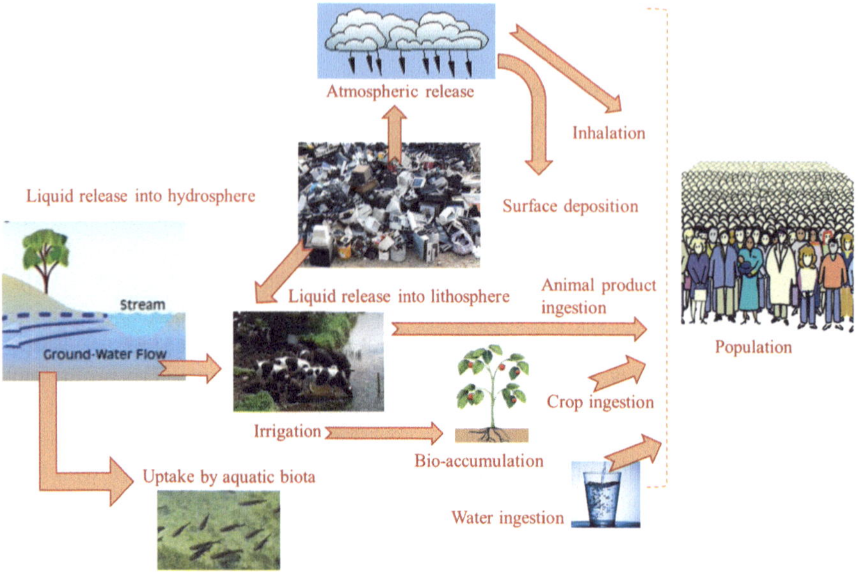

Fig. 14.2 Exposure pathways of toxic substances contained in e-waste

disproportionately large e-waste problem, although there are global mechanisms that work to solve this trade in toxic substances.

Another environmental exposure practice is burning plastic. Long-time exposure through contaminated water and soil elevates chemical loadings of pastures, potentially resulting in high levels of uptake of toxic chemicals in plant foods; moreover, contamination of pastures may give rise to significant contamination of some foods of animal origin (ruminant's milk). Agricultural and animal-rearing activities, both within and immediately surrounding the e-waste sites or downstream to contaminated waterways, may be a cause for general population exposure (Fig. 14.2).

14.5 Toxicological Effects

Toxic waste is a waste material, often in chemical form, which can cause death or damage in a living being. Toxic waste is usually generated as a product of industrial production, but it may also occur as a consequence of activities in urban, agricultural, military, healthcare, radioactive sources and light industrial areas such as chemical cleaning. Toxins are usually released in water, air, or soil. The potential adverse health effects of e-waste long-time exposure have recently been reviewed and can result in some abnormal changes, such as lung dysfunction, problems with the thyroid gland, hormone expression, childhood growth rate disabilities, problems with mental health, etc. Research has been intensively carried out in recent decades

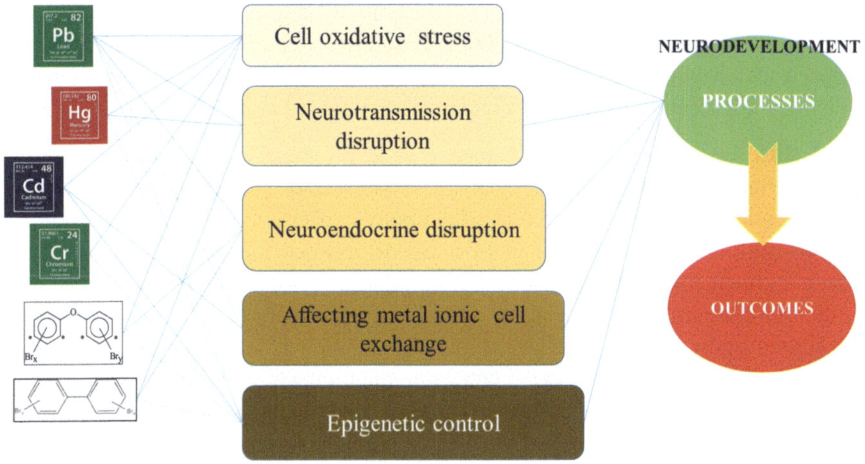

Fig. 14.3 Generalized toxicological processes indicated from toxicants released from e-waste

in order to determine the potential of e-waste for causing carcinogenic changes in biological organisms.

Most frequently, the disposed electronics may release a multitude of potential hazards, such as lead, mercury, chromium, cadmium, zinc, copper, nickel, and other metallic and nonmetallic chemical elements. Organic toxins reveal much significant health effects, but in electronics can contribute to the introduction of polyethylene terephthalate esters, polyvinyl chloride, acrylonitrile-butadiene-styrene polymers, polyethylene, polystyrene, polypropylene, brominated flame retardants, and a dozen more in each part of the environment.

In general, the main and most critical toxicological processes indicated from toxicants released from e-waste are given in Fig. 14.3.

14.5.1 Effects of Heavy Metals

Metals are one of the oldest toxins known to mankind. Metals such as cadmium, chromium, and nickel gain importance with their use in the modern era. Metals are often present in their surroundings, and their presence may be due to a natural occurrence or as a result of anthropogenic activities. Geographical distribution and the concentrations of metals present in various environmental media such as air, water, or soil are important factors that contribute to the degree of exposure (Guo et al. 2009; Tchounwou et al. 2012).

In heavy metals, they usually include antimony, arsenic, bismuth, cadmium, chromium, cobalt, copper, gold, lead, magnesium, mercury, nickel, silver, thallium, tin, molybdenum, zinc, and others. Heavy metals are toxic and tend to be retained long in the body and in the organs and tissues in which they have affinity, for example, in bones, kidneys, the liver, and the brain (Chang et al. 1996; Wang and Shi 2001; Beyersmann and Hartwig 2008; Tchounwou et al. 2012). Heavy metal-induced toxicity and carcinogenicity involves many mechanistic aspects, some of which are not clearly elucidated or understood (Leonard et al. 2004).

Lead (Pb) Lead belongs to the category of heavy metals and therefore will not naturally function in the human organism. It can seriously damage the soft tissues, kidneys, and nervous system, reduces lead body capacity to metabolize vitamin D, and also affects the establishment of a signal between the nerve cells (neurons) disrupting their function. Lead can be accumulated in bones and teeth and can enter in the bloodstream disrupting in the formation of normal red cells as well as their oxygen transport capacity (Yedjou et al. 2010).

Although lead is a danger to the general population, children up to 6 years of age are particularly susceptible to extensive nervous system development and bone structure. Recent studies show that children exposed to lead accumulation remain with a long-term toxicological condition of lead poisoning (Kaul et al. 1999; Litvak et al. 1998; US EPA 1998, 2002; Leonard et al. 2004; Pilsner et al. 2009; Yedjou et al. 2010; Tchounwou et al. 2012).

Apart from children, lead can also cause a variety of disorders in adults: elevated blood pressure, cardiovascular development diseases, discharge problems, kidney damage, pain in stomach and joints, variable mood, reduction of fertility in men, and so on. A special risk group are pregnant women because lead passes easily through the placenta to the fetus, and also a child is at risk during lactation through breast milk (Hertz-Picciotto 2000; Leonard et al. 2004; Tchounwou et al. 2012).

Within the skeleton, lead can be incorporated into the mineral in place of calcium. One of the major mechanisms by which lead exerts its toxic effect is through biochemical processes that include lead's ability to inhibit or mimic the actions of calcium and to interact with proteins (Flora et al. 2006; Sanders et al. 2009; Tchounwou et al. 2012).

Inorganic lead compounds can be introduced into the body orally or thought inhalation of lead vapor or powder, while for the organic compounds of lead, a percutaneous pathway is also possible. Absorption of lead from the lungs depends on its concentration in the air and the size of the lead particles, the largest being the absorption of the particles with a diameter less than 1 micron (Pilsner et al. 2009; Tchounwou et al. 2012).

In normal conditions, the body absorbs about 100 micrograms of lead daily through the air, food, and water for drinking. Absorbed lead in the blood of the poisoned person can be found in highest content in red cells (90 to 95%). Hence, it is distributed through the whole organism, where one part is deposited in the bones by binding for hydroxyapatite crystals. Elimination of lead from the body is mainly through the feces (80–90%) and, to a lesser degree, urine (about 10%). The half-life of lead in the blood and soft tissues is 24–40 days, while the half-life in the bones is 20–30 years (US EPA 1998, 2002).

Lead poisoning can be acute and more commonly chronic. The method of introducing lead in the body can be through oral or inhalation poisoning. Poisonous substances are usually rarely found in acidic form (e.g., lead-coated devices, which in acidic environment convert to lead-soluble compounds). The toxic effect of lead is due to its binding to free sulfhydryl groups of cell enzymes and their enzyme inactivation. At high concentrations in living cells, lead disrupts the tertiary structure of cellular proteins, leading to their denaturation and can cause cell death. The toxic dose for lead acetate is 5–6 g (0.9–1.0 mg /L urine, ie 80–100 mg/100 ml blood). Lethal dose of lead acetate is 20–50 g.

Mercury (Hg) Inorganic mercury compounds originating from electronic waste are mercuric chloride or calomel, mercachloride or sublimate, mercuric sulfide, mercuric iodide, mercury nitrate, etc. They are used as catalytic agents in the production of plastic mass, in the processing of fur and wool, in medicine such as medicaments (e.g., calomel as purgative), etc. Organic mercury compounds that originate from electronic waste are mercury alkyl compounds, phenyl mercury, and others. Some of them are found in the paint industry, agriculture, very small quantities in pesticides, cosmetics, and medicine, such as diuretics and fats (Sarkar 2005).

Elemental mercury in the form of vapor poultry is well absorbed by lungs, while resorption from GIT is insignificant. The absorbed mercury as liposoluble can be distributed throughout the body, can pass the blood–brain barrier and placental sheath, eventually eliminating it from the body through urine and feces. The half-life of elemental mercury in the body is about 60 days.

Inorganic poultry compounds are well absorbed by the GIT and through the skin. When present in large quantities, they can cause corrosive changes with subsequent increased resorption. After resorption inorganic mercury is most concentrated in the kidneys it has in the liver, erythrocytes, bone marrow, white lungs and skin. Excretion is via urine and feces, and the half-life of inorganic mercury in the body is about 40 days. Organic mercury is well absorbed by the GIT and through the skin. Organic mercury due to high liposolubleness can easily be transferred to kidneys, liver, and CNS. Organic mercury is excreted from the body mainly by urine. Half-life in the body is about 70 days (Tchounwou et al. 2003; Zahir et al. 2005; Holmes et al. 2009; Tchounwou et al. 2012).

Mercury poisons can be acute and chronic. According to the pathway of accumulation, the occurrence of mercury in humans is usually accidental. Characteristic clinical manifestation have poisonings with the various forms of mercury. Elemental mercury poisoning, acute and chronic mercury poisoning are most commonly caused by inhalation. Poisoning with inorganic mercury (acute and chronic) are usually with oral intake. Acute poisoning with inorganic mercury can be professional, accidental and suicidal, while chronic inorganic mercury poisoning is most common professional.

Poisoning with organic mercury, acute and chronic is usually oral. According to the manner of occurrence they can be professional and accidental (eg massive fish poisoning contaminated with organic mercury in the Minima Bay in Japan). Chronic poisoning with Organic mercury is usually professional. The mechanism of the toxic effect of mercury in the body is due to the direct corrosive action of the affected skin and mucous membranes, as well systemic action due to the blocking of sulfhydryl groups by cell enzymes (Clarkson and Magos 2006; Tchounwou et al. 2012).

Acute poisonings with elemental mercury occur with the inhalation of poultry vapour. Poisonings manifest with symptoms such as pharyngitis, laryngitis, tracheitis, bronchitis, rhinorrhea, scratching of the throat, hoarseness, cough, and chest pain. During the inhalation of higher concentrations of mercury componds, a toxic pulmonary edema occurs that manifests itself with dyspnoea, chest tightness, plentiful cough, cyanosis, hypotension, tachycardia, and oliguria. Resorbed mercury causes excitability to CNS with the occurrence of psychomotor restlessness and tremor. In chronic poisonings with elemental mercury, it primarily damages the CNS. Initial symptoms are non-specific: obesity, fatigue, anorexia, and weight loss. Characteristic neurological manifestations and changes occur as a result of brain damage. When chronic mercury poisoning occurs, it is usually manifested by shivering. The shivering can be manifested by trembling of the limbs, with varied intensity in different parts of the body. Because of the tremor in the tongue, speech becomes difficult to understand, and the movements of the legs and arms are dysfunctional. Characteristic psychological damage due to mercury poisoning is manifested by irritability, forgetfulness, difficulty in communicating with the environment, and so on (Valko et al. 2004, 2006; Leaner et al. 2007; Pinheiro et al. 2008).

Acute inorganic mercury poisoning is characterized by degenerative alterations of the GIT (gastrointestinal tract) and serious kidney damage. The initial manifestation is severe gastroenteritis syndrome with vomiting, hematemesis, severe pain along the length of the GIT, and bloody diarrhea that leads to severe dehydration and electrolyte imbalance. Decreased blood volume can cause kidney failure. Chronic inorganic mercury poisoning is manifested by the characteristic changes in CNS, i.e., tremor, accompanied by skin changes, changes in GIT, kidneys, and bone marrow. Manifestations of skin changes include generalized rashes, strong pruritus, and hyperkeratosis of the palms and soles. GIT damage is manifested by the occurrence of chronic stomatitis, with the characteristic graying of the gums and loss of teeth. In some cases of chronic inorganic mercury poisoning, kidney damage with proteinuria develops, as well as bone marrow damage leading to anemia and leukopenia. Acute mercury poisoning is characterized clinically by toxic encephalopathy, headache, paraesthesia, muscle pain, and muscle weakness. In the case of serious mercury poisoning, additional adverse events are paresis, paralysis of the limb muscles, and coma disorder. Prenatal organ poisoning mercury causes cerebral palsy of the newborn, due to damage to the cortex from the large and cerebellum. Chronic organic mercury poisoning is characterized by gradual development of a mild form of toxic encephalopathy. For Hg vapor exposure from dental amalgam, two recent large clinical trials did not find adverse effects on cognitive function in children (Bellinger et al. 2006; DeRouen et al. 2006; Dietrich 2010).

Chromium (Cr) Major factors governing the toxicity of chromium compounds are oxidation state and solubility. Cr(VI) compounds, which are powerful oxidizing agents and thus tend to be irritating and corrosive, appear to be much more toxic systemically than Cr(III) compounds, given the similar amount and solubility (Pellerin and Booker 2000). Although the mechanisms of biological interaction are uncertain, the variation in toxicity may be related to the ease with which Cr(VI) can pass through cell membranes and its subsequent intracellular reduction to reactive intermediates. Since Cr(III) is poorly absorbed by any route, the toxicity of chromium is mainly attributable to the Cr(VI) form. It can be absorbed by the lung and gastrointestinal tract and even to a certain extent by intact skin. The reduction of Cr (VI) is considered as being a detoxification process when it occurs at a distance from the target site for toxic or genotoxic effect, while reduction of Cr(VI) may serve to activate chromium toxicity if it takes place in or near the cell nucleus of target organs (Dayan and Paine 2001). If Cr(VI) is reduced to Cr(III) extracellularly, this form of the metal is not readily transported into cells and so toxicity is not observed. The balance that exists between extracellular Cr(VI) and intracellular Cr(III) is what ultimately dictates the amount and rate at which Cr(VI) can enter cells and impart its toxic effects (Velma et al. 2009; Tchounwou et al. 2012).

Cr(VI) enters many types of cells and under physiological conditions can be reduced by hydrogen peroxide (H_2O_2), glutathione (GSH) reductase, ascorbic acid, and GSH to produce reactive intermediates, including Cr(V), Cr(IV) radicals, hydroxyl radicals, and, ultimately, Cr(III). Any of these species could attack DNA, proteins, and membrane lipids, thereby disrupting cellular integrity and functions (Chen et al. 2009; Tchounwou et al. 2012).

Adverse health effects induced by Cr (VI) have also been reported in humans. Epidemiological investigations have reported respiratory cancers in workers occupationally exposed to Cr (VI)-containing compounds (Dayan and Paine 2001; De Mattia et al. 2004; Tchounwou et al. 2012).

Various hypotheses have been proposed to explain the carcinogenicity of chromium and its salts. Some inherent difficulties exist when discussing metal carcinogenesis. A metal cannot be classified as carcinogenic per se since its different compounds may have different potencies. Because of the multiple chemical exposures in industrial establishments, it is difficult from an epidemiological standpoint to relate the carcinogenic effect to a single compound. Thus, the carcinogenic risk must often be related to a process or to a group of metal compounds rather than to a single substance. Differences in carcinogenic potential are related not only to different chemical forms of the same metal but also to the particle size of the inhaled aerosol and to physical characteristics of the particle such as surface charge and crystal modification (De Flora et al. 1990).

Nickel (Ni) Nickel is used in many industries in elemental form or as a mixture with other metals. Therefore nickel poisoning is quite common. The most important mixtures of nickel are nickel oxide and hydroxide, which are used for making enamel, painted ceramics and glass, and then nickel sulfate, nickel carbonyl, and nickel cyanide. For medicine the most important is nickel carbonyl which is a colorless unstable liquid, in contact with oxidants it relieves carbon monoxide. Nickel salts are highly toxic if they reach the circulation. However, nickel salts rarely cause systemic reactions, if they are orally taken in trace amounts (Tchounwou et al. 2012).

Nickel is the best known of the pathogenic properties of nickel allergy. This allergy starts with rash nails followed by itching, pain, burning and sometimes swelling of affected areas. These symptoms can also be passed on to your fingers, forearm, and even chest and face. In extreme cases it can to come to pustulin and ulceration. This allergy is considered as it often attacks people with a brighter complexion. Today is a very common allergy, because people are in constant contact with various items made of nickel. It is usually bracelets, necklaces, earrings, brooches, metal part of belts, and watches which in contact with the skin causes contact dermatitis. This kind of allergy is known as contact allergy. The clinical picture of this type of allergy occurs in the form of a primary lesion that always occurs only locally, at the place where nickel is in contact with the skin. Secondary lesions are not associated with direct contact with nickel but are much more important than primary ones. The mechanism of occurrence is not yet sufficiently clear and usually is manifested with auto-sensitization or neuro-dermatitis. The lesion is usually exudative, often with edema, and skin itching can also occur (Tchounwou et al. 2012).

Cadmium (Cd) Cadmium is a relatively rare metal and only occurs as cadmium sulfide. The whole cadmium is obtained as a by-product at getting zinc. In the form of cadmium oxide, it penetrates the body through the respiratory tract. Its compounds are very toxic and the most toxic is cadmium chloride that is very soluble. Poisoning occurs mainly in the melting of zinc ores containing cadmium, galvanization, and welding (Byrne et al. 2009). Cadmium intake of cigarettes is of great importance because in this way, only one additive is added 0.1–0.2 μg of cadmium. With gastrointestinal resorption, the human body accumulates about 8% of cadmium, and resorption can be increased if the food that is ingested has a lower calcium, iron, and protein content. Cadmium is transported through the blood bound to erythrocytes and large-mass proteins such as albumin. Cadmium's half-life is approximately 30 years. Acute poisoning occurs by inhalation of higher concentrations of cadmium oxide. During the first few hours, there is tingling in nose and eyes and slight irritation of the upper respiratory tract. Further coughing, chest pain, elevated temperature, fever and sweating can also occurs. In severe cases, edema occurs on the lungs that can cause death. Chronic poisoning occurs during prolonged exposure (4–10 years). A common feature of all faces is the appearance of a yellow ring in the middle of the teeth as a result of cadmium deposition (Shaikh et al. 1999; Nordberg 2009; Tchounwou et al. 2012).

Cadmium is excreted through the kidneys and its presence in the urine is evidence of poisoning. Toxic effects are directed toward the renal tubules that are manifested by proteinuria in the urine. There is a condition where cadmium accumulation is possible without developing toxic effects on the kidneys, as this element builds up a compound with metallothionein (a low molecular weight protein). In toxic doses, it affects the metabolism of the calcium, i.e., it increases its excretion through the kidneys. Because of this bone changes in the form of osteoporosis and other deformations of the bone system can occurs as well (Tchounwou et al. 2012). Furthermore, cadmium toxicity can reveal a significant DNA damage and disrupts the synthesis of nucleic acids and proteins (Zheng et al. 2011; Julander et al. 2014).

14.5.2 Effects of Organic Pollutants

Complex molecules, unlike other pollutants, are defined as a function of their chemical nature, starting from four properties that are:

- Toxicity (one or more effects on human health)
- Persistence in the environment (molecules resistant to natural decay)
- Bioaccumulation in tissues of living organisms and increase in the concentration in the food chain (bio-amplification)
- Mobility (transfer of persistent pollutants to a greater distance)

Persistent organic pollutants are a group of organic chemicals that are produced by humans and have certain chemical properties that make them very dangerous, and because of which they are easily spread in nature. Their chemical composition does not change for a long period of time, and therefore they are found even in places where they have never been manufactured or used, such as the Antarctic.

Industrial chemicals were also mainly used until 2004, they are banned, and they are no longer used, but still exist in certain products: in households and landfills in larger quantities. Inadvertently produced pollutants are chemicals that are produced during certain industrial processes or in deliberate or unintentional combustion of open space.

The main consequences that arise from exposure to persistent organic pollutants are various types of cancer, causing serious immune system disorders, and the neurological system, causing various diseases of the endocrine system, resulting in consequences of early puberty in young people. Also, some of these substances can cause disorders even in the young development of children, for example, heart disease and different mental retardations.

The waste of electrical and electronic equipment contains a large number of rare metals, but it also includes the so-called POPs (persistent/non-removable organic compounds) chemicals that are hazardous to chemicals and remain unchanged in the environment in the long run, geographically easily distributed, accumulate in the fat tissue of people and animals and exposure to very low doses of some of them can cause cancer, damage to the central and peripheral nervous system, diseases of the immune system, reproductive system disorders, disturbance of the development of infants and children.

But this does not mean that electrical and electronic equipment is dangerous for everyday use. Improper handling throughout the life cycle of this equipment, especially when managing its waste (e-waste), can lead to POPs chemicals emissions and their accumulation in the human body, in animals, and in the environment (Behnisch et al. 2001, 2003; Van den Berg et al. 2006; Sjodin et al. 2008; Chen et al. 2011).

Interesting about our discussion are POPs chemicals (that have been widely added as substances that slow down burning, in consumer goods like plastic for electronics packaging, polyurethane car tires (90–95% of c-pentaBDE), in furniture and textiles since 1970. The main use of c-octaBDE (95%) was in the ABS polymer used in electronic and electrical equipment housings, especially in cathode ray tube equipment and office equipment such as photocopier machines and printers, common office equipment, and business machines (Sjodin et al. 2008; Chen et al. 2011).

The largest quantities of all chemicals from the POPs-PBDE group were produced between 1970 and 2005 in quantity approximately estimated from 1.3 to 1.5 Mt. (Hutzinger and Thoma 1987; Behnisch et al. 2001, 2003; Van den Berg et al. 2006; Sjodin et al. 2008; Chen et al. 2011). Production in the EU ceased in 1997, while in the United States in 2004, but products containing hazardous POPs-PBDE chemicals are still in use and one day they will become waste (e-waste) and will need to be properly removed from our homes. Electronic and electrical equipment containing these hazardous chemicals is found in industrial facilities, and small- and medium-sized companies and commercial buildings and companies need to manage e-waste in a safe and sustainable manner.

The main challenges regarding the management of new industrial chemicals POPs are:

(a) A lack of a system for informing and delivering data from households, public administration and the commercial sector, waste collectors, carriers and operators for e-waste management, and the transport sector
(b) Low stakeholder awareness and capacity for the content of the components in consumer products/vehicles/electrical and electronic equipment, as well as the potential risks of POPs-PBDE for human health and environment
(c) Lack of collection centers and processing and treatment of vehicles at the end of their lives and collection centers and treatment and processing of e-waste

14.6 Perspectives

Complex and the dangerous component of electrical products makes the recycling process very demanding in terms of safety and price. Safe recycling requires a system that separates poisonous waste from other species in the waste. This process must keep e-waste away from landfills and handle it e-waste. Well-organized management has to coordinate such ventures. Currently, local communities are bearing the entire burden of the e-waste problem on their shoulders. Establishing a system for security recycling e-waste is just one part of the problem. The process of reducing the negative effects of electronics is very difficult, especially if you continue to generate a huge amount of unnecessary waste. To get out of the way with the negative effects of electronics, it is also necessary, first of all, to find a way to reduce the amount of existing waste and toxic substances it contains.

One way to do this is to draw attention to the production problem and product design. Manufacturers of electrical appliances are the cause of the problem, but can this can be their solution: by taking responsibility for their own goods during the whole lifetime of the product—from the date of production to the period when the production stops and the products still exist. In order to prevent problems with e-waste treatment and to protect the environment and human health, manufacturers need to improve three important characteristics of the equipment, reduce the content of hazardous substances, make the equipment last longer, and after the end of the functionality period, manufacturers have to collect the non-functional equipment.

1. *Make products cleaner.* Manufacturers of electronic equipment should produce better products which are safe and easy to recycle, and after the working life will not cause a hazard effect to the environment. This means that they have to exclude poisons from their design. They must include in the design a smaller number of different materials and components in general, as well as the use of recycled materials and reused parts.
2. *Make products that have a longer life.* Manufacturers of electronic equipment must design and promote products which have a longer life and service life. That means they have to produce products that are easy to upgrade and repair, easily decompose to make the components replaced or reused, and easy to recycle. This would include providing information to recyclers firms and repair shops for hazardous substances in products and proper techniques to handle them. It would also mean giving information to the vendors about how to extend the lifetime of their own goods. This item would include details from the process of ordinary and general repairs and the ability to reuse unwanted appliances.
3. Social communities should not suffer the adverse effects on the environment and health as a result of poor e-waste management, and taxpayers should not bear the cost of recycling old electronic equipment. Manufacturers of electronic equipment need to take responsibility for their products throughout their lifecycle. After the equipment has been discontinued, the manufacturer must safely retrieve and recycle the equipment. This means that the handling standards of e-waste must be very high (e.g., e-waste should not be dumped at landfills, crushing or burning, but only quality recycling is considered etc.) and illegal landfills should be removed and controlled.

The production and functionality of electronic equipment include extraction of raw materials, through its transportation, design and production of the product itself, sale and distribution, use, repair, and reuse. All this process involves miners, producers, sellers, consumers, municipalities, recycling centers, and others. All these roles can affect the type of impact that the electronics will have on the population and the planet. However, the best way to create functional and environmentally friendly electronic equipment is to reduce the negative impact on the environment and the human health risk. Manufacturers of electronic equipment need to have some responsibility in the process of recycling electronic waste. In addition, manufacturers are most competent to make changes to products (in order to reduce toxic substances during the technological process). Furthermore, manufacturers of electronic devices know best the technical specifications of their products and have the greatest influence on their design and production process.

14.7 Conclusions

The main danger, in addition to the composition of the landfill, stems from the fact that it is in a densely populated urban suburbs. The state can seriously work to reduce air pollution, which is also associated with the emission of persistent organic pollutants. Much of the composition of air pollution is the consequence of pollution from these chemicals, which are emitted through combustion processes of open air.

As a very important segment that the state can do, it is also a serious improvement in the selection and treatment of waste, which will reduce the burning of landfills and reduce the amount of waste that ends up in landfills and is already contaminated with some of these chemicals. Something that can be done together with the civil sector is raising the public awareness of the citizens and the educators about the health consequences of these polluters, with the citizens being able to contribute, thus not using unsuitable fuels for heating the households.

Extended producer responsibility is widespread in waste management strategies, especially for products that require special handling and end-of-life treatment for your own toxic content. This makes the producer's responsibility suitable for management e-waste. Increased amounts of toxic e-waste and the difficulties that occur in recycling require that we look into it problem at the very root—manufacture and production. Extended producer responsibility is recognized as advocating for a good solution in order to minimize the impact of electronic devices, especially when they become waste.

References

Behnisch PA, Hosoe K, Sakai S (2001) Bioanalytical screening methods for dioxins and dioxin-like compounds a review of bioassay/biomarker technology. Environ Int 27:413–439

Behnisch PA, Hosoe K, Sakai S (2003) Brominated dioxin-like compounds: in vitro assessment in comparison to classical dioxin-like compounds and other polyaromatic compounds. Environ Int 29:861–877

Bellinger DC, Trachtenberg F, Barregard L, Tavares M, Cernichiari E, Daniel D et al (2006) Neuropsychological and renal effects of dental amalgam in children: a randomized clinical trial. JAMA 295:1775–1783

Beyersmann D, Hartwig A (2008) Carcinogenic metal compounds: recent insight into molecular and cellular mechanisms. Arch Toxicol 82:493–512

Byrne C, Divekar SD, Storchan GB, Parodi DA, Martin MB (2009) Cadmium—a metallohormone? Toxicol Appl Pharmacol 238:266–271

Chang LW, Magos L, Suzuki T (1996) Toxicology of metals. CRC, Boca Raton. FL, USA

Chen TL, Wise SS, Kraus S, Shaffiey F, Levine K, Thompson DW, Romano T, O'Hara T, Wise JP (2009) Particulate hexavalent chromium is cytotoxic and genotoxic to the North Atlantic right whale (*Eubalaena glacialis*) lung and skin fibroblasts. Environ Mol Mutagenesis 50:387–393

Chen A, Dietrich KN, Huo X, Ho S (2011) Developmental neurotoxicants in e-waste: an emerging health concern. Environ Health Perspect 119(4):431–438

Clarkson TW, Magos L (2006) The toxicology of mercury and its chemical compounds. Crit Rev Toxicol 36:609–662

Dayan AD, Paine AJ (2001) Mechanisms of chromium toxicity, carcinogenicity and allergenicity: review of the literature from 1985 to 2000. Hum Exp Toxicol 20(9):439–451

De Flora S, Bagnasco M, Serra D, Zanacchi P (1990) Genotoxicity of chromium compounds: a review. Mutat Res 238:99–172

De Mattia G, Bravi MC, Laurenti O, De Luca O, Palmeri A, Sabatucci A, Mendico G, Ghiselli A (2004) Impairment of cell and plasma redox state in subjects professionally exposed to chromium. Am J Ind Med 46(2):120–125

DeRouen TA, Martin MD, Leroux BG, Townes BD, Woods JS, Leitao J et al (2006) Neurobehavioral effects of dental amalgam in children: a randomized clinical trial. JAMA 295:1784–1792

Dietrich KN (2010) Environmental toxicants. In: Yeates KO, Ris MD, Taylor HG, Pennington BF (eds) Pediatric neuropsychology, 2nd edn. Guilford Press, New York, pp 211–264

Flora SJS, Flora GJS, Saxena G (2006) Environmental occurrence, health effects and management of lead poisoning. In: Cascas SB, Sordo J (eds) Lead: chemistry, analytical aspects, environmental impacts and health effects. Elsevier Publication, Netherlands

Frazzoli C, Orisakwe OE, Dragone R, Mantovani A (2010) Diagnostic health risk assessment of electronic waste on the general population in developing countries scenarios. Environ Impact Assess Rev 30:388–399

Grant K, Goldizen FC, Sly PD, Brune M-N, Neira M, van den Berg M, Norman RE (2013) Health consequences of exposure to e-waste: a systematic review. Lancet Glob Health 1:350–361

Guo Y, Huang C, Zhang H, Dong Q (2009) Heavy metal contamination from electronic waste recycling at Guiyu, Southeastern China. J Environ Qual 38(4):1617–1626

Hertz-Picciotto I (2000) The evidence that lead increases the risk for spontaneous abortion. Am J Ind Med 38:300–309

Holmes P, Hames KAF, Levy LS (2009) Is low-level mercury exposure of concern to human health? Sci Total Environ 408:171–182

Hutzinger O, Thoma H (1987) Polybrominated dibenzo-p-dioxins and dibenzofurans: the flame retardant issue. Chemosphere 16:1877–1880

Julander A, Lundgren L, Skare L, Grandér M, Palm B, Vahter M, Lidén C (2014) Formal recycling of e-waste leads to increased exposure to toxic metals: an occupational exposure study from Sweden. Environ Int 73:243–251

Kaul B, Sandhu RS, Depratt C, Reyes F (1999) Follow-up screening of lead-poisoned children near an auto battery recycling plant, Haina, Dominican Republic. Environ Health Perspect 107 (11):917–920

Leaner VD, Donninger H, Birrer MJ (2007) Transcription factors as targets for cancer therapy: AP-1 a potential therapeutic target. Curr Cancer Therap Rev 3:1–6

Leonard SS, Harris GK, Shi X (2004) Metal-induced oxidative stress and signal transduction. Free Radic Biol Med 37:1921–1942

Litvak P, Slavkovich V, Liu X, Popovac D, Preteni E, Capuni-Paracka S, Hadzialjevic S, Lekic V, Lolacono N, Kline J, Graziano J (1998) Hyperproduction of erythropoietin in nonanemic lead-exposed children. Environ Health Perspect 106(6):361–364

Nordberg GF (2009) Historical perspectives on cadmium toxicology. Toxicol Appl Pharmocol 238 (3):192–200

Pellerin C, Booker SM (2000) Reflections on hexavalent chromium: health hazards of an industrial heavyweight. Environ Health Perspect 108:A402–A407

Pilsner JR, Hu H, Ettinger A, Sanchez BN, Wright RO, Cantonwine D et al (2009) Influence of prenatal lead exposure on genomic methylation of cord blood DNA. Environ Health Perspect 117:1466–1471

Pinheiro MCN, Macchi BM, Vieira JLF, Oikawa T, Amoras WW, Santos EO (2008) Mercury exposure and antioxidant defenses in women: a comparative study in the Amazon. Environ Res 107:53–59

Ramesh BB, Parande AK, Ahmed BC (2007) Electrical and electronic waste: a global environmental problem. Waste Manag Res 25:307–318

Robinson BH (2009) E-waste: an assessment of global production and environmental impacts. Sci Total Environ 408:183–191

Sanders T, Liu Y, Buchner V, Tchounwou PB (2009) Neurotoxic effects and biomarkers of lead exposure: a review. Rev Environ Health 24:15–45

Sarkar BA (2005) Mercury in the environment: effects on health and reproduction. Rev Environ Health 20:39–56

Shaikh ZA, Vu TI, Zaman K (1999) Oxidative stress as a mechanism of chronic cadmium-induced hepatoxicity and renal toxicity and protection by antioxidants. Toxicol Appl Pharmacol 154:256–263

Sjodin A, Wong LY, Jones RS, Park A, Zhang Y, Hodge C et al (2008) Serum concentrations of polybrominated diphenyl ethers (PBDEs) and polybrominated biphenyl (PBB) in the United States population: 2003–2004. Environ Sci Technol 42:1377–1384

Sthiannopkao S, Wong MH (2013) Handling e-waste in developed and developing countries: initiatives, practices, and consequences. Sci Total Environ 463–464:1147–1153

Tchounwou PB, Ayensu WK, Ninashvilli N, Sutton D (2003) Environmental exposures to mercury and its toxicopathologic implications for public health. Environ Toxicol 18:149–175

Tchounwou PB, Yedjou CG, Patlolla AK, Sutton DJ (2012) Heavy metal toxicity and the environment. In: Luch A (ed) Molecular, clinical and environmental toxicology. Experientia Supplementum, vol 101. Springer, Basel

UN Environment Programme (2007) E-waste, volume 1: Inventory assessment manual. UN Environment Programme, Nairobi

United States Environmental Protection Agency (2002) Lead compounds. Technology transfer network-air toxics website. http://www.epa.gov/cgi-bin/epaprintonly.cgi

US EPA (1998) Guidelines for ecological risk assessment. EPA/630/R-95/002F, final report, Washington, DC

Valko M, Izakovic M, Mazur M, Rhodes CJ, Tesler J (2004) Role of oxygen radicals in DNA damage and cancer incidence. Mol Cell Biochem 266:79–110

Valko M, Rhodes CJ, Monocol J, Izakovic-Mazur M (2006) Free radicals, metals and antioxidants in oxidative stress-induced cancer. Chem Biol Interact 160:1–40

Van den Berg M, Birnbaum LS, Denison M et al (2006) The 2005 World Health Organization reevaluation of human and mammalian toxic equivalency factors for dioxins and dioxin-like compounds. Toxicol Sci 93(2):223–241

Velma V, Vutukuru SS, Tchounwou PB (2009) Ecotoxicology of hexavalent chromium in freshwater fish: a critical review. Rev Environ Health 24(2):129–145

Wang S, Shi X (2001) Molecular mechanisms of metal toxicity and carcinogenesis. Mol Cell Biochem 222:3–9

Widmera R, Oswald-Krapf H, Sinha-Khetriwalb D, Schnellmannc M, Bonia H (2005) Global perspectives on e-waste. Environ Impact Assess Rev 25:436–458

Yedjou CG, Milner J, Howard C, Tchounwou PB (2010) Basic apoptotic mechanisms of lead toxicity in human leukemia (HL-60) cells. Int J Environ Res Public Health 7(5):2008–2017

Zahir A, Rizwi SJ, Haq SK, Khan RH (2005) Low dose mercury toxicity and human health. Environ Toxicol Pharmacol 20:351–360

Zheng J, Luo XJ, Yuan JG, He LY, Zhou YH, Luo Y et al (2011) Heavy metals in hair of residents in an e-waste recycling area, south China: contents and assessment of bodily state. Arch Environ Contam Toxicol 61(4):696–703

Chapter 15
Toxicity, Eco-toxicity, and Phytoremediation of E-waste

Bushra Rehman, Mazhar Iqbal, and Ismat Nawaz

Abstract E-waste represents any discarded complete electronic product or in components form. Due to the rapid technological advancement, electronics are available easily in the markets as well as online services thus originated the issue of e-waste worldwide. Discarded electronics consisting of mainly cellphones, computers, refrigerators, monitors, and televisions weigh approximately equal to 200 Empire State buildings. Rate of production is very high for Electrical and Electronic Equipment (EEE) throughout the world due to increase in use of these equipments; advancement in the fields of electronics, IT, and telecom industries; and improved living standards of consumers. The economic growth and increase in technological innovations day by day lead to the faster deposition of such products as outdated ones; thus EEE can be designated as having short life span nowadays.

Keywords Toxicity · Eco-toxicity · Phytoremediation · E-waste

15.1 Composition of E-waste

E-waste mainly consists of mixtures of plastic, various metals, and other types of synthetic materials. In our societies it also includes electronic products like televisions, mobile phones, computers, printers, photo state machines, and electronic toys. Components of e-waste can be divided on the basis of their quantity: large, small, and trace. E-waste consists of over 60% metals (large amount) like iron, copper, aluminum, and gold, 30% plastics (small amount), and 2.7% hazardous pollutants

B. Rehman · M. Iqbal (✉)
Department of Environmental Sciences, Faculty of Biological Sciences, Quaid-i-Azam University, Islamabad, Pakistan
e-mail: bushrarehman@bs.qau.edu.pk; miqbal@qau.edu.pk

I. Nawaz
Department of Biotechnology, COMSATS University Islamabad, Abbottabad, Pakistan
e-mail: ismatnawaz@cuiatd.edu.pk

M. Z. Hashmi, A. Varma (eds.), *Electronic Waste Pollution*, Soil Biology 57,
https://doi.org/10.1007/978-3-030-26615-8_15

(trace amounts) like flame retardants such as polybrominated biphenyls (PBB) and polybrominated diphenyl-ethers (PBDEs) (Devi et al. 2004).

15.2 Production and Recycling of E-waste at Global Scale

With electronic industry being the fastest-growing industry in the world, developed countries produce, use, and discard devices at a highest amount and thus generate lot of e-waste at a rapid rate. Mostly, they export these products to developing countries where local buyer processes these items to reuse or recover the valuable elements out of them. Recovery rate ranges from 70 to 80%, while reuse and recycling rate is usually 50–75% (Kaya 2007).

At the global level, e-waste production is estimated to be 20–25 million tons per year, with a high percentage of e-waste being produced in Europe, China, America, and Australia. Advanced technologies such as miniaturization and fast cloud computing networks on the Internet from far-flung areas may account for addition in e-waste generation. In developed countries, disposed e-waste is either incinerated or processed to recover the valuable materials without harming the environment (Table 15.1) (Fig. 15.1).

15.3 Developing Countries and E-waste

Several developing countries including Pakistan, the Philippines, China, Nigeria, India, Ghana, and Vietnam acquire e-wastes by two means: from domestic production and from smuggling. Thus, there is a great challenge for management of the e-waste (Nnorom and Osibanjo 2008). Common disposal activities for the e-waste in several poor countries with low to middle income include unsanitary, uncontrolled landfill sites and recycling of e-waste informally. Copper is recovered by open-air burning in order to remove plastic. Valuable metals like gold, platinum, palladium, and silver are extracted from e-wastes such as printed circuit boards by using acids. The e-waste is broken down and opened up by local labor who are unaware of the hazards and the safe handling techniques (SEPA 2011).

According to scientific research and development, people are exposed to chemicals from e-waste and the recycling process. Health effects include abnormal thyroid development and function, neural and behavioral disorders, miscarriages, prematurity, low birth weight, and genetic mutations (Robinson 2009).

Table 15.1 E-waste components and toxicity

Products →	 Mobile phone	 Printed circuit board	 Cathode ray tube	 Rechargeable batteries	 Lighting screens display	 USB data storage	 Wires and cables
Hazardous elements	Pb, hg, Cd, and Be, Brominated flame retardant, PVC plastic	Cu, Fe, Al brominated flame retardants	Al, Zn, Ni Cs, Sr, high quantity of Pb	Cd in rechargeable batteries Switches and old CRTs	Hg in lighting devices, flat screen display	Hexavalent chromium	Polyvinyl chloride (PVC) Chlorinated dioxins and furans
Toxicity	Carcinogenic and endocrine disruption	Impaired learning and memory functions Behavioral problems due to womb exposure, interfere with thyroid and estrogen hormone	Cause intellectual impairment in children Damage the nervous system Blood and reproductive systems in adults	Affects kidneys and bones Osteoporosis, lung disease	Damage of the brain and CNS during early development, kidney damage	Highly toxic, disrupt balance between ROS and antioxidants, carcinogenic, ulcers DNA adducts	Toxic even in very low concentrations
Eco-toxicity	Highly persistent in the environment	Persistent and bio-accumulative	Pb is toxic to aquatic species Cs effected badly to fish and mammals	Bioaccumulate in the environment	Adverse effects on soil invertebrates	Adverse effects to aquatic life	Highly persistent in the environment

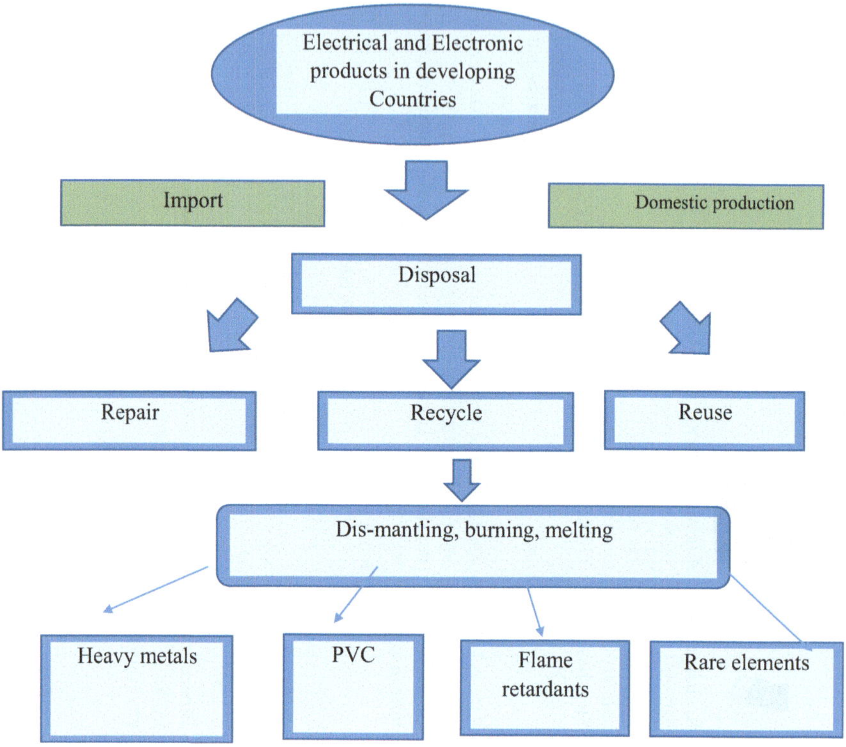

Fig. 15.1 Ultimate fate of e-waste

15.4 Pakistan Is an E-waste Dumping Ground

With a great population involved in import of e-waste, Pakistan has near to become a major e-waste destination. Increase in e-waste production was 5% from 2012 to 2014, whereas the population expansion rate was just 1.6%. In the coming years, Pakistan is going to become one of the top-listed countries for generation of e-waste. Handling of such situation needs some regulatory measures. Recycling activities in Karachi release flame retardants whose inhalation or soil absorption are very low. However, the dermal contact and ingestion through food need to be assessed for a variety of generated toxic compounds including PCBs, dioxins, PAHs, and various heavy metals (Iqbal et al. 2017).

The major cities of Pakistan indulged in the e-waste business are Rawalpindi, Islamabad, Peshawar, Gujranwala, Karachi, Lahore, and Faisalabad, where Karachi is ranked first with its largest and busiest deep-water seaport. The containers of e-waste from other countries are received and unloaded here. Seaport authorities focus on the usual routine of custom clearance and transportation of e-waste to the warehouses. From where, some of the "scrap dealers" buy these items by weight. After which, e-wastes are dismantled, burnt, or dumped, depending upon the composition of the e-waste. A large group of people from all ages including teenagers and children voluntarily dismantle, burn, and/or separate valuable items from the e-waste. These volunteers perform this work in order to make a living and to support their families (Saeed 2013).

15.5 E-waste Toxicity and Eco-toxicity in Developing Countries

In the recycling sites, workers involved in the dismantling activities are exposed to the hazards according to the nature of the job and persons involved. Duration of hazardous effects is not fully investigated; however, studies are conducted for assessment of metals and other chemicals in human-derived biological samples (Wen et al. 2008).

Heavy metals released by e-waste dust can affect humans through several routes including ingestion, inhalation, and dermal absorption. In polluted environments, according to an estimate, adults may ingest up to 100 mg dust day^{-1} (Leung et al. 2008).

Heavy metals are also absorbed and accumulated in the edible plant species in a large amount. Consumption of such unsafe food which contains high quantities of heavy metals may lead to the accumulation of heavy metal in the kidneys and liver of humans that disturbs the body metabolism leading to cardiovascular, nervous, nephrological, and bone diseases (Järup 2003). Heavy metals are toxic because they are bio-accumulated in organisms with the passage of time. The level of toxicity for every heavy metal differs among each other and depends upon the concentration, time of exposure, surroundings, nutritional condition, species, and type of gender. Several types of toxic effects are observed due to chronic exposure of heavy metals such as mutations, cancers, DNA damage, damage to immunity, general health, reproduction, and development (Ali et al. 2013).

Our environment is greatly affected due to the fast-growing electronic waste in the recent years. The best solution to this problem is Reverse Supply Chain (RSC) method. This method is based on reuse and recover. Several models are designed for RSC operation such as collection, transportation, and treatment risks, but most of them do not include the risks associated with this operation linked to the total cost of e-waste treatment in that particular designed RSC model. Another model is designed for RSC that includes risk costs and based on mixed integer linear programming. The results of such models are used to indicate the locations of facilities and show the flow of materials. Illustrative examples are also used to evaluate the effectiveness of the model. Such models are very useful for the decision makers of e-waste management (Linh et al. 2018).

15.6 Major E-waste Pollutants in Pakistan

Mainly, organic flame retardants (FRs) released into environment due to recycling activities of e-waste in Pakistan, China, India, and Africa pose serious effects to the environment and humans. During analysis of soil samples taken from Shershah and Lyari site of Karachi (well-known e-waste recycling areas), Dechlorane Plus (DP) and "novel" brominated flame retardants (NBFRs) were found in high concentrations. Additionally, organo-phosphorus-based FRs (OPFRs) were also found in high quantity among different air samples. Among other substances, decabromodiphenyl ether (BDE-209) (726 ng g^{-1}), decabromodiphenyl ethane (DBDPE) (551 ng g^{-1}), 1,2-bis(2,4,6-tribromophenoxy)ethane (BTBPE) (362 ng g^{-1}), and triphenyl-phosphate (∑TPP) (296 ng g^{-1}) were found to be prevalent in soils, while in air samples of OPFR congeners (5903–24,986 ng m^{-3}) were also present. Presence of high-quantity FRs is an indication of informal e-waste recycling practices. Thus, average emissions for these FRs in air and soil are directly proportional to the labor's working hours which they spent at the recycling sites (Iqbal et al. 2017). E-waste workers suffer health defects in their respiratory tract and skin (Robinson 2009).

15.7 Measurement of Heavy Metals Toxicity

Most of the e-waste is dumped into the landfill sites. Toxicity Characteristic Leaching Procedure (TCLP) and Synthetic Precipitation Leaching Procedures (SPLP) are tests which are used to check the level of toxicity. A study was conducted in order to test the heavy metal toxicity due to waste printed circuit boards and they showed that mean concentration of leaching was greatest in case of copper in both tests but lead concentration was 6.2 mg L^{-1}, which exceed the Chinese EPA regulatory or TCLP limit (5 mg L^{-1}). However, the other compounds such as PBDEs and TBBPA were not detected in both types of tests. Leaching trends of BFRs were further tested in landfill sites and reported to be the main cause of contamination (Zhou et al. 2013).

"Chemical cocktail" as leachate, produced due to deposition of e-waste and detected in different toxicity test, was found to be toxic for aquatic life as well. Scientists performed "aquatic toxicity assays" which include acute and chronic growth inhibition assay (only for PCBs) and MetPlate test for acute heavy metal toxicity. They discovered that 51 out of 56 compounds from the discarded electronic devices were producing toxic leachate (Dagan et al. 2007).

15.8 Toxicity Tests for Flame Retardants on Daphnia

Three polybrominated FRs, i.e., polybrominated and polychlorinated biphenyls (PBB-153 and PCB-153) and polybrominated diphenyl ether (PBDE)-153, were tested for acute and reproductive toxicity tests on *Daphnia magna*. In 24-hour acute toxicity test, concentration ranges between 0 μg L^{-1} and 210 μg L^{-1} of these chemicals were not found to be toxic to *Daphnia*. However, according to reproductive tests, concentration ranges between 50 and 100 μg L^{-1} of these FRs were found to be lethal as animals died at these concentrations. The most toxic was PCB-153 which accumulated 3 times more compared to others (particularly at a concentration of 12.5 and 25 μg L^{-1}) and thus effected severely on reproduction. Least toxic was PBDE-153 but induced a high rate of mortality than the control group. This difference in toxic effect on reproduction of *Daphnia* may be due to different chemical structure of these RFs. The tested compounds also accumulated in a different way into *Daphnia* (Nakari and Huhtala 2008).

15.9 Effects on Soil Microbiota on E-waste Recycling Sites

Conventional e-waste recycling practices release heavy metals and high quantity of organic pollutants into the surrounding environment. Soil microbial community is also affected by these contaminants during pollutant remediation. Contaminated soil samples from e-waste recycling sites of Pakistan and neighboring country of China were analyzed for indigenous microbial communities at whole and core levels. Severe effects on nutrition (N, K, and total organic carbon) were established at both levels due to heavy metals (Pb, Cu, and Zn), PAHs, and PBDEs. Tested pollutants affected 75% of groups which are created on the basis of operational taxonomic units (OTUs) sensitively. Major OTUs tested include lactic acid bacteria, *Proteobacteria*, and *Nitrospirae* in the ecological network (Jiang et al. 2017).

15.10 E-waste Treatment with Biochar

Recycling sites of e-waste especially in urban areas need special attention for remediation. There are many studies that documented the positive effect of application of biochar (BC) on the remediation of contaminated sites. Scientists had studied the changes in soil morphology by the application of dairy-manure-derived biochar and mobility of metals, in order to gather information regarding the immobilization of metals, which leads to the restoration of vegetation of the landscape on contaminated e-waste recycling sites. Biochar ratio and incubation time were the main factors which determine positive effect produced by the amended soil on the bioavailability of metals via reducing the mobility of Cu, Cd, Pb, and Zn. Heavy metals were transformed to more stable fraction, ultimately decreasing their toxic concentrations. Pb and Cu leaching was significantly reduced. Other effects of BC applications were increase in the soil pH, cation exchange capacity, and phosphorous availability which indicates that biochar is an effective amendment material for abandoned e-waste recycling sites for the restoration of vegetation on any landscape. It is concluded that, after the alkaline treatment the BC showed higher efficiency (probably due to increase in porosity), surface area and functional groups (Chen et al. 2017).

15.11 Biodegradation of Brominated Flame Retardants (BFRs) in Soil

"Readily biodegradable" tests are simple and quick methods. In this case the test bottle having the test compound with an aqueous solution of sludge is incubated for 28 days by measuring the biological oxygen demand (BOD). If BOD is more than 60% of the theoretical oxygen demand, the test compound is considered readily degradable (Sedykh and Klopman 2007).

BFRs can be stored in the soil. Some of the studies to check their persistence indicated that brominated diphenyl ethers (BDE 28 and BDE 209) degraded in both types of aerobic and anaerobic conditions. On the other hand, the compound 246-brph and TBECH showed fast degradation in all conditions. Earthworms also accumulated BFRs from soil while zebra fish also tested for accumulation that showed presence of various metabolites as degradation products of BFRs in the body of fish (Nyholm 2009).

15.12 Phytoremediation: Alternate Option for E-waste-Contaminated Sites

Use of plants for the uptake, degradation, immobilization, volatilization, or conversion into harmless compounds of heavy metals and organic contaminants (PCBs and PAHs) is referred as phytoremediation (Greipsson 2011). Phytoremediation technique can be employed for the detoxification, degradation, or removal of heavy metals and other environmental pollutants like PCBs and pesticides from the media such as soil or water. This approach is novel, cost-effective, eco-friendly, and easy to apply (Clemens 2001).

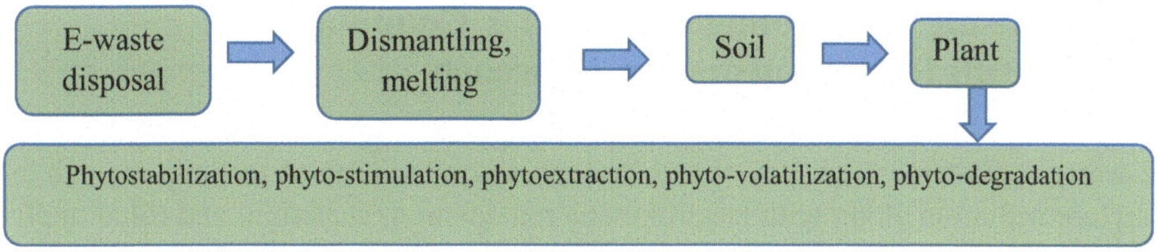

Fig. 15.2 Phytoremediation techniques for e-waste

15.13 Techniques of Phytoremediation

Several techniques of phytoremediation have been explained in Fig.15.2 (Schwitzguebel 2000).

15.14 Plant Selection for Phytoremediation

There are many cellular mechanisms in plants which enable them to tolerate and detoxify the higher concentrations of heavy metals. Plants that can accumulate large quantities of heavy metals inside them are called hyperaccumulators. This is a genetic determinant which makes them suitable for phytoremediation purposes. More than 500 species of plants fall under this category, while only 2% of the angiosperms are categorized as hyperaccumulators (Sarma 2011).

Types of vegetation/plants generally used for this purpose include poplar tree; agronomic plants, i.e., sunflower, white radish, and carrot; aquatic plants like alfalfa, blue-green algae, water lilies (yellow or white), and weeds, i.e., duck weeds; and brake fern, arrowroot, and periwinkle. Grasses including vetiver grass, Sudan grass, rye grass, Bermuda grass, alpine blue grass, and switch grass have been documented as phytoremediation agents (Vasavi et al. 2010).

15.15 Wetland's Phytoremediation

Under natural wetland conditions, a dense population zone of microbes being provided with conditions for reducing contaminant loads of petroleum hydrocarbons, additives, pesticides, organo-halogens, and volatile organic compounds was found to be degraded effectively and thus used as an efficient technique for remediation. Laboratory microcosm and mesocosm technologies are used within and between these two systems to study wetland phytoremediation. Results are positive for phytoextraction and rhizospheric degradation (Williams 2002).

15.16 Using Low Voltage and EDTA Combined with Rhizoremediation Approach

E-waste decontamination was tested through a combination of techniques such as phytoremediation along with electrokinetic remediation, chelation, and foliar application of plant hormone for different metal-contaminated soils using *E. globulus*. EDTA treatment showed a significant increase in decontamination ability of Cd, Pb, and Cu metals per plant with 0.9–11.5 times more metals leaching. Only low- and moderate-voltage (2 V and 4 V) electric fields supported the plant growth for the removal of heavy metal; however, high voltage (10 V) enhanced the metal accumulation. Voltage increased from 0 to 10 V resulted in leachate volume reduction from 1224 to 134 mL probably due to electro-osmosis and electrolysis. Cytokinin foliar application accounted 56% more biomass and 2.5 times more leachate versus individual phytoremediation because of high transpiration rate. Greater biomass production and metal accumulation was found in combination. Multi-technique approach for the decontamination of Cd, Pb, and Cu from soil takes 2–10-fold less time than other treatments which take more time with very low leachate produced. This approach is suggested to be most suitable (Luo et al. 2017).

15.17 Conclusions

E-waste is increasing day by day in the developed countries, and its proper management is necessary. Because of the import and recycling business of e-waste in the developing countries, it needs more attention toward the recycling procedure and their management practices. We should design the policies to limit the import of such items, at first, and use technologies for the remediation of hazardous contaminants (produced due to e-waste) like heavy metals, PCBs, and flame retardants from soil, air, and water. Toxicity tests suggest that these pollutants are very harmful for living organisms (human, animal, plant, and beneficial microorganisms). Among many decontamination strategies, phytoremediation can be applied in-situ or ex-situ to a variety of contaminants. It is an innovative, cost-effective, and environment-friendly technology, thus providing a good alternative to conventional cleanup methods. Many plants are reported to detoxify harmful compounds like heavy metals, PBB, PBDE, and PCB thus convert them into less harmful or harmless compounds. We can also use biochar as a chelator, for additional support, along with phytoremediation plant(s).

References

Ali H, Khan E, Sajad MA (2013) Phytoremediation of heavy metals concepts and applications. Chemosphere 91:869–881

Chen Z, Zhang J, Liu M, Wu Y, Yuan Z (2017) Immobilization of metals in contaminated soil from E-waste recycling site by dairy-manure-derived biochar. Environ Technol 9:2801–2809

Clemens S (2001) Developing tools for phytoremediation towards a molecular understanding of plant metal tolerance and accumulation. Int J Occup Med Environ Health 14:235–239

Dagan R, Dubey B, Bitton G, Townsend T (2007) Aquatic toxicity of leachates generated from electronic devices. Arch Environ Contam Toxicol 53:168–173

Devi B, Shobha S, Kamble R (2004) E-waste: the hidden harm of technological revolution. J Indian Assoc Environ Manag 31:196–205

Greipsson S (2011) Phytoremediation. Nat Edu Knowl 3:7

Iqbal M, Syed JH, Breivik K, Chaudhry MJI, Li J, Zhang G, Malik RN (2017) E-waste driven pollution in Pakistan: the first evidence of environmental and human exposure to flame retardants (FRs) in Karachi City. Environ Sci Technol 51:13895–13905

Järup L (2003) Hazards of heavy metal contamination. Br Med Bull 68:167–182

Jiang L, Cheng Z, Zhang D, Song M, Wang Y, Luo C, Zhang G (2017) The influence of e-waste recycling on the molecular ecological network of soil microbial communities in Pakistan and China. Environ Pollut 231:173–181

Kaya M (2007) Electronic waste disposal, reuse and recycle technologies. In: Proceedings of European metallurgical conference, Düsseldorf, Germany, pp 1613–1631

Leung AO, Duzgoren-Aydin NS, Cheung K, Wong MH (2008) Heavy metals concentrations of surface dust from e-waste recycling and its human health implications in Southeast China. Environ Sci Technol 42:2674–2680

Linh DTT, Yousef A, Lee S-HL, Phan NKP (2018) Development of a reverse supply chain odel for electronic waste incorporating transportation risk. IPTEK J Proc Series:3. https://doi.org/10.12962/j23546026.y2018i3.3703

Luo J, Cai L, Qi S, Wu J, Gu XS (2017) A multi-technique phytoremediation approach to purify metals contaminated soil from e-waste recycling site. J Environ Manag 204:17–22

Nakari T, Huhtala S (2008) Comparison of toxicity of congener-153 of PCB, PBB, and PBDE to Daphnia magna. Ecotoxicol Environ Saf 71:514–518

Nnorom IC, Osibanjo O (2008) Overview of electronic waste (e-waste) management practices and legislations and their poor applications in the developing countries. Resour Conserv Recycl 52:843–858

Nyholm RJ (2009) Persistency, bioaccumulation and toxicity assessment of selected brominated flame retardants. Umeå University, Umea, Sweden, p 22

Robinson BH (2009) E-waste: an assessment of global production and environmental impacts. Sci Total Environ 408:183–191

Saeed A (2013) Dangerous labour: the little known e-waste phenomenon. Dawn 2013. http://www.dawn.com/news/1032967

Sarma H (2011) Metal hyperaccumulation in plants: a review focusing on phytoremediation technology. J Environ Sci Technol 4:118–138

Schwitzguebel J (2000) Potential of phytoremediation, an emerging green technology. In: Ecosystem service and sustainable watershed management in North China international conference, Beijing, vol 3, pp 210–220

Sedykh A, Klopman G (2007) Data analysis and alternative modelling of MITI-I aerobic biodegradation. SAR QSAR Environ Res 18:693–709

SEPA (2011) Recycling and disposal of electronic waste: health hazards and environmental impacts Stockholm. ISBN 978–91–620-6417-4

Vasavi A, Usha R, Swamy P (2010) Phytoremediation-an overview review. J Indust Pollut Cont 26:83–88

Wen S, Yang F, Gong Y, Zhang X, Hui Y, Li J (2008) Elevated levels of urinary 8-hydroxy-2′-deoxyguanosine in male electrical and electronic equipment dismantling workers exposed to high concentrations of polychlorinated dibenzo-p-dioxins and dibenzofurans, polybrominated diphenyl ethers and polychlorinated biphenyls. Environ Sci Technol 42:4202–4207

Williams JB (2002) Phytoremediation in wetland ecosystems: progress, problems, and potential. Crit Rev Plant Sci 21:607–635

Zhou X, Guo J, Lin K, Huang K, Deng J (2013) Leaching characteristics of heavy metals and brominated flame retardants from waste printed circuit boards. J Hazard Mater 246:96–102

Chapter 16
Phytoremediation of Electronic Waste: A Mechanistic Overview and Role of Plant Secondary Metabolites

Mubarak Ali Khan, Nazif Ullah, Tariq Khan, Muhsin Jamal, Naseer Ali Shah, and Huma Ali

Abstract The escalating economic growth, urbanization and globalization over the last three decades have resulted in the huge production and consumption of electronic devices and appliances all over the world. This has caused an alarming situation of the disposition of electronic waste (e-waste) from the used and discarded electronic products to the environment, which can adversely affect the ecosystem and health of the humans. Management, treatment and recycling of e-waste become crucial to prevent the serious environmental complications and diseases. Among the several methods for treatment of e-waste, phytoremediation is of vital importance, which involves the application of plants and vegetation for the remediation of e-waste contaminants. Phytoremediation technology is a cost-effective green technology known for its optimal results on-site and is considered as environment-friendly and generally socially acceptable. The success of phytoremediation technology is by virtue of some unique plants which possess selective capabilities such as uptake of the metals by roots, translocation through stem and bioaccumulation in the leaves.

In this chapter, we have described in detail the process of phytoremediation as a suitable and sustainable method for remediation of e-waste contaminants including

M. A. Khan (✉) · N. Ullah
Department of Biotechnology, Faculty of Chemical and Life Sciences, Abdul Wali Khan University Mardan (AWKUM), Mardan, Pakistan
e-mail: makhan@awkum.edu.pk

T. Khan
Department of Biotechnology, University of Malakand, Chakdara Dir Lower, Pakistan

M. Jamal
Department of Microbiology, Abdul Wali Khan University Mardan (AWKUM), Mardan, Pakistan

N. A. Shah
Department of Biosciences, COMSATS Institute of Information Technology, Islamabad, Pakistan

H. Ali
Department of Biotechnology, Bacha Khan University, Charsadda, KP, Pakistan

M. Z. Hashmi, A. Varma (eds.), *Electronic Waste Pollution*, Soil Biology 57,
https://doi.org/10.1007/978-3-030-26615-8_16

heavy metals and other hazardous substances. Further, a mechanistic overview of the process of phytoremediation technology for treatment of e-waste has been elucidated to highlight the functional role of phytochemicals of plants in contaminants removal through phytoremediation.

Keywords Phytoremediation · Electronic waste · Mechanistic overview · Plant secondary metabolites

16.1 Introduction

Electronic waste (e-waste) refers to the used and unwanted material left from the different electronic devices; those have completed their normal shelf life in many household applications. E-waste is mainly composed of the metals used in the manufacturing of electronic appliances. For instance, gold, silver, copper, nickel, mercury and other hazardous metals can be extracted from the discarded materials of various electronic devices (Robinson 2009; Tsydenova and Bengtsson 2011). The primary sources of e-waste are the discarded household electronic products such as radios, TVs, photocopier, printers, CD players, batteries, calculators, tonners, cell phones, etc. However, the industrial, military and laboratory electronic machineries such as fans, washing machines, air conditioners, fridge, oven, heater, iron, grinder, keyboards, etc. also contribute in the major production of e-waste to the environment (Heacock et al. 2016). These electronic products produce large quantities of different types of e-waste. Recent analytical technologies have shown that e-waste from many electronic products is composed of hazardous substances such as polyvinyl chlorides (PVCs and PCBs), epoxy resins, thermosetting plastics, fibreglass, cadmium, germanium, lead, mercury, carbon and iron and the elements such as tin, gallium, thallium, copper, gold, beryllium, silicon, silver, vanadium, aluminium, selenium and indium (Muszyńska et al. 2015).

The escalating economic growth in the last 30 years has substantially increased the consumption and production of electronic products and equipment worldwide (Babu et al. 2007). In fact, the new electric appliances have drastically influenced every aspect of the modern-day life by providing more comfort, health and ease in acquisition and exchange of information (Sinha 2007). Further, the urbanization and globalization have forcefully uplifted the economic and technological growth through digital revolution worldwide. People all over the world are now using more electronic products than their forefathers. This has caused an alarming situation of the disposition of a variety of e-waste to the environment, which can adversely affect the ecosystem and of course the health of the humans. It is harder and challenging to dismantle and recycle e-waste due to the complex nature of its composition. Since a lot of the electronic products are made up of different materials, such as glass, plastics and some coating or colouring chemicals, this mixed composition of e-waste is the major bottleneck in its effective treatment, management and low-cost recycling (Yu et al. 2010). The mixed composition also makes e-waste to cause negative impacts on human health. The hazardous materials present in e-waste are reported to cause many serious ailments including bronchitis, skin disease, reproductive complaints and cancer (Yu et al. 2010; Robinson 2009; Li et al. 2009). Many lethal and poisonous substances from e-waste enter the human food chain through the soil-crop-food pathway and cause severe health complications. According to a recent survey, the annual global production of e-waste is estimated at approximately 40 million metric tons, and 13% of that is reported to be recycled mostly in the developing world (Laurent et al. 2014). Further, the combustion and burning of e-waste on the site results in the production of fine particulate matter that is the main pollutant responsible for cardiac and pulmonary diseases. The largest e-waste recycling unit in the world is located in Guiyu city of China. People of this city are suffering from many diseases including digestive complaints, respiratory and neurological disorders. For instance, 80% of the infants and children in Guiyu city are at risk of respiratory diseases and poisoning due to lead (Leung et al. 2008). Combustion from burning e-waste creates fine particulate matter, which is linked to the pulmonary and cardiovascular disease. While the health implications of e-waste are difficult to isolate due to the informal working conditions, poverty and poor sanitation. The worrisome thing is that majority of the chemicals from e-waste are not biodegradable and they persist in the environment for long periods of time, increasing exposure risk (Ogunseitan et al. 2009). Several methods including physical, thermal and chemical are currently employed for the treatments, management and recycling of e-waste. However, these methods have some limitations. Most of these methods are costly and do not provide optimal performance on-site (Kofoworola 2007). One of the several biological methods for remediation of e-waste metals is phytoremediation, which involves the application of plants and vegetation for the treatment of contaminated soil. Phytoremediation technology is economically justified and is considered as least environmentally invasive and generally socially acceptable. The extensive and deep root system of the hyperaccumulator plants enables purification of groundwater environment from the pollution caused by disposition of e-waste into the environment. That is why this technology of remediation of chemical contaminants is also called as green technology (Tangahu et al. 2011).

16.2 Available Methods for E-waste Management and Treatment

E-waste contains many hazardous substances, which are harmful not only to the environment but for humans as well. Thus, it is crucial to search out for the effective and efficient techniques for the clean-up and removal of e-waste. Several thermal and chemical methods are already being used for the management, treatment and recycling of e-waste. These methods include (1) the recycling treatment and metal recovery; (2) shredding, pulverization and crushing; (3) sustainable disposal; (4) landfilling; (5) incineration; (6) open field dumping; and (7) open burning (Kofoworola 2007). As discussed in details in other chapters of this book, each method has its own prospects and limitation. Most of these methods either are costly or do not provide optimal performance on-site. Therefore, biological methods for treatment of e-waste, for example, phytoremediation, are preferred for their potential in clean-up of the environment and recovery of valuable metals. Besides, these methods are cost-effective, eco-friendly and sustainable in optimal treatment of e-waste (Brandl et al. 2001).

16.3 Phytoremediation: An Efficient Technology for Treatment of Electronic Waste

Phytoremediation is the biological method for the remediation of e-waste through plants. This process can be used for the removal of hazardous substances from e-waste in contaminated soil, sediments and water. There are many exceptional plant species which possess the natural potential of accumulating e-waste metals and are known as hyperaccumulators (Tangahu et al. 2011). As described in Table 16.1, a substantial number of research studies are available in literatures, showing the potential of many plant species in the remediation of several types of e-waste including heavy metals and other hazardous substances. Through the distinct, unique and selective capabilities in the entire plant system such as uptake of the metals by roots, translocation through stem and bioaccumulation in the leaves, phytoremediation takes the first line in the bioremediation of e-waste. Exploitation of phytoremediation technology by using green plants and vegetation has successfully accomplished the in situ treatment of soil, sediment and water, which were highly contaminated by polychlorinated biphenyls (PCBs) and other organic hazardous substances of e-waste (Brandl et al. 2001). In Guiyu city of China, a very large portion of soil was contaminated by different types of e-waste including polycyclic aromatic hydrocarbons (PAHs), brominated diphenyl ethers (BDEs) and deca-BDE. It is worth mentioning that the area for rice fields near burning sites was less than e-waste open burning sites. Further, it was observed that e-waste open burning sites in the soil possessed fairly higher concentrations of total PCBs, polybrominated diphenyl ethers (PBDEs) and polychlorinated dibenzodioxins (PCDDs). The non-e-waste open burning sites were highly contaminated by all persistent toxic substances (PTSs), with 5–50 times more concentrations than the PTSs in the rice field. This was the very first detailed research analysis on the PTS contamination in soils which was due to open burning of e-waste. Throughout the phytoremediation technology using alfalfa plants, the soil enzyme and microbial community were enhanced for removal of polychlorinated biphenyls (PCBs) in the contaminated soil field. For remediation of polycyclic aromatic hydrocarbons (PAHs), a multi-component phytoremediation system includes PAHs degrading

Table 16.1 Application of different plant species used in the phytoremediation of e-waste metals/contaminants and their phytochemical composition

E-waste contaminants	Plant species used in phytoremediation	References	Bioactive metabolites in the plant
Antimony (Sb)	*Achillea wilhelmsii*	Hajiani et al. (2015)	Terpenoids, phenolics, flavonoids
	Matthiola farinosa	Hajiani et al. (2015)	Flavonoids
	Pteris fauriei	Feng et al. (2015)	Flavonoids, phenols, tannins
	Pteris vittata	Müller et al. (2013)	Flavonoids, titerpenoids, phenolics
	Pteris cretica	Feng et al. (2011)	Alkaloids, flavonoids, saponins
Arsenic (As)	*Azolla caroliniana*	Zhang et al. (2008)	Alkaloids, terpenoids, steroids
	Populus alba	Vamerali et al. (2009)	Flavonoids, polyphenols
	Daucus carota	Helgesen and Larsen (1998)	Phenolic compounds, ascorbic acid
	Oryza sativa	Heitkemper et al. (2001)	Saponins, terpenoids, tannins
	Malus domestica	Caruso et al. (2001)	Quercetin, chlorogenic acid
Barium (Ba)	*Helianthus annuus*	Sampaio Junior et al. (2015)	Tannins, saponins, flavonoids
	Brassica juncea Czern. (mustard)	Coscione and Berton (2009)	Flavonoids, n-octacosane, linolenic, oleic acid
	Lactuca sativa L.	Lamb et al. (2013)	Carotenoids, phenolic acids
	Medicago sativa L.	Gardea-Torresdey et al. (1999)	Phenols, terpenoids, flavonoids
	Calotropis procera	Gardea-Torresdey et al. (1999)	Tannins, saponins, flavonoids
Beryllium (Be)	*Brassica napus* L.	Ali et al. (2018)	Caffeic acid, chlorogenic acid, quercetin, kaempferol
Cadmium (Cd)	*Oryza sativa* L.	Liu et al. (2007)	Saponins, terpenoids, tannins
	Sorghum bicolor L.	Muranyi and Ködöböcz (2008)	Tannins, saponins, flavonoids
	Pyxine cocoes	Muranyi and Ködöböcz (2008)	Tannins, saponins, flavonoids
	Hordeum vulgare	Peralta-Videa et al. (2009)	Phenolics, flavonoids, tannins

(continued)

Table 16.1 (continued)

E-waste contaminants	Plant species used in phytoremediation	References	Bioactive metabolites in the plant
	Spinacia oleracea (spinach)	Intawongse and Dean (2006)	Rutin, quercitin, gallic acid
	Brassica juncea L.	Peralta-Videa et al. (2009)	Phytoanticipins, phytoprotectants
	Eucaliptus camaldulenses Dehnh	Pence et al. (2000)	Tannins, saponins, flavonoids
Hexavalent chromium/chromium VI (Cr VI)	*Nicotiana tabacum*	Kim et al. (2006)	Alkaloids, steroids, phenols
	Convolvulus arvensis	Montes-Holguin et al. (2006)	Phenolic compounds, ascorbic acid
	Brassica oleracea var. *botrytis (cauliflower)*	Peralta-Videa et al. (2009)	Phenolic compounds, ascorbic acid
	Lycopersicon esculentum L.	Peralta-Videa et al. (2009)	Rutin, quercitin, gallic acid
	Calotropis procera	Kim et al. (2006)	Cardenolides, flavonoids, saponins
Lead (Pb)	*Alternanthera philoxeroides*	Cho-Ruk et al. (2006)	Phenols, cardiac glycosides
	Amaranthus hybridus L.	Tangahu et al. (2011)	Flavonoids, steroids, terpenoids
	Brassica campestris L.	Tangahu et al. (2011)	Flavonoids, anthocyanins
	Brassica juncea (L.) Czern.	Van Ginneken et al. (2007)	Phytoanticipins, phytoprotectants
	Brassica nigra (L.) Koch	Cho-Ruk et al. (2006)	Phenolics, flavonoids, tannins
Mercury (Hg)	*Brassica juncea* L.	Van Ginneken et al. (2007)	Phytoanticipins, phytoprotectants
	Colocasia esculenta	Skinner et al. (2007)	Phenolics, flavonoids, tannins
	Eichornia crassipes	Skinner et al. (2007)	Phenolics, flavonoids, tannins
	Helianthus tuberosus	Sas-Nowosielska et al. (2008)	Chlorogenic acids, phenolic compounds
	Oryza sativa L.	Liu et al. (2007)	Saponins, terpenoids, tannins
Nickel (Ni)	*Salix viminalis*	Watson et al. (2003)	Phenolics, flavonoids, tannins
	Sorghum bicolor L.	Muranyi and Ködöböcz (2008)	Phenolics, flavonoids, tannins
	Hypogymnia physodes	Muranyi and Ködöböcz (2008)	Tranorin, chloroatranorin, usnic acid

(continued)

Table 16.1 (continued)

E-waste contaminants	Plant species used in phytoremediation	References	Bioactive metabolites in the plant
	Canna indica L.	Subhashini and Swamy (2014)	Phenolic compounds, tannin, saponins
	Vetiveria Zizanioides L.	Muranyi and Ködöböcz (2008)	Phenolics, flavonoids, tannins
Polychlorinated biphenyls (PCBs)	*Medicago sativa*	Petruzzelli et al. (2012)	Protchaechenic acid, caffate, kaempherol
	Lespedeza cuneata	Petruzzelli et al. (2012)	Phenolics, flavonoids, tannins
	Panicum clandestinum	Petruzzelli et al. (2012)	Alkaloids, tannins, saponins, flavonoids
	Phalaris arundinacea L.	Petruzzelli et al. (2012)	Alkaloids
	Panicum variegatum L.	Petruzzelli et al. (2012)	Alkaloids, tannins, saponins, flavonoids
Selenium (Se)	*Oryza sativa*	Dhillon and Dhillon (2009)	Saponins, terpenoids, tannins
	Brassica juncea L.	Schiavon and Pilon-Smits (2017)	Phytoanticipins, phytoprotectants
	Hibiscus cannabinus L.	Parker et al. (2003)	Phenolics, flavonoids, tannins
	Pteris vittata	Parker et al. (2003)	Phenolics, flavonoids, tannins
	Typha angustifolia L.	Srivastava et al. (2005)	Phenolics, flavonoids, tannins

Data of the bioactive compounds present in the respective plant species was obtained from the plant metabolites database (http://pmn.plantcyc.org/)

bacteria (*Acinetobacter* sp.), carbuncular mycorrhizal fungus (AMF; *Glomus mosseae*) and ryegrass (*Lolium multiflorum*). The application of AMF considerably ($p < 0.05$) enhanced the growth of ryegrass. The cultivation of ryegrass subsequently improved the growth of PAH-degrading bacteria and which consequently enhanced the peroxidase activities in soil. Similarly, the interactions of ryegrass with PAH-degrading bacteria or AMF considerably ($p < 0.05$) enhanced the dissipation of phenanthrene (PHE) and PYR (pyrene) from the soil. Using rhizobox experimentation system, a decreasing dissipation gradient of PHE and PYR was revealed along the radial direction of maize (*Zea mays* L.) root, in which the highest dissipation rates were observed in rhizosphere zone followed by near rhizosphere zone and bulk soil zone in outer sections. The results revealed that there is a great potential for the development of a multi-component phytoremediation system for PAH-contaminated soil such as PAH-degrading bacteria, plants and AMF (Xiezhi 2008). In another study, Lin et al. (2003) described the importance of phytoremediation phenomenon for treatment of PCB-contaminated soils from e-waste recycling zone. The study was targeted to compare the capabilities of four different plants including alfalfa, rice, tall fescue and ryegrass for phytoremediation of PCBs. The plants were applied to remediate PCB-contaminated soil of Taizhou city, which is one of the largest e-waste recycling centres in China. They recorded optimal results of PCBs remediation by the cultivated plants in the soil after 120 days, as compared with the unplanted soil.

16.3.1 Phytoextraction

Phytoextraction also called as phytoaccumulation is the process of removal or movement of e-waste metals from the contaminated soil through plant roots into stem and leaves (Jutsz and Gnida 2015). The plants which exhibit the natural potential of accumulating higher levels of metals are called hyperaccumulators. Thus through the phenomenon of hyperaccumulation during phytoextraction, some plants can take up a variety of e-waste metals in enormous concentration from the contaminated soil. These hyperaccumulators can carry and accumulate the obnoxious metals in different above-the-ground organs such as stem, branches and leaves in concentrations from 100 to 1000 times higher than the normal plants without being affected by any visible phytotoxic effects (Figs. 16.1 and 16.2). Further, these plants are generally observed to grow abundantly in the areas contaminated with the continuous disposition of e-waste and produce greater biomass that can be easily handled for harvesting and recovery of different valuable metals. Therefore, such plants are suitable for the process of phytoremediation (Rascio and Navari-Izzo 2011). On the basis of the quantity of metals in dried foliage such as Cd 100; Co, Cu, and Cr 300; Pb and Ni 1000; and Zn 3000 μg/g, respectively, a large number of plants (about 500 taxa) have been recognized as hyperaccumulators of important metals (Van der Ent et al. 2013). A list of such is given in Table 16.1. These plants belong to different diverse families such as *Brassicaceae*, *Violaceae*, *Cunouniaceae*, *Lamiaceae*, *Asteraceae*, *Poaceae*, *Euphorbiaceae*, *Caryophyllaceae*, *Cyperaceae*, *Fabaceae*, *Caryophyllaceae* and *Flacourtiaceae* (Muszynska and Hanus-Fajerska 2015). These plants are exclusively different from other plants due to the following characteristics: (1) a higher capability to sequester heavy metals from the contaminated soils, (2) better root to shoot transportation of metal ions, (3) a superior capability to detoxify and collect/bin tremendously huge quantity of heavy metals in the shoots, (4) fast-growing ability and accumulation capability of heavy metals anions in the shoots, and (5) a well-developed and plentiful root system (Jabeen et al. 2009; Rascio and Navari-Izzo 2011).

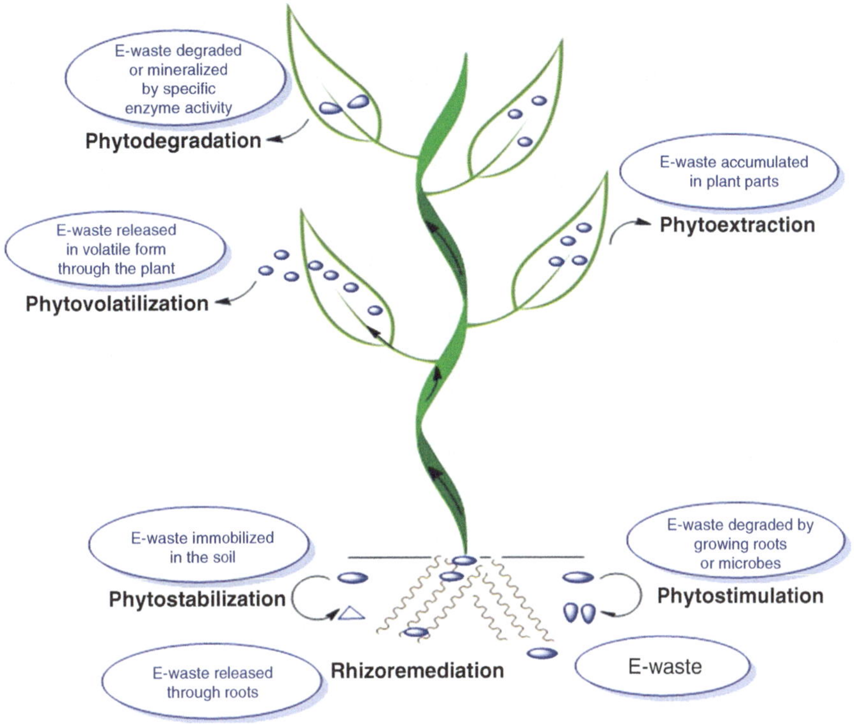

Fig. 16.1 Schematic representation of the different processes involved in the phytoremediation of e-waste in the contaminated soil

16.3.2 Phytofiltration

Phytofiltration is the use of roots, seeds and plants to adsorb or precipitate toxic metal ions from the aqueous medium (Ali et al. 2013). As illustrated in Figs. 16.1 and 16.2, phytofiltration is classified further to the following three forms, i.e. (1) rhizofiltration is the application of the plant roots for extraction of heavy metals, (2) blastofiltration is the application of the plant seedlings for extraction of heavy metals and (3) caulofiltration is the application of the cut/excised plant shoots for extraction of heavy metals. Overall, it is the remediation of heavy metals from the polluted and contaminated sites by using plant roots or seedlings (Chen et al. 2015).

16.3.3 Phytostimulation

Phytostimulation is the process of enhancing plant capability to degrade/detoxify organic wastes by stimulation of microbial enzyme activities for decontamination

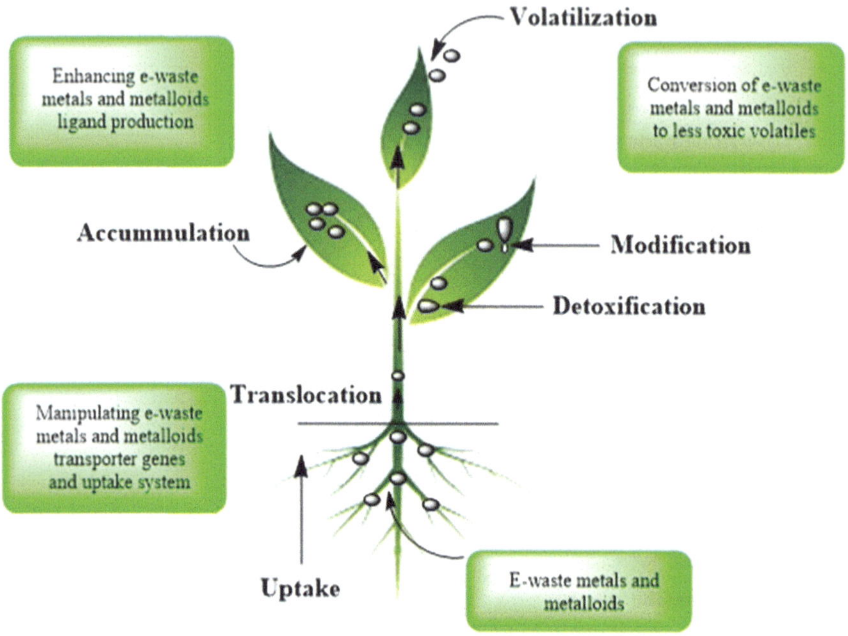

Fig. 16.2 Mechanistic overview of phytoremediation of e-waste contaminants by using hyperaccumulator plants

(Fig. 16.1). Certain plant extracts/exudates secretions from roots of the plants can be enhanced by certain microbes. For example, ethylene (a plant hormone) on one hand can stimulate elongation of roots at very low concentration and on the other hand at a higher concentration can inhibit cell division and DNA synthesis. However, this effect can be stopped by reducing ethylene concentration in plants. The reduction in ethylene can be obtained by some specific enzyme such as 1-aminocyclopropane-1-carboxylase deaminase which regulates ethylene biosynthesis by balancing ethylene-level production in plants (Donot et al. 2012; Gaiero et al. 2013). This enzyme is made by plant growth-promoting rhizobacteria (PGPR) linked with plant roots using exudates/extract released by plants as carbon and energy sources to cause degradation of e-waste metal contaminants (Tak et al. 2013).

16.3.4 *Phytostabilization*

This process of phytoremediation refers to the application of plant roots to absorb pollutants from the soil and to retain them within the rhizosphere (Figs. 16.1 and 16.2). By this process, e-waste contaminants, especially hazardous heavy metals, are separated from the source and stabilized, limiting this contaminant from spreading to other places in the environment (Lone et al. 2008). The metals are reduced by the root system of the plants through precipitation, absorption, complication and valence reduction in the region around plant roots and thus the access and mobility of contaminant to the environment are restricted (Choudhary and Varma 2016). The quantity of heavy metals found in the rhizosphere soil around a plant indicates the efficiency and success of phytostabilization in restricting the mobility of heavy within the plant (Rajkumar et al. 2012). Plants capable of the phytostabilization process should have a broad metals recognition and the tendency for low mobilization of metals from roots to shoots (Islam et al. 2013). The phytostabilization capacity of plants can be improved by changing the physicochemical conditions such as pH and organic matter contents. These conditions can be changed by adding some substances such as biochar or compost which will increase the yield of plants and also immobilize the metals. Phytostabilization is a superior substitute to other techniques because of its higher potential of capturing hazardous metals in-situ. The contaminants are not taken up into other tissues of the plant and therefore do not disperse into the environment. It focuses primarily on heavy metal sequestration only within the rhizosphere (Tak et al. 2013).

16.3.5 Phytovolatilization

Phytovolatilization is a remediation process which uses plants for the elimination of soil contaminants which are readily changed into vapours and so are released into the environment (Ali et al. 2013). Some plants such as tobacco plants have the good capability towards the accumulation of extremely toxic methylmercury from Hg-contaminated sites and convert it to the less toxic elemental Hg in a volatile form that releases through the leaves of plants to the environment (Mukhopadhyay and Maiti 2010). This conversion of the volatile form of contaminants during phytovolatilization is due to plants' metabolic potential in combination with microbes living inside the rhizosphere (Tak et al. 2013).

16.3.6 Phytodegradation

It is the degradation of toxic organic contaminants into less or non-hazardous chemicals through plant enzymes (Ali et al. 2013). Some enzymes such as nitroreductases and dehalogenases are plant-specific enzymes which are involved in the degradation of organic contaminants (Favas et al. 2014). There should be optimum conditions such as pH and temperature for these enzymes to cause effective contaminants degradation. The process of conversion of hazards toxic organic pollutants can be improved in the soil by applying rhizospheric microbes through the process of rhizodegradation (Ogunmayowa et al. 2015). This effective conversion occurs because the rhizospheric region of the plants contains a higher amount of nutrients released from the roots. These nutrients attract more bacteria to improve the conversion of the contaminants compared to the bulk soil which has little organic compounds and would contain less population of microbes (Babalola 2010). However, phytodegradation is mainly limited to the elimination of organic pollutants since heavy metals such as Cu, Ag, Hg and Au are non-biodegradable.

16.3.7 Rhizofiltration

Rhizofiltration is the process of removing toxic substances/chemicals or pollutants from groundwater through filtration using the roots of plants. This process depends on the mechanism of rhizospheric accumulation by plants (Figs. 16.1 and 16.2). Among plants, the terrestrial plants are more proficient for the rhizofiltration of toxic chemicals compared to other aquatic plants because the former plants have special natural solar-driven pumps to sequester particular elements from the nearby environment (López-Chuken 2012). The plants that have the potential of translocation and resistance towards high amounts of toxic heavy metals such as hyperaccumulators are highly fit for the process of rhizofiltration. Addition of PGPR to an e-waste contaminated site results in the decrease of heavy metal toxicity by raising the capability of plants to become free from heavy metal contamination and safe from environmental stress (Tak et al. 2013). However, there are certain limitations of the phytoremediation technology which include: reduce the rate at which remediation take place which normally becomes inadequate when there are a large number of pollutants at the contaminated area and also low accumulation and storage of pollutants in the plant materials (Ma et al. 2011).

16.4 Silencing Mechanisms Involved in Phytoremediation of E-waste Metals

Metals at excess level hinder the metabolic processes of plant and thus stop normal plant functioning. The harm to plants is caused in various processes such as the generation of reactive oxygen species (ROS) and/or the dislocation of amino acids through the formation of bonds between these heavy metals and –SH groups of the amino acids (Emamverdian et al. 2015; Krumova et al. 2016). ROS damages the cell membrane in a way that they hinder the functional groups of important molecules in the cell which results in abnormal functioning of enzymes and pigments. In addition to these, the heavy metals suppress photosynthesis, respiration and other enzymatic activities of the plant (Emamverdian et al. 2015; Pence et al. 2000). Among the metals there are those which can undergo oxido-reduction that is redox reaction and are classified as Redox-active metals such as Chromium (Cr), Copper (Cu), Manganese (Mn), Iron (Fe), Lead (Pb) and those which are non-redox active metals such as Cadmium (Cd), Nickle (Ni), Mercury (Hg), Zinc (Zn) and Aluminium

(Al) (Bücker-Neto et al. 2017). Redox-active metals directly produce ROS and thus generate oxidative stress in cells causing disruption to DNA structure and function, chloroplast and other pigments eventually destroying the cell (Singh et al. 2016). On the other hand, the non-redox active metals activate ROS-producing enzymes and restrain antioxidant system thus causing the damage (Emamverdian et al. 2015). In any case, ROS is generated, and the ultimate damage is caused by the excess oxidation of membranes and biomolecules. Plants protect itself from these metals by hindering the uptake through physical barriers such as thick cuticle, cell walls, and tissues such as trichomes. However, as we are studying the phytoremediation of these metals, we will discuss the system in which the plant modify/detoxify these metals for its own good and thus protect the environment too.

Plants protect itself from oxidation through its defence system primarily run by the secondary metabolites classified as phytochemicals. Once the metals surpass the barriers and enter the tissues and cells of the plant, different defence mechanisms in the cell are initiated to alleviate the damaging effects of the heavy metal (Silva and Matos 2016). One mechanism is the activation of antioxidant-generating enzymes such as superoxide dismutase, catalase and glutathione reductase and non-enzymatic antioxidants such as phenolic compounds, ascorbate, glutathione, alkaloids and tocopherols that remove the free radicals (Sharma et al. 2012). As an example, the detoxification of metals by phenolic compounds is detailed later. Apart from the role of secondary metabolites, one example of the defence process used by the plants is the production of the enzyme phytochelatin synthase that binds to heavy metals (Gupta et al. 2013). Phytochelatin synthase results in the formation of phytochelatins (PCs) which are short-chain thiol-rich repetitions of peptides of low-molecular-weight and are used as biomarkers for detecting the level of metals (Saba et al. 2013). Other than PCs, plant synthesize metallothioneins (MTs) which are also low-molecular-weight proteins rich in cysteine and having affinity for metals such as Cu, Zn, Cd and As (Guo et al. 2013). These are among the many different mechanisms used by the plants to silence heavy metals. However, our main aim is to focus on how secondary metabolites plant their role in remediation of toxic metals of e-waste.

16.5 Role of Plant Secondary Metabolites in the Phytoremediation of E-waste Metals

Controlling soil contamination such as that from e-waste through phytoremediation has been in the limelight since recent. Although there are many different schemes of the exact mechanism of contaminant removal through phytoremediation, the role of secondary metabolites in plants cannot be undermined. Secondary metabolites are phytochemicals produced as a product of secondary metabolism which is not directly involved in the growth and development of plants. Secondary metabolites generally play an important role in plant interactions and defence system. There is no sharp rule for classification of secondary metabolites; however, the phytochemicals which play a role in plant defence response are categorized into the following classes: alkaloids, flavonoids, glycosides, phenols and terpenoids broadly. Every category encompasses a variety of types of metabolites, every class playing a role in different defence processes of the plant (Bourgaud et al. 2001).

Many different studies have hinted towards the role of phytochemicals and antioxidants in the detoxification of heavy metals accumulated at the sites of e-waste disposal. For instance, results from studies by Agwaramgbo (2005) suggested that the phytoremediation capabilities of the plants tested had a direct correlation with the antioxidant potential. They further concluded that the plants having the highest amount of phytochemicals such as carotene and vitamin C were having the highest antioxidant potential and thus were able to remediate 2,4,6-trinitrotoluene effectively.

Similarly, another study has suggested that the concentration of important phytochemicals such as flavonoid, alkaloid, tannin, saponin and steroid was detected to be higher in leafy vegetables which has the high accumulation of metals such as Copper (Cu), nickel (Ni), zinc (Zn), lead (Pb), cadmium (Cd) and arsenic (As) (Ogoko 2015). In another study, Smeets et al. (2005) observed an increased accumulation of phenolic compounds in *Phaseolus vulgaris* when exposed to cadmium. This suggests the direct involvement of phenolic compounds in heavy metal detoxification. Similarly, studies on leaves of *Phyllanthus tenellus* sprayed with copper sulfate reported an elevated level of phenolic compounds compared to the control plants (Michalak 2006). Other plants such as wheat also induced an increased shift towards phenolic compound biosynthesis pathway (shikimate pathway) in response to nickel toxicity (Díaz et al. 2001).

Although the majority of the studies so far have reported the role of enzymes and other biomolecules in the process of quenching the toxic metals in the plant during phytoremediation, the role of secondary metabolites is emerging as an important area of consideration. There is a diverse variety of secondary metabolites in plants and thus there may be many different possible mechanisms for the detoxification of heavy metals in the plants. Taking the example of phenolic compounds for detoxification of lead, which is one of the harmful components accumulated via e-wastes, can be a case for the involvement of secondary metabolites in phytoremediation.

A general sketch of the silencing of metals through phenolic compounds is given (Figs. 16.3 and 16.4). Metals accumulated in soil from e-waste are toxic to plants. These metals upon uptake by the plants generate reactive oxygen species (ROS) in the plant. ROS because of quick and high rates of oxidation is lethal for the plant as it damages the membranes. The plants cope with ROS through different defence systems, the secondary metabolite system being notable. This system employs the antioxidant action of secondary metabolites such as phenolic compounds. Through this antioxidant action, the phenolic compounds chelate metals such as lead from e-waste. Phenolics possess hydroxyl and carboxyl groups, able to bind particularly metals, and thus act as a suitable chelating agent for the intoxicating lead. This may inactivate lead and thus suppress its ROS forming capacity. For instance, direct chelation, or binding to polyphenols, was observed with methanol extracts of

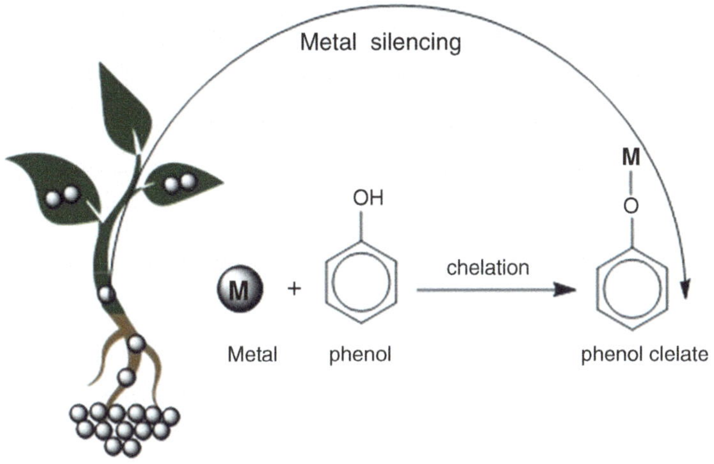

Fig. 16.3 The proposed mechanism of action of e-waste metal detoxification by plant phenolic compounds

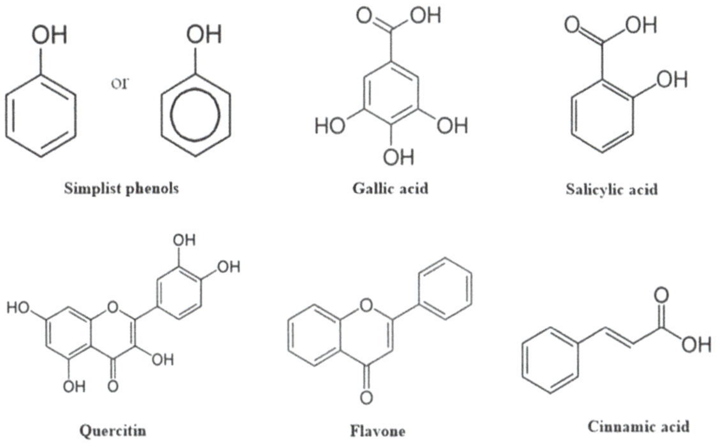

Fig. 16.4 Representative structures of important phenolic compounds found in plants. Data was taken from NCBI PubChem and structures were constructed through ChemDraw Ultra

rhizome polyphenols from *Nympheae* for Pb and other toxic metals such as chromium and mercury (Lavid et al. 2001).

Going further, other than chelation, the antioxidant ability of phenolic compounds is harnessed through another mechanism too. Metal ions decompose lipid hydroperoxide (lOOH) by the hemolytic cleavage of the O-O bond and give lipid alkoxyl radicals, which initiate free radical chain oxidation. Phenolic antioxidants inhibit lipid peroxidation by trapping the lipid alkoxyl radical and thus fight the damaging effect of the metals by unarming them (Dinis et al. 1994).

In short, the phytochemicals specifically secondary metabolites play an inevitable role in detoxifying metals accumulated via e-waste disposal. To be able to fully explore the metal remediation potential of plants, the response of secondary metabolism to metal accumulation needs to be understood. In addition to this, the transformation of plants in terms of metabolic engineering can be a nice area to enhance the generation of certain metabolites that play important role in metal detoxification in specific plants. A major benefit of this technology will be in the application of these plants to remove metals at the very entry point thus inhibiting its bioaccumulation and the chances of ultimate release in the environment through processes such as phytovolatilization.

16.6 Conclusions

Controlling and management of e-waste contamination through phytoremediation have got the global attraction in recent years. Though many different routes are involved in the mechanism of remediation of e-waste contaminants through phytoremediation, the role of secondary metabolites in plants cannot be undermined. Secondary metabolites generally play an important role in plant interactions and defence system in the overall process of phytoremediation of contaminants. To get more and more advantages from the phytoremediation technology, more research studies need to be done to explore the existing plants potential to more effectively remediate the heavy metals from the contaminated environments. Additionally, further studies are needed to find some more plants with promising characters such as plants with more aggressive nature towards metal extraction and accumulation. Besides, genetic engineering techniques can be used to engineer new plant varieties for efficient phytoremediation of heavy metal contaminants.

References

Agwaramgbo L (2005) Screening plants for antioxidant & phytoremediation potential. Dillard University, New Orleans, LA

Ali H, Khan E, Sajad MA (2013) Phytoremediation of heavy metals—concepts and applications. Chemosphere 91(7):869–881

Ali S, Jin R, Gill RA, Mwamba TM, Zhang N, Islam F, Ali S, Zhou W (2018) Beryllium stress-induced modifications in antioxidant machinery and plant ultrastructure in the seedlings of black and yellow seeded oilseed rape. Biomed Res Int. https://doi.org/10.1155/2018/1615968

Babalola OO (2010) Beneficial bacteria of agricultural importance. Biotechnol Lett 32 (11):1559–1570

Babu BR, Parande AK, Basha CA (2007) Electrical and electronic waste: a global environmental problem. Waste Manag Res 25(4):307–318

Bourgaud F, Gravot A, Milesi S, Gontier E (2001) Production of plant secondary metabolites: a historical perspective. Plant Sci 161(5):839–851

Brandl H, Bosshard R, Wegmann M (2001) Computer-munching microbes: metal leaching from electronic scrap by bacteria and fungi. Hydrometallurgy 59(2–3):319–326

Bücker-Neto L, Paiva ALS, Machado RD, Arenhart RA, Margis-Pinheiro M (2017) Interactions between plant hormones and heavy metals responses. Genet Mol Biol 40(1):373–386

Caruso JA, Heitkemper DT, B'Hymer C (2001) An evaluation of extraction techniques for arsenic species from freeze-dried apple samples. Analyst 126(2):136–140

Chen G, Liu X, Brookes PC, Xu J (2015) Opportunities for phytoremediation and bioindication of arsenic contaminated water using a submerged aquatic plant: *Vallisneria natans* (Lour.) Hara. Int J Phytoremediation 17(3):249–255

Cho-Ruk K, Kurukote J, Supprung P, Vetayasuporn S (2006) Perennial plants in the phytoremediation of lead-contaminated soils. Biotechnology 5(1):1–4

Choudhary DK, Varma A (2016) Microbial-mediated induced systemic resistance in plants. Springer, New York

Coscione AR, Berton RS (2009) Barium extraction potential by mustard, sunflower and castor bean. Sci Agric 66(1):59–63

Dhillon S, Dhillon K (2009) Phytoremediation of selenium-contaminated soils: the efficiency of different cropping systems. Soil Use Manag 25(4):441–453

Díaz J, Bernal A, Pomar F, Merino F (2001) Induction of shikimate dehydrogenase and peroxidase in pepper (*Capsicum annuum* L.) seedlings in response to copper stress and its relation to lignification. Plant Sci 161(1):179–188

Dinis TC, Madeira VM, Almeida LM (1994) Action of phenolic derivatives (acetaminophen, salicylate, and 5-aminosalicylate) as inhibitors of membrane lipid peroxidation and as peroxyl radical scavengers. Arch Biochem Biophys 315(1):161–169

Donot F, Fontana A, Baccou J, Schorr-Galindo S (2012) Microbial exopolysaccharides: main examples of synthesis, excretion, genetics and extraction. Carbohydr Polym 87(2):951–962

Emamverdian A, Ding Y, Mokhberdoran F, Xie Y (2015) Heavy metal stress and some mechanisms of plant defense response. Sci World J. https://doi.org/10.1155/2015/756120

Favas PJ, Pratas J, Varun M, D'Souza R, Paul MS (2014) Phytoremediation of soils contaminated with metals and metalloids at mining areas: potential of native flora. In: Environmental risk assessment of soil contamination. InTech

Feng R, Wei C, Tu S, Tang S, Wu F (2011) Simultaneous hyperaccumulation of arsenic and antimony in Cretan brake fern: evidence of plant uptake and subcellular distributions. Microchem J 97(1):38–43

Feng R, Wang X, Wei C, Tu S (2015) The accumulation and subcellular distribution of arsenic and antimony in four fern plants. Int J Phytoremediation 17(4):348–354

Gaiero JR, McCall CA, Thompson KA, Day NJ, Best AS, Dunfield KE (2013) Inside the root microbiome: bacterial root endophytes and plant growth promotion. Am J Bot 100 (9):1738–1750

Gardea-Torresdey J, Tiemann K, Polette L, Chianelli R, Pingitore N, Mackay W (1999) Phytoremediation of heavy metals with creosote plants. Google Patents

Guo J, Xu L, Su Y, Wang H, Gao S, Xu J, Que Y (2013) ScMT2-1-3, a metallothionein gene of sugarcane, plays an important role in the regulation of heavy metal tolerance/accumulation. Biomed Res Int 2013. https://doi.org/10.1155/2013/904769

Gupta D, Huang H, Corpas F (2013) Lead tolerance in plants: strategies for phytoremediation. Environ Sci Pollut Res 20(4):2150–2161

Hajiani NJ, Ghaderian SM, Karimi N, Schat H (2015) A comparative study of antimony accumulation in plants growing in two mining areas in Iran, Moghanlo, and Patyar. Environ Sci Pollut Res 22(21):16542–16553

Heacock M, Kelly CB, Asante KA, Birnbaum LS, Bergman ÅL, Bruné M-N, Buka I, Carpenter DO, Chen A, Huo X (2016) E-waste and harm to vulnerable populations: a growing global problem. Environ Health Perspect 124(5):550

Heitkemper DT, Vela NP, Stewart KR, Westphal CS (2001) Determination of total and speciated arsenic in rice by ion chromatography and inductively coupled plasma mass spectrometry. J Anal At Spectrom 16(4):299–306
Helgesen H, Larsen EH (1998) Bioavailability and speciation of arsenic in carrots grown in contaminated soil. Analyst 123(5):791–796
Intawongse M, Dean JR (2006) Uptake of heavy metals by vegetable plants grown on contaminated soil and their bioavailability in the human gastrointestinal tract. Food Addit Contam 23 (1):36–48
Islam MS, Ueno Y, Sikder MT, Kurasaki M (2013) Phytofiltration of arsenic and cadmium from the water environment using *Micranthemum umbrosum* (JF Gmel) SF Blake as a hyperaccumulator. Int J Phytoremediation 15(10):1010–1021
Jabeen R, Ahmad A, Iqbal M (2009) Phytoremediation of heavy metals: physiological and molecular mechanisms. Bot Rev 75(4):339–364
Jutsz AM, Gnida A (2015) Mechanisms of stress avoidance and tolerance by plants used in phytoremediation of heavy metals. Arch Environ Protection 41(4):104–114
Kim YJ, Kim JH, Lee CE, Mok YG, Choi JS, Shin HS, Hwang S (2006) Expression of yeast transcriptional activator MSN1 promotes accumulation of chromium and sulfur by enhancing sulfate transporter level in plants. FEBS Lett 580(1):206–210
Kofoworola O (2007) Recovery and recycling practices in municipal solid waste management in Lagos, Nigeria. Waste Manag 27(9):1139–1143
Krumova E, Kostadinova N, Miteva-Staleva J, Gryshko V, Angelova M (2016) Cellular response to Cu-and Zn-induced oxidative stress in *Aspergillus fumigatus* isolated from polluted soils in Bulgaria. CLEAN Soil Air Water 44(6):657–666
Lamb DT, Matanitobua VP, Palanisami T, Megharaj M, Naidu R (2013) Bioavailability of barium to plants and invertebrates in soils contaminated by barite. Environ Sci Technol 47 (9):4670–4676
Laurent A, Bakas I, Clavreul J, Bernstad A, Niero M, Gentil E, Hauschild MZ, Christensen TH (2014) Review of LCA studies of solid waste management systems–part I: lessons learned and perspectives. Waste Manag 34(3):573–588
Lavid N, Schwartz A, Yarden O, Tel-Or E (2001) The involvement of polyphenols and peroxidase activities in heavy-metal accumulation by epidermal glands of the waterlily (Nymphaeaceae). Planta 212(3):323–331
Leung AO, Duzgoren-Aydin NS, Cheung K, Wong MH (2008) Heavy metals concentrations of surface dust from e-waste recycling and its human health implications in southeast China. Environ Sci Technol 42(7):2674–2680
Li Y, Richardson JB, Bricka RM, Niu X, Yang H, Li L, Jimenez A (2009) Leaching of heavy metals from E-waste in simulated landfill columns. Waste Manag 29(7):2147–2150
Lin J, Jiang W, Liu D (2003) Accumulation of copper by roots, hypocotyls, cotyledons and leaves of sunflower (*Helianthus annuus* L.). Bioresour Technol 86(2):151–155
Liu W-X, Shen L-F, Liu J-W, Wang Y-W, Li S-R (2007) Uptake of toxic heavy metals by rice (*Oryza sativa* L.) cultivated in the agricultural soil near Zhengzhou City, People's Republic of China. Bull Environ Contam Toxicol 79(2):209–213
Lone MI, He Z-l, Stoffella PJ, Yang X-E (2008) Phytoremediation of heavy metal polluted soils and water: progresses and perspectives. J Zhejiang Univ Sci B 9(3):210–220
López-Chuken UJ (2012) Hydroponics and environmental clean-up. In: Hydroponics—a standard methodology for plant biological researches. InTech
Ma Y, Prasad M, Rajkumar M, Freitas H (2011) Plant growth promoting rhizobacteria and endophytes accelerate phytoremediation of metalliferous soils. Biotechnol Adv 29(2):248–258
Michalak A (2006) Phenolic compounds and their antioxidant activity in plants growing under heavy metal stress. Pol J Environ Stud 15(4):523–530
Montes-Holguin MO, Peralta-Videa JR, Meitzner G, Martinez-Martinez A, de la Rosa G, Castillo-Michel HA, Gardea-Torresdey JL (2006) Biochemical and spectroscopic studies of the response

of *Convolvulus arvensis* L. to chromium (III) and chromium (VI) stress. Environ Toxicol Chem 25(1):220–226
Mukhopadhyay S, Maiti SK (2010) Phytoremediation of metal mine waste. Appl Ecol Environ Res 8(3):207–222
Müller K, Daus B, Mattusch J, Vetterlein D, Merbach I, Wennrich R (2013) Impact of arsenic on uptake and bio-accumulation of antimony by arsenic hyperaccumulator *Pteris vittata*. Environ Pollut 174:128–133
Muranyi A, Ködöböcz L (2008) Heavy metal uptake by plants in different phytoremediation treatments. Cereal Res Commun 36:387–390
Muszynska E, Hanus-Fajerska E (2015) Why are heavy metal hyperaccumulating plants so amazing? BioTechnologia 96(4). https://doi.org/10.5114/bta.2015.57730
Muszyńska B, Rojowski J, Dobosz K, Opoka W (2015) Biological and physico-chemical properties of thallium. Medicina Internacia Revuo 26(105):180–185
Ogoko E (2015) Accumulation of heavy metal in soil and their transfer to leafy vegetables with phytoremediation potential. Am J Chem 5(5):125–131
Ogunmayowa OT, Dzantor K, Adeleke E (2015) Coupling bio/phytoremediation with switchgrass to biofuel feedstock production in mixed-contaminant soils. PhD Thesis
Ogunseitan OA, Schoenung JM, Saphores J-DM, Shapiro AA (2009) The electronics revolution: from e-wonderland to e-wasteland. Science 326(5953):670–671
Parker DR, Feist LJ, Varvel TW, Thomason DN, Zhang Y (2003) Selenium phytoremediation potential of *Stanleya pinnata*. Plant Soil 249(1):157–165
Pence NS, Larsen PB, Ebbs SD, Letham DL, Lasat MM, Garvin DF, Eide D, Kochian LV (2000) The molecular physiology of heavy metal transport in the Zn/Cd hyperaccumulator Thlaspi caerulescens. Proc Natl Acad Sci 97(9):4956–4960
Peralta-Videa JR, Lopez ML, Narayan M, Saupe G, Gardea-Torresdey J (2009) The biochemistry of environmental heavy metal uptake by plants: implications for the food chain. Int J Biochem Cell Biol 41(8–9):1665–1677
Petruzzelli G, Pedron F, Rosellini I, Tassi E, Gorini F, Barbafieri M (2012) Integrating bioremediation and phytoremediation to clean up polychlorinated biphenyls contaminated soils. World Acad Sci Eng Technol 6(6):357–360
Rajkumar M, Sandhya S, Prasad M, Freitas H (2012) Perspectives of plant-associated microbes in heavy metal phytoremediation. Biotechnol Adv 30(6):1562–1574
Rascio N, Navari-Izzo F (2011) Heavy metal hyperaccumulating plants: how and why do they do it? And what makes them so interesting? Plant Sci 180(2):169–181
Robinson BH (2009) E-waste: an assessment of global production and environmental impacts. Sci Total Environ 408(2):183–191
Saba H, Jyoti P, Neha S (2013) Mycorrhizae and phytochelators as remedy in heavy metal contaminated land remediation. Intl Res J Environ Sci 2(1):74–78
Sampaio Junior J, Amaral N, Zonta E, Magalhães MO (2015) Barium and sodium in sunflower plants cultivated in soil treated with wastes of drilling of oil well. Rev Bras Eng Agríc Ambient 19(11):1100–1106
Sas-Nowosielska A, Galimska-Stypa R, Kucharski R, Zielonka U, Małkowski E, Gray L (2008) Remediation aspect of microbial changes of plant rhizosphere in mercury contaminated soil. Environ Monit Assess 137(1–3):101–109
Schiavon M, Pilon-Smits EA (2017) Selenium biofortification and phytoremediation phytotechnologies: a review. J Environ Qual 46(1):10–19
Sharma P, Jha AB, Dubey RS, Pessarakli M (2012) Reactive oxygen species, oxidative damage, and antioxidative defense mechanism in plants under stressful conditions. J Bot 2012. https://doi.org/10.1155/2012/217037
Silva P, Matos M (2016) Assessment of the impact of aluminum on germination, early growth and free proline content in *Lactuca sativa* L. Ecotoxicol Environ Saf 131:151–156
Singh A, Prasad SM, Singh S, Singh M (2016) Phytoremediation potential of weed plants' oxidative biomarker and antioxidant responses. Chem Ecol 32(7):684–706

Sinha S (2007) Downside of the digital revolution. Toxics Link 28
Skinner K, Wright N, Porter-Goff E (2007) Mercury uptake and accumulation by four species of aquatic plants. Environ Pollut 145(1):234–237
Smeets K, Cuypers A, Lambrechts A, Semane B, Hoet P, Van Laere A, Vangronsveld J (2005) Induction of oxidative stress and antioxidative mechanisms in *Phaseolus vulgaris* after Cd application. Plant Physiol Biochem 43(5):437–444
Srivastava M, Ma LQ, Cotruvo JA (2005) Uptake and distribution of selenium in different fern species. Int J Phytoremediation 7(1):33–42
Subhashini V, Swamy A (2014) Phytoremediation of metal (Pb, Ni, Zn, Cd and Cr) contaminated soils using *Canna indica*. Curr World Environ 9(3):780
Tak HI, Ahmad F, Babalola OO (2013) Advances in the application of plant growth-promoting rhizobacteria in phytoremediation of heavy metals. In: Reviews of environmental contamination and toxicology, vol 223. Springer, New York, pp 33–52
Tangahu BV, Abdullah S, Rozaimah S, Basri H, Idris M, Anuar N, Mukhlisin M (2011) A review on heavy metals (As, Pb, and Hg) uptake by plants through phytoremediation. Int J Chem Eng 2011. https://doi.org/10.1155/2011/939161
Tsydenova O, Bengtsson M (2011) Chemical hazards associated with treatment of waste electrical and electronic equipment. Waste Manag 31(1):45–58
Vamerali T, Bandiera M, Coletto L, Zanetti F, Dickinson NM, Mosca G (2009) Phytoremediation trials on metal-and arsenic-contaminated pyrite wastes (Torviscosa, Italy). Environ Pollut 157 (3):887–894
Van der Ent A, Baker AJ, Reeves RD, Pollard AJ, Schat H (2013) Hyperaccumulators of metal and metalloid trace elements: facts and fiction. Plant Soil 362(1–2):319–334
Van Ginneken L, Meers E, Guisson R, Ruttens A, Elst K, Tack FM, Vangronsveld J, Diels L, Dejonghe W (2007) Phytoremediation for heavy metal-contaminated soils combined with bioenergy production. J Environ Eng Landsc Manag 15(4):227–236
Watson C, Pulford I, Riddell-Black D (2003) Screening of willow species for resistance to heavy metals: comparison of performance in a hydroponics system and field trials. Int J Phytoremediation 5(4):351–365
Xiezhi Y (2008) Assessment and bioremediation of soils contaminated by uncontrolled recycling of electronic-waste at Guiyu, SE China. Hong Kong Baptist University, Hong Kong
Yu J, Williams E, Ju M, Shao C (2010) Managing e-waste in China: policies, pilot projects and alternative approaches. Resour Conserv Recycl 54(11):991–999
Zhang X, Lin A-J, Zhao F-J, Xu G-Z, Duan G-L, Zhu Y-G (2008) Arsenic accumulation by the aquatic fern Azolla: comparison of arsenate uptake, speciation and efflux by *A. caroliniana* and *A. filiculoides*. Environ Pollut 156(3):1149–1155

Chapter 17
Aspects of E-waste Management in India

Rajiv Ganguly

Abstract Increased urbanization and globalization have led to marked increase in use of consumer products with the use of electronics being the most prevalent. Wherein in one aspect it has led to global connectivity and social inclusivity, adversely it has led to generation of e-waste in massive proportions. In principle, e-waste comprises of precious heavy metals which are highly toxic in nature and thereby have severe human and environmental effects if disposed of unscientifically. The chapter deals with the different aspects of e-waste generation in India, its potential environmental health and environmental impacts, and associated legislative measures and discusses sustainable methods for effective disposal of generated e-waste including the commercial viability of such waste.

Keywords E-waste management · Environmental effects · E-waste management handling rules

17.1 Introduction

Increased urbanization and globalization have led to rapid urban development leading to rise of use of electronic goods and appliances in our daily lives. These electronic goods and appliances enhance our living standards but have a definitive shelf life. The problem arises when it comes to the disposal of such electronics after the completion of shelf lives of such electronics. This is primarily because such electronics comprise of precious and toxic metals which have severe human health and environmental effects if they are not disposed of carefully. The increased generation of such electronic wastes (e-waste) is a rising concern, particularly in the context of presence of both toxic and precious metals (Babu et al. 2007). The production of e-waste is expected to rise continually with an expected production of

R. Ganguly (✉)
Department of Civil Engineering, Jaypee University of Information Technology, District Solan, Waknaghat, Himachal Pradesh, India
e-mail: rajiv.ganguly@juit.ac.in

M. Z. Hashmi, A. Varma (eds.), *Electronic Waste Pollution*, Soil Biology 57,
https://doi.org/10.1007/978-3-030-26615-8_17

about 72 million tons in 2017 (Garlapati 2016) with China and the United States being the major producers (Nisa 2014). Recent reports (Bhutta et al. 2011) suggest that annual generation of electronic waste worldwide varies from 20 to 50 million tons having the potential to cause significant human health and environmental damage. In the context of developing countries, earlier reports suggest that such electronic waste would comprise 1% of the total municipal solid waste generated and will rise to 2% by 2010 (Chaturvedi et al. 2007) with the total generation between the periods 2007 to 2011 being estimated at 2.5 metric tons (Dwivedy and Mittal 2010) (Fig. 17.1). Recent studies conducted by Associated Chambers of Commerce of India (ASSOCHAM) reveal that the growth rate of e-waste in India is about 25% with an expected generation quantity of 15 metric tons by 2015 (Nisa 2014). From the existing statistics, it is quite evident that the usage of such products will continue to increase, and the fact that electronic industry is one of the fastest-expanding industries in present context leads to more manufacture and generation of electronic waste (Garlapati 2016). Figure 17.1 shows the increasing trend of generation of e-waste in India.

The major drawback associated with the safe disposal and recycling of such electronic wastes is its harmful human health and environmental impacts. For example, older televisions and computer monitors have toxic metals like lead, cadmium, and mercury as their major constituents, while other electronic circuitry have nickel, beryllium, and zinc as primary constituents (European Union 2012). Hence, it is imperative to have an effective waste management system for such e-wastes which pose significant challenges to both government and common public as it comprises both precious and toxic elements. In this context, special e-waste handling rules have been formulated for safe handling practices of such wastes.

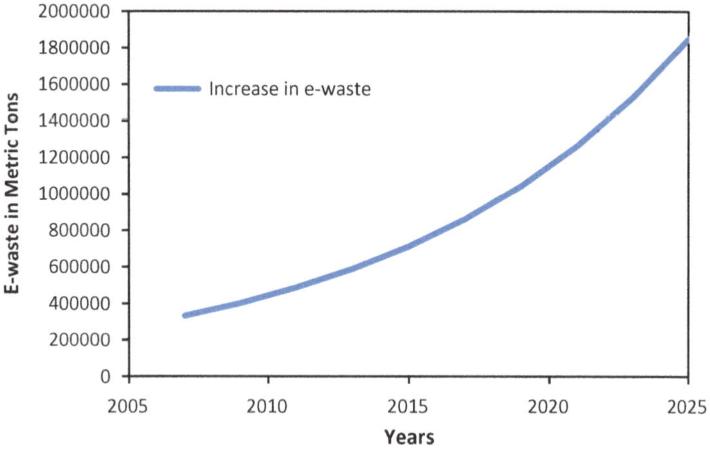

Fig. 17.1 Increasing trend of generation of e-waste in India

17.2 Definition and Classification of E-waste

E-waste or waste electrical and electronic equipment (WEEE) is a terminology that is used to define all electronic and electrical products achieving the end of shelf life. It also incorporates discarded appliances or electronic goods that are rejected during the manufacturing, refurbishing, and the repair process (Ganguly 2016). In practice, e-waste includes a broad denomination with both personal and household electronic equipments varying from television, mobiles, refrigerator, microwave, juicer grinder, mixer, and personal computer all falling within its purview (Puckett et al. 2002). Since the term e-waste encompasses a broad category, certain definitive bodies and researchers have given the definition of e-waste, the most important of which are highlighted in Table 17.1.

As per the definition of electronic waste by EU WEEE directives 2002/96/EC and 2012/19/EU, there exist ten different classifications of e-waste which have been summarized in Table 17.2 along with their respective percentages.

The constituents of e-waste is widely varied as they not only include metals (both ferrous and non-ferrous) but also ceramics, plastics, and different other combinations (APME 2004). The specifications of the electronics depend on the different constituents present in the manufactured product being categorized as both hazardous and non-hazardous waste. Classification of the metal components is primarily done on the basis of presence of ferrous and non-ferrous metals, whereas other components include plastics, ceramics, glass woods, and other inert and non-reactive materials (Garlapati 2016). In the category of non-ferrous metals, metals like aluminium, copper, silver, platinum, etc. are considered precious metals and often considered for recovery. Presence of other metals like mercury, lead, arsenic, cadmium, selenium, and chromium classifies the e-waste as hazardous in nature (Garlapati 2016).

Table 17.1 Definition of E-waste as per different conventions

EU WEEE Directive (EU 2002)	'Electrical or electronic equipment which is waste including all components, sub-assemblies and consumables, which are part of the product at the time of discarding.' Directive 75/442/EEC, Article 1 (a) defines 'waste' as 'any substance or object which the holder disposes of or is required to dispose of pursuant to the provisions of national law in force'
Basel Action Network	'E-waste encompasses a broad and growing range of electronic devices ranging from large household devices such as refrigerators, air-conditioners, cell phones, personal stereos, and consumer electronics to computers which have been discarded by their users'
OECD (2001)	'Any appliance using an electric power supply that has reached its end of life'
Sinha (2008)	'An electrically powered appliance that no longer satisfies the current owner for its original purpose'
Gregory (2009)	E-waste refers to 'the reverse supply chain which collects products no longer desired by a given consumer and refurbishes for other consumers recycles or otherwise processes wastes'

Table 17.2 Definition of e-waste as per different conventions

Sr. no.	Waste category	Label	Contribution (%)
1.	Major household gadgets	Large HH	42.1
2.	Minor household gadgets	Small HH	4.7
3.	IT and telecommunication gadgets	ICT	33.9
4.	User gadgets	CE	13.7
5.	Illumination gadgets	Lighting	1.4
6.	Electrical and electronics appliances	E&E tools	1.4
7.	Toys, leisure, and sports gadgets	Toys	0.2
8.	Medical device	Medical equipment	1.9
9.	Monitoring and control instruments	M&C	0.1
10.	Automatic dispensers	Dispensers	0.7

17.3 E-waste Generation in Global and Indian Context

Recent studies mention that the proportion of e-waste generated from developing countries will double within the next decade. Further by 2030, it is expected that developed and developing countries will have obsolete computer wastes of 200–300 and 400–700 million, respectively (Duan et al. 2015). The rapid increase in generation of e-waste can be attributed to the discontinuance of the electronic goods. Primarily in developed countries, the lifespan of such electric products is only 2 years after which therein they are discarded or are then exported to developing countries as there is a demand for such used electronic goods (Ganguly 2016). In the context of production of such voluminous quantities of e-waste, it is imperative to have proper regulations for handling of such wastes. As per United Nations Environmental Program (UNEP), developing countries like China, India, Mexico, and Brazil would encounter the maxim of the environmental and health effects of such wastes.

The problem is more compounded for developing countries wherein there exist almost no governmental records on the production of e-waste in the country. Available data from studies conducted by Non-Governmental organizations (NGO) and other sources are reported. An earlier study conducted by the Comptroller and Auditor General (CAG) of India suggested the presence of 1.5 MT of plastic waste, 7.5 MT of industrial wastes, 1.7 MT of medical waste, and about 4 lakh tones of electronic waste being generated in the country (Agarwal 2010).

Another primary concern is the import of used electronics goods from developed countries to developing countries for usage and dumping which is of serious concern. There exists no official figures for import of such wastes, and they are often passed into developing nations on basis of false pretences like goodwill, reusable category, and charity purposes (Garlapati 2016). Such increased burden of e-waste severely affects the local population and the surrounding environment.

17.4 Health and Environmental Effects of E-waste

The hazardous components of the e-waste generated have the propensity to cause severe environmental damage. Primary emissions involve release of PCB and heavy metals (selenium, cadmium, chromium, etc.), whereas dioxins and furans are generated from secondary sources. Secondary emissions arise from untreated or improper treatment of the e-waste, whereas tertiary treatment involves emissions from those hazardous chemicals involved in treatment of e-waste (Gupta 2014). In simpler terms, electronic appliances can be classified into white category (household electronics), brown category (televisions, camera), and grey category (PC and other associated subsidiaries), and earlier studies have reported that the grey category is more hazardous than the other two categories (Sheng and Etsell 2007; Jomova et al. 2011). The practice of unscientific and lay methods in handling leads to hazardous emissions from such wastes.

Human health effects associated with exposure to emissions from such e-wastes lead to adverse functioning of reproductive system, thyroid, nervous system, lung functioning, and stunted growth (Grant et al. 2013). Workers associated with handling of e-waste without protective equipments can develop further ailments like skin cancer, tumours, anaemia, and related diseases (Janagam and Jeymani 2011). In this context, it is important to mention that exposure to e-waste is highly detrimental to human health. In particular, the effects are more profound for developing countries like India wherein there is a lack of proper implementation of handling of such e-wastes. The worst affected population groups are those living nearby the landfill dumpsites and the personnel involved in the handling of these wastes (Ganguly 2016).

17.5 Existing Practices of Handling of E-waste

Handling of e-waste is highly complex and requires special expertise primarily due to its constituents which are toxic in nature to both the environment and the human health. Generated e-wastes from LCDs, CRTs, capacitors, electrical circuitry, copier, and cartridges are highly toxic due to presence of heavy metals like mercury, selenium, lead, and chromium. Further, incineration (secondary emission source) of circuitry boards, plastic casings, and PVC materials leads to generation of dioxins and furans.

In developing countries, most of the e-wastes are generally disposed of in open landfills. For example, land filling of such burnt wastes can lead to severe groundwater pollution if hydrographical and soil conditions are favourable in nature. Cadmium (from mobile phones) alone has the potential for polluting 600 m^3 of water and has the potential for long-lasting effects (Garlapati 2016). Further, land filling of sites with e-waste can produce highly toxic nature of leachate which can severely contaminate the groundwater (Rana et al. 2018). Additionally, acid treatment of dissolving computer chips can cause acidification of soil and the groundwater (Milovantseva and Saphores 2013).

Similarly, incineration of such wastes leads to air pollution. For example, flammable components of electrical equipment when burnt (incinerated in dump sites) releases brominated flame retardants causing extensive damage to human health and surrounding environment. Hence, continual methods of open dumping and incineration have the potential to cause severe and long-lasting health and environmental impacts (Osibanjo and Nnorom 2007; Chen et al. 2011; Milovantseva and Saphores 2013; Rana et al. 2018).

17.6 Strategies for Management of E-waste

17.6.1 Responsibility and Role of Industries Including Sustainable Design Philosophy

It is given that management of e-waste is imperative for electronic goods and other appliances stored in warehouses which account for almost 75–80% (Garlapati 2016). It is imperative that if the generation of e-waste is controlled at the manufacturing phase, it will be highly beneficial. This can be achieved by reducing the volume of hazardous waste in the manufacturing process and also reducing the stock quantity of such process will reduce the generation of e-waste. Further the implementation of reuse and recycle policy could significantly reduce associated disposal costs and raw material procurement and also boost resale of such electronics. In principle, both on- and off-site recovery of wastes can be carried out along with the possibility of exploring inter-industry exchange (Garlapati 2016). These may involve inventory management, post-production, modification, and salvaging of such wastes (Freeman 1997).

The concept of managing inter-industry exchange of such wastes can be easily maintained by an updated inventory management system. A proper inventory system will regulate the use of such hazardous materials in the industry and will also help in alleviating the generation of excess wastes. This is primarily achieved by reducing the use of hazardous and surplus raw materials, verification of material purchase, and evaluation and regulation of control procedures (Garlapati 2016). In general, the first step involves using a detailed methodology for identifying the process and than followed by material procurement. Alternatively, a procedure of ordering required volume of materials may be implemented followed by detailed inventorisation to reduce purchase of any surplus hazardous material (Gaidajis et al. 2010).

Further to the above, reduction in e-waste generation can be controlled by using accomplished and natural materials in the manufacturing process and by altering the production process thereby conceptualizing new product design. This can be achieved by material alteration and amendment in manufacturing process, better working measures, and more endurable end products. For example, plant and bio-based plastics and plant-based polymers may be used in the design system instead of those made from petrochemicals. Similarly, natural (bio-based) inks and toners glue could be used more periodically. Proper implementation of abovementioned schemes can help in optimizing the system in utilization of such hazardous materials. Finally, the personnel involved in such activities should be provided training and made aware of process of optimizations, safety guidelines, etc. in handling of such wastes as they play a significant role in this aspect (Gaidajis et al. 2010).

Similarly, the use of natural and non-hazardous materials for production of electronic goods can significantly reduce the generation of e-waste, thereby reducing disposal costs in the future (Gaidajis et al. 2010). This can be successfully carried out by source segregation and waste concentration methods. The former involves the separation of metal components and application of treatment methods for recovery of such precious metals and is one of the most economical options. Treatment techniques generally involve reverse osmosis, bioleaching, ultrafiltration, and other pressure-based filtration techniques (Dasgupta et al. 2017). This is also termed as 'reverse production' (Ganguly 2016). Possible recovery of such metals leads to a closed-loop system minimizing the potential health and environmental impacts (Dasgupta et al. 2017). Labelling on packaging should be done to make aware of potential environmental hazards by the industries.

17.6.2 Extended Producer Responsibility (EPR) and Producer Responsibility Organization (PRO)

The EPR is the newer concept wherein the control of e-waste generation focuses primarily on the system instead of the production facilities, i.e., the manufacturer takes the responsibility for post-consumer stage of the product's life cycle, including its final disposal. In this aspect, the manufacturer bears the full responsibility of the manufacture, performance, and the care needed after the shelf life of the equipment. This in turn induces the industry in promoting and using sustainable product design and technology thereby preventing pollution. The main aim of EPR is promoting sustainable development via environmentally friendly product development and recovery.

EPR has been successfully implemented in developing countries but is still in its infancy in developing countries. For example, Germany was the first country to implement EPR policies in spirit followed by other European countries in facilitating legislative policies for implementation of EPR. In the United States, it is the state's prerogative for implementing EPR wherein the state directives dictate policies like disposal fees and refund system and compulsory take-back of old electronics. Similar policies are also implemented in Japan, but the regulations allow charging the consumers for the benefit.

17.6.3 Responsibility of Citizens

Another important aspect in controlling e-waste generation is proper management of such electronic products used by the consumers. In this context, consumers should be advised to use the electronic products till the end of its shelf life with proper servicing schedules to increase their life spans. Further old and unwanted electronic products should be given away for refurbishing and reusable purposes. Further, the consumers should be made aware of environmentally friendly products including their maintenance, recycling, and disposal methods.

17.6.4 Responsibility of Government

It is the government's prerogative in setting up regulatory bodies for integrating policy decisions of different authorities for management of hazardous wastes. In principle, the government should lay down regulations, laws, administration, and penalties for management of such wastes. The regulations should entrust the agency in monitoring the activities of the government. It should provide information on raw materials procured and used by the different manufacturers and supervisory authority on checking the roster of the industry (Skinner et al. 2011).

Further, relevant information on the toxicity of the materials and the results of their testing for adverse health and environmental effects should be made available to the regulatory authority. Policies related to risk management should be clearly mentioned for the manufacture, processing, distribution, use, and disposal of electronic wastes. The government should promote the concept of reuse of e-waste and discuss potential business avenues for the same. Similarly, they should impart education to the consumers on the environmental and health effects of such wastes and suggest precautionary measures for such cases (Skinner et al. 2011). Finally, the government should use the endeavours of NGOs in managing such voluminous quantities of waste and can explore partnerships with them for promoting reuse and recycling of the wastes.

17.7 E-waste Legislation in India

In the context of management of health effects of such wastes, the government has highlighted the latest E-waste Management and Handling Rules, 2011. It was approved in May 2011 and came into effect in May 2012. A year's time was given to the industries to make them aware of the rules and policies so as to help them abide and consider the products they manufacture. Some of the important excerpts from the rule are as follows: (Garlapati 2016)

- As per Schedule-I, regulations will be applicable to producer, purchaser, collection centres, and recycling of the wastes. The CPCB and SPCB will be the main regulatory authorities.
- The Battery Management and Handling Act passed in 2001 will encompass all industries producing lead batteries, whereas the Enterprises Development Act will be responsible for handling all micro- and small-scale industries. Similarly, the Atomic Energy Act of 1962 will cover all radioactive wastes.
- Schedule-I will be applicable to all manufacturing equipment's and raw products involved in the generation of the electronics and the electrical appliances and will be functional part of the product till the shelf life of the goods.
- The regulations also impose stringent measures on the hazardous materials used in the production and enables reduced use of such materials in electronic goods and appliances. This suggests that the manufacturers have to abide by the specified ranges of heavy metals used for the manufacture of such goods.
- The MOEF/CPCB has the final authority in deciding the use of any hazardous waste material which does not conform to Schedule-I components after comparison.
- The manufacturing unit is beyond the scope of Schedule-I components; the wastes generated from such manufacturing processes have to be transferred to a recycling facility and ensured that complete recycling takes place.

In the context of management of health effects of such wastes, the government has highlighted the 'E-waste Management and Handling Rules, 2011' which was approved in May 2011. The potential drawbacks of the legislation have been discussed in detail elsewhere (Ganguly 2016). However, to summarize, one of the major potential flaws of these regulations is subdivision of enforcement policy between state and the centre. This could possibly induce 'spillover' from those states which diligently implement the regulations to other states which might be slightly lax in imposition. This would also involve additional administrative costs to the states in keeping detailed roster about the details of formal recycling units within their jurisdiction (Skinner et al. 2011). The regulations also discuss about bringing the informal sector within the purview of the governmental monitoring but do not discuss in detail how the process will be implemented and the fate of the workers associated with the informal sector. Further, the regulations also fail to describe how awareness about the hazards of e-waste would be distributed to the consumers. This is particularly important as manufacturers do not describe the potential pitfalls of electronic wastes on the packaging as a result of which consumers are unaware about safe disposal of these products after their life span.

17.8 Recycling of E-waste

Recycling of hazardous wastes is one of the most economical methods for preventing environmental problems associated with them and is even commercially viable. However, setting up of recycling of such wastes involves detailed logistical planning and requirement as it involves collection, transportation, segregation, treatment, recovery, and disposal of such wastes. In principle, such logistical functioning should be supervised by the pertaining regulatory authority and incentives should be provided for improving the functioning of the entire system. As mentioned earlier, the NGOs at national and state levels may be involved for smooth functioning of such systems (Garlapati 2016).

Further, skilled personnel should be involved to ensure environmentally friendly practices of recycling as they help in separating out toxic materials and recover materials from the complex electronic system. Recycling facilities should have control equipment for capturing air emissions released from the toxic pollutants during the recovery process. Proper control equipment and safety measures should be provided to the personnel involved during the recovery process. In this context, methods like bioleaching which involves use of microbes for recovery of precious metals are more favourable as it prevents exposure to personnel and is an environmentally friendly method.

This is particularly important in Indian context as majority of the recycling operations are carried out by the unorganized (informal) sector. This sector is responsible for about 90–95% of all recycling activities in India with an efficiency of about 30% in recovery of precious metals. The sector mainly comprises of unskilled labourers who reside from city slum areas and use pristine methods to carry out such recycling methods, exposing them to such harmful toxics like acidic solutions, fumes, burnt residual ashes, and other agents and in turn causing asthma and respiratory, reproductive, and physical injuries. Long-term exposure may also lead to cancer. Since the informal sector falls outside the purview of the government, they are not guided by governmental healthcare policies and hence suffer from such repeated exposures.

In contrast, the organized sector (under governmental regulations) accounts for only 10–15% of the recycling process carried out in the country. The problems experienced by the formal sector of recycling are the competition faced from the informal sector in the collection and disposal of e-waste leading to financial losses. Further, many of the collected e-waste are transferred for refurbishing or resold with only a small proportion accounting for the recycling process. The treatment process used by the formal recycling sector is an environmentally friendly process without the use of toxic chemicals or open incinerations. Primarily, the clients of such formal recycling units are surrounding industries and even MNCs.

17.9 Economic Potential from E-waste

E-waste is a potential income-generating source for industries and also has the potential for creating a new job market in India. This is primarily because a large proportion of e-waste consists of precious material in sufficient quantity to be harnessed for potential economic viability. Further, there exists scope for reusing some materials from older e-wastes to newly built electronic equipments like circuit boards, capacitors plastics, and other components. Further, thermo-chemical treatment is a viable option for treating plastic waste generated from e-waste as it is an environmentally friendly process (Garlapati 2016). This treatment process can lead to production of synthetic fuel from plastics which can be used to power diesel generators (Kantarelis et al. 2011). The plastic waste generated from the e-waste can also be used as a raw feed for production of hydrogen using two-stage pyrolysis-gasification systems (Acomb et al. 2013). Other possible recovery from e-waste includes use of LCD computer screens as electrodes for electricity generation from Microbial Fuel Cell (MFC) (Gangadharan et al. 2015). Such business initiatives can possibly reduce dumping of e-waste.

17.10 Conclusions

E-waste is one of the fastest-growing wastes in both developed and developing countries due to rapid obsolescence of present-day modern electronics. The chapter discusses about the existing condition of e-waste generation both globally and in the Indian context and the need for quantification of overall wastes generated. Discussion on health effects of exposure to such hazardous wastes has also been presented in the chapter. The chapter further discusses some of the E-waste management strategies in India including the specific roles of the government, manufacturers, and the consumers in reducing generation and impacts of e-waste. This includes important tools like EPR, implementation of legislations, and establishment of recycling facilities to ensure safe disposal of such wastes in an environmentally friendly manner. The chapter also discusses in detail the importance of recycling such e-wastes including the roles of both the formal and informal sectors involved in the process. The chapter also discusses the existing legislation of handling such e-wastes in India and some of the potential drawbacks associated with the existing legislation. A discussion on potential business initiatives from the generated e-wastes has been presented.

References

Acomb JC, Nahil MA, Williams PT (2013) Thermal processing of plastics from waste electrical and electronic equipment for hydrogen production. J Anal Appl Pyrol 103:320–327

Agarwal R (2010, May 4) A policy? Rubbish. The Hindustan Times. https://www.hindustantimes.com/india/a-policy-rubbish/story-sjKDRiMXlKQ1IjcVeQk5pM.html. Accessed 13 September 2018

Association of Plastics Manufacturers in Europe (APME) (2004) Plastics – a material of choice for the electrical and electronic industry-plastics consumption and recovery in Western Europe 1995. APME, Brussels, 1

Babu R, Parande AK, Basha AC (2007) Electrical and electronic waste: a global environmental problem. Waste Manag Res 25:307–318

Bhutta KS, Adnam O, Xia Ozhe Y (2011) Electronic waste: a growing concern in today's environment. Econ Res Int 474230. https://www.hindawi.com/journals/ecri/2011/474230/. Accessed 31 Aug 2018

Chaturvedi A, Arora R, Khatter V, Kaur J (2007) E-waste assessment in India – specific focus on Delhi. MAIT-GTZ study. http://www.nswai.com/nswaiadmin/Pdfs/insertPdf/i_2015/i_Nov15/e-Waste%20Assessment%20in%20India%20-%20Specific%20Focus%20on%20Delhi.pdf. Accessed 7 Sept 2018

Chen A, Dietrich KN, Huo X, Ho S (2011) Developmental neurotoxicants in e-waste: an emerging health concern. Environ Health Perspect 119(4):431–438

Dasgupta D, Debsarkar A, Hazra T, Bala BK, Gangopadhyay A, Chatterjee D (2017) Scenario of future e-waste generation and recycle-reuse-landfill-based disposal pattern in India: a system dynamics approach. Environ Dev Sustain 19(4):1473–1487

Duan H, Wang J, Huang Q (2015) Encouraging the environmentally sound management of C&D waste in China: an integrative review and research agenda. Renew Sustain Energ Rev 43:611–620

Dwivedy M, Mittal RK (2010) Estimation of future outflows in India. Waste Manag 30(3):483–491

EU (2002) Directive 2002/96/EC of the European parliament and of the council of 27 January 2003 on waste electrical and electronic equipment (WEEE) – joint declaration of the European parliament, the council and the commission relating to article 9. Official Journal L037:0024–39 (13/02/2003; 2002 http://europa.eu.int/eur-lex/en/)

European Parliament (2012) Directive 2012/19/EU of the European parliament and of the Council of 4 July 2012 on waste electrical and electronic equipment (WEEE). Off J Eur Union L197:38–71

Freeman MH (1997) Standard handbook of hazardous waste treatment and disposal. McGraw-Hill, New York

Gaidajis G, Angelakoglou K, Aktsoglou D (2010) E-waste: environmental problems and current management. J Eng Sci Technol Rev 3(1):193–199

Gangadharan P, Nambii M, Jaganathan S (2015) Liquid crystal polaroid glass electrode from e-waste for synchronized removal/recovery of Cr(+6) from waste water by microbial fuel cell. Bioresour Technol 195:96–101

Ganguly R (2016) E-Waste management in India – an overview. Int J Earth Sci Eng 9(2):574–588

Garlapati VK (2016) E-waste in India and developed countries: management, recycling, business and biotechnological initiatives. Renew Sustain Energy Rev 54:874–881

Grant K, Goldizen FC, Sly PD, Brune MN, Neira M, van den Berg M, Norman RE (2013) Health consequences of exposure to e-waste: a systematic review. Lancet Glob Health 1(6): 350–361. https://www.thelancet.com/action/showPdf?pii=S2214-109X%2813%2970101-3. Accessed 13 Sept 2018

Gregory J, Magalini F, Kuehr R, Huisman J (2009) E-waste take-back system design and policy approaches. Solving the e-Waste Problem (StEP), White Paper

Gupta M (2014) Management of hazardous electronics waste. Int J Comput Appl 90(1):11–14

Janagam D, Jeymani M (2011) E-Waste—a major threat to environment and health. Ind J Sci Tech 4(3):313–317

Jomova K, Jenisova Z, Feszterova M, Baros S, Liska J, Hudecova D, Rhodes CJ, Valko M (2011) Arsenic: toxicity, oxidative stress and human disease. J Appl Toxicol 31(2):95–107

Kantarelis E, Yang W, Blasiak W, Forsgren C, Zabaniotou A (2011) Thermo chemical treatment of e-waste from small household appliances using highly pre-heated nitrogen-thermo gravimetric investigation and pyrolysis kinetics. Appl Energy 88(3):922–929

Milovantseva N, Saphores JD (2013) E-waste bans and U.S. household's preferences for disposing of their waste. J Environ Manag 124:8–16

Nisa M (2014) E-waste management. J Nanosci Nanotechnol 2(1):766–768

Organization for Economic Co-operation and Development (OECD) (2001) Extended producer responsibility – a guidance manual for governments. OECD, Paris

Osibanjo O, Nnorom IC (2007) The challenge of electronic waste (e-waste) management in developing countries. Waste Manag Res 25(6):489–501

Puckett J, Byster L, Westervelt S, Gutierrez R, Davis S, Hussain A, Dutta M (2002) Exporting harm: the high-tech trashing of Asia, the Basel action network (BAN) and silicon valley toxics coalition (SVTC). SVTC report. http://svtc.org/wp-content/uploads/technotrash.pdf. Accessed 10 Aug 2018

Rana R, Ganguly R, Gupta AK (2018) Indexing method for assessment of pollution potential of leachate from non-engineered landfill sites and its effect on ground water quality. Environ Monit Assess 190(1):46. https://link.springer.com/content/pdf/10.1007%2Fs10661-017-6417-1.pdf. Accessed 15 Sept 2018

Sheng PP, Etsell TH (2007) Recovery of gold from computer circuit board scrap using aqua regia. Waste Manag Res 25:380–383

Sinha S (2008) Dark shadows of digitization on Indian horizon. In: Johri R (ed) E-waste: implications, regulations, and management in India. The Energy and Resource Institute, New Delhi, pp 23–44

Skinner A, Dinter Y, Lloyd A, Strothmann P (2011) The challenges of e-waste management in India: can India draw lessons from the EU and the USA? ASIEN 117:S7–S26

Chapter 18
Bioremediation Approaches for E-waste Management: A Step Toward Sustainable Environment

Toqeer Ahmed, Irfan Liaqat, Rabbia Murtaza, and Anjum Rasheed

Abstract Electronic waste commonly known as e-waste is becoming an important concern all over the world and has life-threatening effects on both human and the environment as it comprises toxic metals and emissions like chlorofluorocarbons. E-waste comprises both hazardous and valued materials demanding distinct management and recycling techniques to evade adverse ecological effect and detrimental impact on human well-being. Inappropriate recycling practices and release of toxic chemicals from the e-waste can damage human health including lung diseases, cancer, and mental and reproductive health. Children are most vulnerable to toxic metals. Transport of hazardous wastes in the form of electronic products from developed to developing countries is increasing health risks, and less attention has been paid on this matter during the past two decades. It is highly recommended to raise awareness about the health hazards related to e-waste. In this chapter, toxicity, health concern, bioremediation approaches for e-waste management (EWM), international regulations, and e-waste issues with reference to Pakistan perspective have been discussed in detail.

Keywords Bioremediation · E-waste management · Sustainable environment

18.1 Introduction

Electronic waste is becoming a prime concern around the globe due to its unsafe disposal and life-threatening health impacts on the biosphere and communal health (UNEP 2007; Song et al. 2014; Balde et al. 2015). The environmental impacts of improper e-waste management have been observed in water, air, and soil because of the release of toxic substances. The extensive industrial processing and chemical handling have raised serious issues for global soil, water, and air through

T. Ahmed (✉) · I. Liaqat · R. Murtaza · A. Rasheed
Centre for Climate Research and Development (CCRD), COMSATS University Islamabad, Islamabad, Pakistan
e-mail: toqeer.ahmed@comsats.edu.pk

M. Z. Hashmi, A. Varma (eds.), *Electronic Waste Pollution*, Soil Biology 57,
https://doi.org/10.1007/978-3-030-26615-8_18

contamination of hazardous substances such as xenobiotics, polyaromatic hydrocarbons (PAHs), insecticides, and heavy metals. These contaminants were recognized as poisonous, oncogenic, and mutagenic for living organisms (Tripathi 2011). During the last decade, modification in our lifestyle has hugely impacted the environment, where the need to execute a supplementary sustainable strategy regarding our consumption behaviors appears of much consequence. This trend specially in commercial sectors that are most likely affecting our environment including electronic manufacturing where the lower half-life and the rapidly evolving technology have resulted in higher electronic-waste volumes across the globe. "E-waste comprises of a varied and emerging variety of electronic devices ranging from bulky everyday appliances, like mobile phones, air-conditioners, freezers, audio systems and expendable electronic items to Personal Computers (PC's) disposed of by customers" (Patel and Kasture 2014).

Most of the electronic-waste components are directed to landfills. Their incomplete recyclability, because of substantial configuration along with the inevitable limitations in landfills, has directed to the growth of retrieval practices for their recovery, indicating the importance of electronic-waste recycling, to retrieve valuable constituents other than waste managing aspect. E-waste ranging from urban to industrial slightly differs in physical and chemical properties. E-waste comprises both hazardous and valued materials demanding distinct handling and recycling techniques to evade adverse ecological effect and detrimental impact on human well-being. Although the valuable and base metal recovery is quite possible by recycling practices, the higher labor cost and the firm environmental regulations have limited these activities (Esteve-Nunez et al. 2001). The classification of electronic-waste components with their sources and exposure routes has been shown in Fig. 18.1.

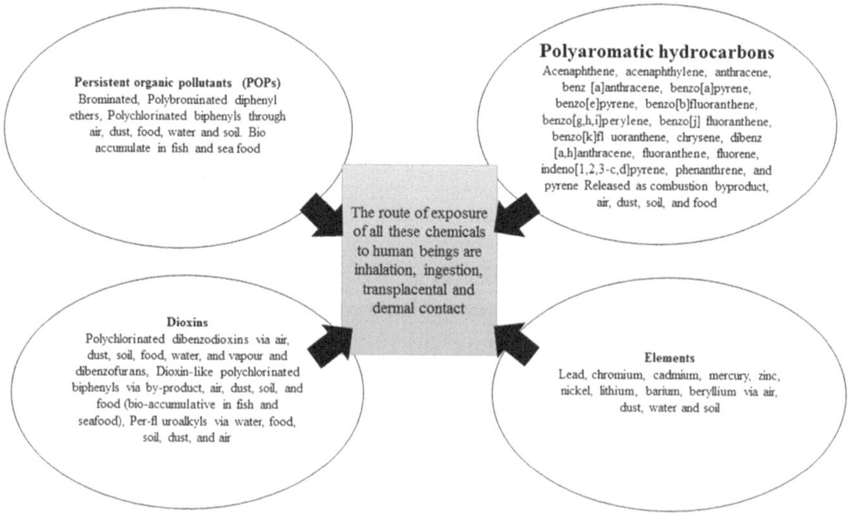

Fig. 18.1 The classification of electronic-waste components with their sources and exposure routes

Many countries have faced immense problems in handling e-waste that is illegally imported or locally produced (Nnorom and Osibanjo 2008). Developed countries ship e-waste to developing countries like Pakistan, India, China, etc. where improper e-waste management techniques are practiced (Sthiannopkao and Wong 2013), while in the Middle East and North African (MENA) countries, 90% e-waste is discarded in landfill and 5% is casually recycled (WAMDA 2018). In low-income countries, e-waste is discarded in unsanitary or uncontrolled sites. These disposal activities of e-waste are generally practiced at Pakistan, India, the Philippines, Nigeria, and China. The wires are burnt to remove copper and plastic, while acid from the batteries is extracted to recover gold, silver, platinum, and palladium from printed circuit boards (PCBs). These e-waste management approaches have been seen in countries, like Pakistan, India, China, Vietnam, Ghana, and the Philippines, to retrieve precious metals where people have scarcity of facilities to protect public health and environment (Leung et al. 2006; SEPA 2011).

In order to treat water and soil from the contamination caused by electronic waste or any other toxic metals or hydrocarbons, persulfate oxidation is used because of its benefits like easy function, stability, no secondary pollution, and greater efficiency (Tsitonaki et al. 2010). Due to its greater solubility, organic compounds are easily oxidized in the aqueous solution and water, thus proving to be a good solution to treat water and soil (Yang and Yeh 2011). The recycling of e-waste generates wastewater issues because it contains toxic metals like Cu, Ni, Cd, Zn, and Pb causing pollution in water, air, and soil (Sojinu et al. 2012; Wong et al. 2007a, b; Deng et al. 2008)

A major challenge in developing approaches for electronic-waste (e-waste) handling is the varying composition of several elements due to quick high-tech advancement, specifically in the electronic components. It is possible to efficiently restore the polluted environment using bioremediation-based methods. Bioremediation usually refers to the degradation of contaminants through microorganisms by acting in parallel or order to completely degrade the hazardous substances that caused risks for the environment and human (Surajit Das 2014). A few microbes are the finest candidates among all microorganisms to eradicate major contaminants in the environment due to their better diversity and adaptability in the environment. Yet, a variety of environmental pollutants have been found to be resistant, due to their partial metabolization or conversion into other xenobiotics by microbes that ultimately again accumulate in the environment. Therefore, it may be more fruitful to find out novel catabolic mechanisms that might result in comprehensive mineralization of these toxic contaminants. Hence, it has become necessary worldwide to take satisfactory actions to regulate the increase in electronic waste as they might increase 50% in the upcoming few years subsequently increasing higher ecological contamination and health risks (Singh and Ward 2004).

18.2 E-waste Composition

Electronic-waste composition depends entirely on the nature and the lifespan of the fragment; telecommunication system and IT scraps comprise a larger quantity of valued metals as compared to scrap from everyday appliances (Tanskanen 2013).

A typical cell phone comprises over 40 components including costly metals such as silver, palladium, and gold; some common metals like tin, copper, and iron; superior metals like lithium, antimony, cobalt, and indium; and remnants like plastic and ceramic (Izatt et al. 2012). The higher use of common metals in electronics like iron is also extensive: around 6 kg iron is utilized commonly in a PC; that means almost 930,000 tons of iron were used for the assembly of personal computers vended in 2007 only. The mutual unit sales in 2007 of cell phones and PCs had further gone up to 3% of the excavated stock of gold (Au), silver (Ag), palladium (Pd), and cobalt (Co) worldwide (UNEP 2009). The worldwide demand of few metals for electronics is almost 80% of indium (for conductive layers in LCD), 50% of antimony (for flame retardants), and above 80% of ruthenium (for magnetic properties). Outdated freezers, refrigerators, and air conditioning elements comprise chlorofluorocarbons (CFCs). The chlorofluorocarbons (CFCs) discharged from objects disposed in landfills contribute majorly to the earth's stratospheric ozone depletion (Scheutz et al. 2004). The harmful contaminants present in e-waste scrap are listed in Table 18.1.

18.3 Toxicity and Public Health Effects of E-waste

E-waste processing has become a universal health problem. Informal and non-standard e-waste processing activities into environment led to malicious chemical leakage. During inappropriate recycling practices, e-waste containing heavy metals are released that results in damaging vegetation, humans, animals, and further environmental resources. E-waste causes the deposition of heavy metals in human tissues, thus posing serious diseases of the liver, lungs, and kidneys (Li et al. 2011). The countries that are utmost affected by inappropriate disposal are India (Bengaluru and Delhi), China (Guiyu and Taizhou), and few African countries (Nigeria, Ghana), in which e-waste has been cast off with minimal or no guiding principle, primarily by means of obsolete technology (de Garbino 2004). Humans are normally unprotected to persistent organic pollutants POPs such as polycyclic-aromatic hydrocarbons (PAHs), polychlorinated biphenyls (PCBs), polybrominated diphenyl ethers (PBDEs), dioxin, polybrominated dibenzo-p-dioxins and dibenzofurans (PBDD), polyvinyl chloride (PVC), polychlorinated dibenzo-p-dioxins and dibenzofurans (PCDD), and heavy metals in water, soil, air, and food reservoirs through numerous ways including ingestion, inhalation, and dermal absorption from ignition, expulsions, and industrial facilities. Heavy metals leaked from e-waste reprocessing cannot be broken down into less dangerous end products, unlike organic pollutants

Table 18.1 Hazardous contaminants contained in electronic waste

Electronics scrap	Hazardous substances	E-waste concentration (mg/kg)	E-waste emission per year worldwide (tons)
Condensers, transformers	Polychlorinated biphenyls (PCB)	14	280
Insulation foam, cooling parts	Chlorofluorocarbon (CFC)	–	–
Combustion products	Polycyclic aromatic hydrocarbons (PAHs)	–	–
Products of lower-temperature (*T*) combustion of plastics such as PVCs and others	Polyhalogenated aromatic hydrocarbons (PHAHs)	–	–
Flame retardants	Polybrominated diphenyl ethers (PBDEs)	–	–
Fire retardants	Polybrominated biphenyls (PBBs)	–	–
	Tetra-bromo-bisphenol-A (TBBPA)	–	–
Smoke detectors	Americium (Am)	–	–
Fire retardants, plastics	Antimony	1700	34,000
Si doping material	Arsenic (As)	–	–
Cathode ray tubes (CRTs) Getters	Barium (Ba)	–	–
Silicon-based rectifiers	Beryllium (Be)	–	–
Plastics, batteries, toners,	Cadmium (Cd)	180	3600
Data tapes and floppy disks	Chromium (Cr)	9900	198,000
Wiring	Copper (Cu)	41,000	820,000
Semiconductors	Gallium (Ga)	–	–
LCD displays	Indium (In)	–	–
Solder CRTs, batteries	Lead (Pb)	2900	58,000
Batteries	Lithium (Li)	–	–
Fluorescent lamps, batteries, switches	Mercury (Hg)	0.68	13.6
Batteries	Nickel (Ni)	10,300	206,000
Rectifiers	Selenium (Se)	–	–
Wiring, switches	Silver (Ag)	–	–
Solder, LCD screens	Tin (Sn)	2400	48,000
Solder, LCD screens	Zinc (Zn)	5100	102,000
CRT screens	Rare earth elements	–	–

Adapted from Morf et al. (2007), Robinson (2009)

that have biodegradability capacity. Bioaccumulation of electronic-waste elements and their end products (heavy metals) in living organisms can adversely affect human health when their concentration exceeds the minimal level over time (Zeng et al. 2016; Premalatha et al. 2014).

E-waste components influence public health through two different routes that include (1) contamination of food chain, entering of toxic materials into food chain that are released from recycling processes and transferring to animals and humans, and (2) direct occupational exposure of worker to toxic substances working in recycling sites. The dust from all the recycling activities is released into the atmosphere. Fang et al. (2013) proposed special masks for screening of PM 2.5 to mitigate the inhalation of pollutants. Multiple studies suggest that plants and microbes can be used to treat the soil. Extracellular polymeric substances (EPS), proteins, and mucopolysaccharides are released through microorganisms into the soil that forms toxic metal complexes, hence reducing the movement of metals in soil (Rajkumar et al. 2012; Seguel et al. 2015). It has been detected that the exposed population to e-waste has higher frequencies of micro-nucleated bi-nucleated cells. Also, the DNA damage is more common to the people exposed to e-waste. It has been reported that polycyclic-aromatic hydrocarbons, polybrominated diphenyl ethers, and perfluoroalkyls cause adverse outcomes in the infants and neonatals. Multiple chronic disorders in life like type 2 diabetes, obesity, hypertension, and cardiovascular diseases are linked with dioxin-like polychlorinated biphenyls, dioxins exposure, perfluoroalkyls, cadmium, and lead. Similarly, lung damage and lung cancer are associated with certain other compounds that are found in e-waste such as hexavalent chromium, polycyclic-aromatic hydrocarbons, lithium, cadmium, nickel, and arsenic.

Pregnant females exposed to dioxins, polycyclic-aromatic hydrocarbons, polychlorinated biphenyls, polybrominated diphenyl ethers, perfluoroalkyls, and cadmium have suffered with impaired fetal growth, low birth length and weight, preterm delivery, and low head circumference. The chemical constituents of e-waste also affect fertility and reproductive development. Lead causes delayed puberty in girls. The sperm quality of males is affected by polychlorinated biphenyls, TCDD, and perfluoroalkyls, while POPs are also the endocrine disruptors. Other impacts include DNA damage, mental, reproductive, and respiratory issues.

Qu et al. (2007) reported, the direct-exposure of PBDEs on labors in e-waste reprocessing sites and detect the increased levels of PBDEs in the sample of serum groups with the maximum concentration level of BDEs is 3436 ng/g lipid dry weight that is the higher level of BDE in humans until now documented. Similarly, Zhao et al. (2008) measured different organic pollutant levels in hair sample groups including PCBs, PBBs, and PBDEs at 181.99, 57.77, and 29.64 ng/g dry weight, respectively, that are much higher in concentration as compared to reference areas (Zhao et al. 2008; Qu et al. 2007). Wang et al. (2009) detected lead (Pb, 49.5 lg/g) and copper (Cu, 39.8 lg/g) concentration in hair scalp samples. Xing et al. (2009) and Chan et al. (2007) measured PCDD and PCBs in samples of milk of human at 9.50 and 21.02 pg/g, respectively. Ha et al. (2009) detected concentration level of Bi, Cu, and Sb in the hair sample of e-waste recycling labors in India that was much greater than the tolerance level and the concentration levels of tri to tetra-chlorinated (PCBs), tetra-BPhs, tri to tetra-chlorinated (OHPCBs), octa-brominated (OH-PBDEs), and PBDEs in the samples of workers serum were higher than those in serum sample that were taken from individuals existing near the shore area (Wang et al. 2009; Chan et al. 2007; Xing et al. 2009; Ha et al. 2009). These results approve that organic pollutants and heavy metal exposure leaked from e-waste handling processes cause substantial public health risk to labors and native residents specially children and women.

Children are highly vulnerable to toxic metal elements in comparison to adults for few reasons: surplus exposure routes including object-to-mouth, hand-to-mouth, breastfeeding, placental exposures; higher surface area in relation to body-weight, larger ventilation rate per minute relative to body-size, their high basal metabolic rate, comparatively increased food uptake, lower toxin removal rates, and they can inhale much toxic substances; and their tissues and organs are under development stage and so extra sensitive to disturbed cellular processes. Increased levels of cadmium (Cd) and lead (Pb) were found in blood samples of children living near e-waste processing sites (Kiddee et al. 2013).

18.4 Bioremediation Approaches for EWM

Bioremediation is a wide-ranging perception that comprises all procedures and activities that happen to biotransform an environment, previously changed by pollutants, to its native state. Even though the processes that are being used to accomplish the required outcomes differ, they still have similar principles, by using microbes or enzymes, either indigenous or activated by the nutrient addition and optimization of environment or planted into the soil. For gaining energy, microorganisms catalyze energy-producing reactions involving breaking chemical reactions that transfer electrons away from the pollutant. The energy that is acquired from electrons transfer is then devoted along with some added electrons and carbon atoms from the pollutant to produce even more cells. The thermal or physiochemical techniques alone are not as effective as the biological techniques, having greater removal efficiency and being more successful. Bioremediation has now turned out to be a commonly accepted opportunity for the cleaning of polluted soils and aquifers (Pramila et al. 2012).

Currently, four major biological techniques for the cleanup of groundwater and soil are present: (1) native microorganisms' stimulation by adding the nutrients, optimization of redox and pH conditions, etc.; (2) site-specific microorganism inoculation with biotransforming capabilities; (3) immobilized enzymes usage; and (4) phytoremediation procedure to remove or convert contaminants. The specific methods that are being used for bioremediating polluted water and soil include intrinsic bioremediation, slurry bioreactor, land farming, and composting (Bollag and Bollag 1995).

18.4.1 EWM Approaches in Water

18.4.1.1 Bioleaching

Microbes are dynamic in the generation and breakdown of numerous organic as well as inorganic material on earth's surface. Bioleaching is known as the self-capability of microorganisms to alter solid metal composites to its solvable and extractable form. In the metal's bioleaching, three major categories of microorganisms are involved, heterotrophic bacteria (*Pseudomonas* sp., *Bacillus* sp.), autotrophic bacteria (*Thiobacillus* sp.), and heterotrophic fungi (*Penicillium* spp., *Aspergillus* sp.) (Schinner and Burgstaller 1989), as explained in Table 18.2.

Common mechanisms that are involved in bioleaching are acidolysis, complexolysis, and redoxolysis. At 40°C or less, microbes have an ability to remove heavy metals from its iron and sulfide covering mineral concentrates and ores. Through microbial oxidation, iron and sulfide ions are oxidized to produce ferric ion and sulfuric acid, respectively; subsequently these substances can transform insoluble metal sulfides such as copper, zinc, and nickel to soluble metal sulfates that may be simply removed from the final solution (Pham and Ting 2009). The mechanisms of bioleaching process are demonstrated in Fig. 18.2.

Suzuki (2001) discovered that through contact (direct) leaching through bacteria and indirect leaching through ferric ion by bacterial oxidation of ferrous ion, mineral solubilization can be accomplished. Other than *Thiobacillus ferrooxidans*, the organism for microbial leaching that is mostly studied; iron and sulfur oxidizing archaea/bacteria are the probable candidates for metal bioleaching at low-pH and high-temperature conditions. Some moderately thermophilic (acidophilic-chemolithotrophic) and heterotrophic bacteria were exploited for leaching of metals from scrap that comprises *Sulfobacillus thermosulfidooxidans* (Ilyas et al. 2007).

18.4.1.2 Biosorption

Biosorption is based on the absorption and binding of solvable pollutants (ionized toxic metals) on the cellular surface. The ability of microbes, e.g., bacteria, algae, fungi, and plant biomass, to eradicate radionuclide and heavy metal ions or to encourage their conversion to less poisonous forms has captivated many researchers and biotechnologist's attention for many years. Thus, several thoughts for elimination of heavy metals from e-waste of polluted environment through microbes are being estimated, and some of them were taken to the industrial level (Volesky et al. 1993).

Biosorption mechanisms can be categorized according to the need of cell's metabolism that are known as metabolism dependent and independent according to the area from where the metal is found: from the removed solution that includes like cell surface sorption/precipitation extracellular accumulation/precipitation, and intracellular accumulation/precipitation (Muyzer and Stams 2008; Neethu et al.

Table 18.2 Use of several microorganisms for heavy metals removal from e-waste through bioleaching

Metals removed	Microorganisms	Temperature (°C)	pH
Copper (Cu)	*Acidithiobacillus ferrooxidans*	45	2–3
	Acidithiobacillus thiooxidans	45	2–3
	Aspergillus niger	30	3–4
	Penicillium simplicissimum	32	2–6
	Gallionella sp.	30	2–6
	Leptospirillum sp.	45	3
Zinc (Zn)	*Acidithiobacillus ferrooxidans*	45	2–3
	Acidithiobacillus thiooxidans	45	3–4
	Aspergillus niger	30	2–6
	Penicillium simplicissimum	32	2–6
Lead (Pb)	Strain of F3-02 *Acidithiobacillus ferrooxidans*	45	2–3
	Acidithiobacillus thiooxidans	45	3–4
	Aspergillus niger	30	2–6
	Penicillium simplicissimum	32	2–6
Aluminum (Al)	*Bacillus circulans* and *B. mucilaginosus*	35	4–5
	Acidithiobacillus thiooxidans	45	3–4
	A. ferrooxidans	45	2–3
	Aspergillus niger	30	2–6
	Penicillium simplicissimum	32	2–6
Nikel (Ni)	*Acidithiobacillus thiooxidans*	45	3–4
	A. ferrooxidans	45	2–3
	Aspergillus niger	30	2–6
	Penicillium simplicissimum	32	2–6
Gold (Au)	*Chromobacterium violaceum*	30–37	3.0
	Pseudomonas fluorescens	25	7–8
	Desulfovibrio desulfuricans	35–43	2.0
	Acidithiobacillus sp.	45	3–4
	Leptospirillum thiooxidans	45	2–3
	Ferromicrobium	45	2–3
	Acidiphilium	45	2–3
Silver (Ag)	*Acidithiobacillus* sp.	45–60	2–3
	Leptospirillum sp.	45	2–3
	Ferromicrobium	45	2–3
	Acidiphilium	45	2–3
Cadmium (Cd)	*Aspergillus niger*	30	2—6

Adapted from Pant et al. (2012)

2015). During metabolism-independent mechanism, metal uptake is based on physiochemical interactions present between the functional groups existing on the bacterial cell surface and metal. As it is not based on cell's metabolism, it depends on chemical sorption, physical adsorption, and ion exchange. Microbial cell wall primarily comprises of carbohydrates, lipids, and proteins that have plentiful metal

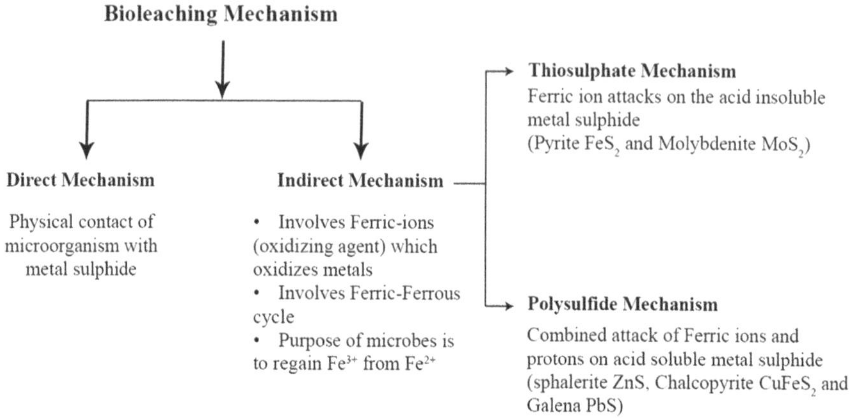

Fig. 18.2 The mechanisms of bioleaching process (Adapted from Pant et al. 2012)

absorption groups such as phosphate, sulfate, amino, and carboxyl groups (Ahalya et al. 2003). The cleansing mechanism of microbes can be classified into:

- Intracellular segregation.
- Exporting the toxic ion away from the cell by changing the membrane transport system that is present in early cell's accumulation.
- Reducing the cell's permeability.
- Extracellular segregation by definite mineral ion absorption. Extracellular decontamination involves the enzymatic transformation of more toxic cations or anions into less toxic form.

The different bacterial, algal, and fugal strains that are used for heavy metals removal are listed in Table 18.3.

The difference between microbially enhanced chemisorption of metals and adsorption is a reaction that takes place between the adsorbate and the surface. In this case, a sequence of reactions take place in which microorganisms first infiltrate a biomineral of non-specific metals (priming deposits) and then they use these priming deposits as nucleation focus for the consequent target metal's deposition (Tabak et al. 2005).

18.4.1.3 Bioaccumulation

Bioaccumulation is based on the absorption of pollutants within the organism, which are changed into a cell biomass inside the cellular organelle and concerted there; this method involves dynamic metabolism. For organic chemicals, sometimes they are catalyzed inside cell cytoplasm to other less toxic contaminants; but, instead, the toxic metals that go inside are sequestered (Hou et al. 2006; Prakash et al. 2012).

Table 18.3 Common species that are used and their biosorption capacity for heavy metal removal

Species	Microorganisms	Metal removed	Biosorption capacity (mg/g)
Bacterial species	*Enterobacter* sp. *J*	Copper (Cu)	32.12
	Arthrobacter sp.	Copper (Cu)	17.87
	Pseudomonas fluorescence	Chromium (Cr)	40.8
	Pseudomonas sp.	Chromium (Cr)	95
	Enterobacter sp. *J*	Cadmium (Cd)	46.2
	E. coli	Nickel (Ni)	6.9
	Pseudomonas fluorescence	Nickel (Ni)	40.8
	Pseudomonas putida	Zinc (Zn)	17.7
	Bacillus jeotgali	Zinc (Zn)	222.2
Fungal species	*Aspergillus niger*	Lead (Pb)	34.4
		Copper (Cu)	28.7
	Saccharomyces cerevisiae	Lead (Pb)	270
		mercury (Hg)	64.2
	Penicillium simplicissimum	Cadmium (Cd)	52.5
		Zinc (Zn)	65.6
	Penicillium chrysogenum	Nickel (Ni)	260
		Lead (Pb)	204
		Copper (Cu)	92
	Penicillium purpurogenum	Chromium vi (Cr)	36.5
Algal species	*Sargassum* sp.	Cadmium (Cd)	84.7
	Chlorella miniata	Chromium (Cr)	34.6
	Spirulina platensis	Copper (Cu)	67.93
	Spirogyra sp.	Lead (Pb)	140
	Sargassum muticum	Zinc (Zn)	34.1
	Ulva lactuca	Cadmium (Cd)	43.02

Adapted from Mustapha and Halimoon (2015)

Vibrio harveyi, a common inhabitant of the saline atmosphere, is testified to have the ability for cadmium bioaccumulation (Cd) up to 23.3 mg Cd^{2+} per g of gasping cells. The other strains that are being used for the bioaccumulation include *Bacillus circulans*, *Bacillus megaterium*, *Deinococcus radiodurans*, and *Micrococcus luteus* for the removal of chromium and uranium metals; and some fungal strains *Aspergillus niger* and *Monodictys pelagica* are reported for the bioaccumulation including chromium and lead removal from the electronic scrap (Patel and Kasture 2014).

18.4.1.4 Biotransformation

Biotransformation approaches have the potential to use the microorganism's natural metabolic capability to accumulate, convert, or destroy the toxic complexes including radionuclides, hydrocarbon compounds, pharmaceutical constituents, and noxious heavy metals. Biotransformation denotes the method in which a substance is transformed by chemical reactions from one chemical form to alternative chemical form. For noxious metals, the oxidation state is altered by the electron's addition and removal; hence, their chemical characteristics are also different (Karigar and Rao 2011).

Rhodobacter sphaeroides and *Rhodobium marinum* have the potential of eradicating toxic metals like zinc (Zn), copper (Cu), lead (Pb), and cadmium (Cd) from the polluted water through biotransformation. Two mechanisms for biotransformation are: (1) direct-enzymatic reduction mechanism, wherein the exterior of the cell multivalent metal ions is lessen by taking electrons from the enzymes, and (2) indirect reduction mechanism, where multivalent metals are reduced and immobilized by the aid of metal-reducing or sulfate-reducing microbes in sedimentary or subsurface environment (Tabak et al. 2005).

Polyaromatic hydrocarbons (PAHs) are abundant in environment and are of much ecological distress because of their perseverance, noxiousness, and mutagenicity in nature. Some new aquatic bacterial species such as *Lutibacterium anuloederans*, *Cycloclasticus spirillensus*, and *Neptunomonas naphthovorans* have also been employed in advanced PAH biodegradation methods in an aquatic environment (Chung and King 2001). Numerous categories of plastics including polyethylene, polystyrene, polyvinyl chloride, polyethylene terephthalate, and polypropylene are being used for packing, fishing, etc. in our surroundings that eventually contaminate our environment. Still, microbes have advanced the mechanism for the breakdown of plastic to harmless form. Recent discoveries proved that *Rhodococcus ruber* can reduce 8% dry weight of plastic in concerted liquid culture (in vitro) in just 30 days. In the same way, bacterial strains from genera *Psychrobacter*, *Moritella*, *Shewanella*, and *Pseudomonas* that are being secluded from Japan's deep seas have the ability to degrade the ε-caprolactone in a much effective manner (Kathiresan 2003).

18.4.1.5 Biomineralization

Biomineralization is a method through which living organisms can produce minerals. The prokaryote mineralization can be grouped into two classes: biological-induced mineralization (BIM) and biological-controlled mineralization (BCM). In the case of BCM, minerals are formed at a definite site inside or on the cell at specific conditions, but in (BIM), the minerals are synthesized extracellularly because of organism's metabolic potential. The extracellular formation of these minerals gives a concept to the scientific community to employ this ability of microorganisms for wastewater treatment purposes (Lowenstam and Weiner 1989; Lowenstam 1981). The use of bacterial-induced carbonate biomineral is becoming progressively prevalent every day; removal of calcium and heavy metals from wastewater, removal of radionucleotides, CO_2 sequestration, contaminant biodegradation, and to the remediation of construction materials.

Biomineralization of contaminant toxic metals and radionuclides into calcite happens as a result of competitive co-precipitation reaction where appropriate divalent cations are fused into the calcite structure.

$$^{90}Sr^{2+} + OH^{-} + HCO^{-3} = SrCO^{3}(s) + H_2O$$

These radionuclides and cations assimilate into the calcite lattice via exchange of calcium (Ca^{2+}) ions in the microenvironment of mineral precipitate, making a low strontium carbonate mineral that has poorer solubility (Fujita et al. 2000). Scaling of channels and reactors occurs due to higher concentration of calcium ions almost 500–1500 mg/L in the wastewater that results in larger calcium formation in the form of carbonates, phosphates, and gypsum. In his study, Hammes and Verstraete 2002 found the ability of eliminating calcium ions (Ca^{2+}) from industrial wastewaters through microbial-induced calcite precipitation (MICP) method (Kumari et al. 2015). Almost 90% calcium ions (Ca^{2+}) was effectively detached from the inspected wastewater by the addition of low urea concentration (0–16 g/L). Likewise, Okwadha and Li (2011) described the capability of calcium for polychlorinated biphenyls (PCB) treatment and urea ureolytic *S. pasteurii* cultures. Bacterial strains like *Cupriavidus metallidurans*, *Bacillus fusiformis*, and *Sporosarcina ginsengisoli* are able to remove cadmium, arsenic, and lead, while fungal strain (*Aspergillus flavus*) can remove lead from e-waste through biomineralization (Patel and Kasture 2014).

18.4.2 Recommended EWM for Air Quality

The main factor in exposure to toxic chemicals during recycling and disposal is taking services of unskilled workers. First and foremost, remedial measure would be to recognize recycling as a formal industry and to enforce establishment of safe and state-of-the-art facilities for recycling and disposal of e-waste. Technical training courses need to be developed for skill development in this sector. Only trained workers should be engaged in recycling sites. There is also essential to improve the technology for recycling and to train the workers in safe extraction, recycling, and disposal systems. Adsorption of metal through adsorbents particularly alumina may be adopted for adsorption of metal fumes (Saini et al. 2017) which could otherwise be emitted into the atmosphere. For disposal of compact fluorescent lamps (CFLs), CFLs should be crushed in closed crushing units to separate and capture mercury in collection unit. This process will help prevent emissions of mercury fumes into atmosphere. Similar metal extraction processes may be opted for safe recycling process.

An example of good governance may be taken from Nigeria wherein the country imposed import duty on imported electronic waste. Recently, Thailand has executed an embargo on import of e-waste for recycling as the country is concerned about the dangers to public health and environment. Such initiatives bring awareness to public about the hazards of recycling which are usually overlooked for economic benefit. Despite being signatory to the Basel Convention, the Government of Thailand took the right action considering lack of safe disposal facilities and practices.

18.5 International Regulations

To regularize the transboundary movement of harmful waste, "Basel Convention on the Control of Transboundary Movements of Hazardous Wastes and Their Disposal" was implemented in 1989 and came into force in 1992. There are 178 participants to the Basel Convention, including almost every country in East Asia and the Pacific. The purpose of the Basel Convention was to provide the importing countries with regulating mechanism of "notice and consent." The purpose of the Basel Convention was to guard human health and the ecosystem from the hazards rising due to recycling and removal of harmful waste. E-waste is also regulated under the Basel Convention. Currently, used products are also exported to developing countries for refurbishment and repair under the Basel Convention. The United States holds an opposing position with the justification that transboundary movement of used electronic products does not establish such material to be waste and hence such items may not fall within the purview of the Basel Convention.

PCBs, commercial penta-bromo-diphenyl ether (c-Penta BDE), constituents of hexabromobiphenyl (HBB), and industrial octa-bromo-diphenyl ether (c-Octa BDE) have been constrained under Stockholm Convention on Persistent Organic Pollutants (POPs) because of their resistance to degradation and bioaccumulation potential (Stockholm Convention 2009a, b) prohibiting their manufacture, consumption, import, and export. However, PBDEs regulated under Stockholm Convention have an exemption of recycling and reuse. Permission to be recycled and reused is not justified due to possible risk of exposure (Stockholm Convention 2010a, b).

Other compounds of major concern are refrigerants released from dumping of refrigerators and air conditioners. These refrigerants include ozone-depleting substances (ODS) and greenhouse gases like HFCs. International trade of ODS is regulated under the Montreal Protocol on Constituents that diminish the ozone layer, whereas HFCs are covered under the Kigali Amendment to the Montreal Protocol and the Kyoto Protocol to the UN Framework Convention on Climate Change. The Kigali Amendment to the Montreal Protocol was implemented on October 15, 2016, in Kigali, Rwanda. It demands high-income developed countries to pledge down hydrofluorocarbons (HFCs), a replacement of CFCs, by 2019 year and low-income developing countries to initiate in the future. The Kigali

Amendment came into power since the higher global warming capabilities of HFCs. Phasing down of HFCs under the Montreal Protocol is projected to evade up to 0.5°C temperature by end of the era.

18.6 EWM Methods Adopted in Developing Countries and Their Impacts

Transboundary movement of almost 80% of the electronic waste from high-income developed countries to low-income developing world has led to increase in the exposure risk of hazardous and toxic chemicals. Developing countries accepting e-waste for recycling include China, India, Pakistan, Thailand, Ghana, Nigeria, Vietnam, and the Philippines. Currently, these countries have very less safe and formal recycling and disposal facilities for electronic waste. Recycling is poorly regulated and is carried out mostly in scrapyards with no occupational, health, and safety measures. In these facilities, hazardous substances are sometimes exploded due to manhandling by non-skilled workers. Increasing burden of electronic waste, and its unsuitable handling and removal through open combustion or in landfills, affects the environment and public health significantly. It also poses some challenges to sustainable development. Electronic waste has raised up some serious problems due to the core constituents present in these which are harmful and are non-biodegradable. According to UNU & ITU report, entire countries in the world collectively produced an overwhelming 44.7 million metric tons (Mt) of electronic waste annually in 2016 or an equivalent of 6.1 kilogram per resident (kg/resident), comparable to the 5.8 kg/resident produced in 2014, that is almost equal to 4500 Eiffel Towers every year. The quantity of e-waste is anticipated to increase to 52.2 million metric tons, or 6.8 kg/citizen, by 2021. 44.7 million metric tons (Mt), nearly 1.7 Mt are dumped as waste in highly developed countries and are possibly to be landfilled or burned. Only 8.9 Mt of electronic waste are said to be collected and recycled worldwide, which is equivalent to 20% of all the electronic waste produced. E-waste statistics are not appropriate only in terms of the ecological effect; there is also a significant economic factor to the discussion. The total worth of all electronic-waste raw materials is projected at nearly 55 billion euros in 2016, which is greater than the gross domestic product (GDP) in 2016 of some countries (Balde et al. 2017). The electronic-waste generation and its collection per continent are listed in Table 18.4.

The electronics business is one where subcontracting of mass-produced products is accompanied by major other companies. Mostly genuine equipment producers have fragmented their straight industrial chains and labor-intensive tasks. Most of labor-intensive manufacturing activities have been migrating to low-income or developing countries, where labor cost is lesser (Plepys 2002).

Table 18.4 E-waste generation per continent and their collection

Regions	Countries in region	Population in region (million)	E-waste generation in region (Mt)	E-waste generation per citizen (kg/citizen)	Documented, collection, and recycled (Mt)	Collection rate (%) in region
Asia	49	4364	18.2	4.2	2.7	15
Europe	40	738	12.3	16.6	4.3	35
America	35	977	11.3	11.6	1.9	17
Africa	53	1174	2.2	1.9	0.004	0
Oceania	13	39	0.7	17.3	0.04	6

Adapted from UNU and ITU (2017)

18.6.1 Landfill Disposal

Regardless of the recent universal change toward zero wastes, the quantity of landfills keeps on multiplying in low-income developing countries. In landfilling, dugouts are built on the plane exteriors, soil is mined from the dugouts, and waste material is concealed in it, which is enclosed by a dense layer of soil. Current practices such as safe landfill have been given with some conveniences like impermeable lining constructed by clay or plastic, leachate collection sink that gathers and handover the further leachate to wastewater treatment plants. Townsend et al. (2004) confirmed 12 diverse categories of electronic devices such as monitors, CPUs, laptops, VCRs, TV sets, remote controls, cellular phones, printers, etc. In most of the cases during experimentation, lead (Pb) concentration level in the leachates exceeds the minimal limit of 5 mg/L, and each device leached lead (Pb) more than its minimal level in at least one assessment. The authors determined that these outcomes give enough evidences that electronic devices scrap that comprise any printer wiring boards or color CRT with lead bearing fuse have the harmful wastes for lead (Pb) metal (Townsend et al. 2004). Contaminants have the capacity to move through groundwater and soils inside and around landfill areas. Organic contaminant in landfill decays and infiltrates into the soil in the form of landfill leachate. Leachates can have higher concentrations of suspended and dissolved inorganic substances, organic materials, and toxic heavy metals. But, the concentration level of contaminated materials from leachate depends on the scrap properties and phases of waste breakdown in a landfill (Qasim and Chiang 1994; Kasassi et al. 2008).

18.6.2 Incineration

Incineration is a controlled and ample burning process, where the waste elements are burned in specifically intended incinerators at higher temperature almost at 900–10000 °C. The e-waste incineration advantages include the reduction of electronic-waste bulk and the consumption of the energy component of ignitable compounds. Secondly, through incineration many environmentally harmful organic materials are transformed into less harmful substances. Stewart and Lemieux performed tests on incineration of the PC's motherboards, keyboards, and cases mixtures by means of pilot-scale rotary kiln incinerator. The vent gas was investigated for volatile and semi-volatile organic compounds of partial burning, halogens, heavy metals, as well as PCDDs/Fs. The detected metal discharge was substantial and comprised of copper (Cu), cadmium (Cd), lead (Pb), mercury (Hg), and antimony (Funcke and Hemminghaus 1997; Stewart and Lemieux 2003). The shortcomings of incineration are the compound discharge to air, the evasion from vent of gas cleaning, and the high number of residues from combustion and gas cleaning. In countries such as Ghana, India, Indonesia, and China, where electronic waste is transported for recycling, there is no advancement of technology; therefore, e-waste scrap is burned in the open environment. All the highly noxious substances escape into the atmosphere. Hence, the inappropriate electronic-waste recycling, specifically e-waste incineration, poses a severe risk to human beings as well as to the whole ecosystem (Pramila et al. 2012).

18.6.3 Recycling

Recycling comprises disassembling, i.e., exclusion of diverse parts of electronic waste covering harmful materials, such as PCB and mercury (Hg), plastic and printed circuit board separation, elimination of CRT, and sequestration of non-ferrous and ferrous metals. Enormous amount of e-waste is now being transported around the globe for recycling in low-income/developing countries by labor-intensive process in the yards of residential areas. E-waste recycling is significant because of the rich sources of raw material, removal of toxic substances, and solid waste management (Asante et al. 2012). The process of e-waste recycling is illustrated in Fig. 18.3.

Although the recycling process is a developing and evolving industry, its hazardous effects are being neglected so far. The extent of harms posed to labors and the atmosphere, thus, differs significantly depending on the particulars of the separate facility maneuvers. The risk related to dismantling phase is the unintentional releases and leakages of harmful materials. For example, mercury (Hg) inside light sources such as fluorescent tubes of photocopiers, scanners, etc. as well as switches could be discharged into the surroundings of a recycling area on rupturing of the shell (Aucott et al. 2003).

In a US electronics recycling facility, evaluation of air quality in the area of electronic-waste shredders has revealed lead (Pb) and cadmium (Cd) concentration level as high as 0.27 and 1.4 lg/m^3, respectively. This outcome shows that there were workplace pollution and a likelihood of constant exposure of labors to these noxious metals (Peters-Michaud et al. 2003). Wang and Guo (2006) measured the considerable concentration level of lead (Pb) in external water downstream of the recycling facility in Guiyu, China. The level of lead (Pb) concentration was high as 0.4 mg/L,

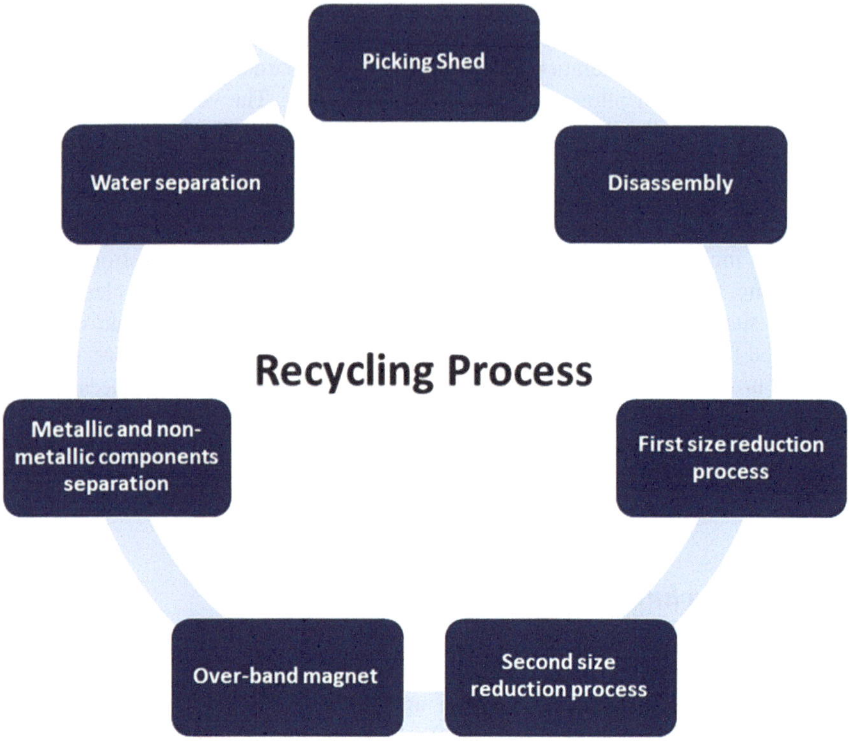

Fig. 18.3 Different steps involved in e-waste recycling process

that is, almost eight times greater than the China drinking water standard. The results from few other studies also validate the drastic pollution of the atmospheric air from brominated and chlorinated substances and toxic metals near e-waste recycling areas in Guiyu, China, and India (Li et al. 2007; Ha et al. 2009).

18.7 E-waste with Reference to Pakistani Perspectives

Most of the electronic waste from developed countries is transported to Pakistan. According to an estimate by the European Union, the e-waste will rise annually by 3–5%, while it also comprises of greater part of municipal waste. Karachi is the hub when it comes to dumping site of electronic waste in Pakistan. Lots of children are engaged in e-waste industry when it comes to reuse and recycling of e-waste products in Pakistan. Metals and plastics are taken out from electronic devices such as mobile phones, computers, televisions, etc. All the management of e-waste is done by adults and children with no personal protective equipment (PPE), but children are more vulnerable at the growing stage. In Layari, the e-waste chemicals

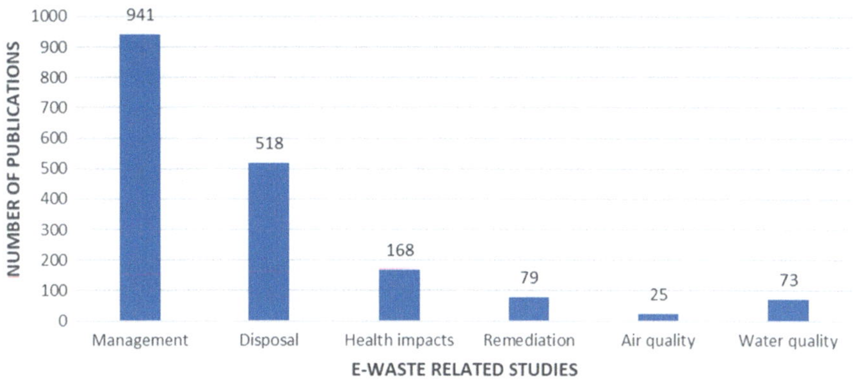

Fig. 18.4 Studies conducted on e-waste during the period between 2000 and mid-2018

from the dumping site seep into the nearby river that turns the river water black (Hai 2013). The same practice has been seen in other cities like Lahore, Rawalpindi, Peshawar, etc. Safe ways for disposal, proper regulations at the government level, and their implementation are recommended in already-published studies to circumvent health-related problems in the country (Iqbal et al. 2015; Sajid et al. 2018; Umair 2015).

18.7.1 E-waste-Related Studies

Studies conducted on different thematic areas of electronic waste such as e-waste management, its disposal, health impacts, and e-waste with reference to air and water quality during the past 17.5 years were analyzed (Fig. 18.4). The information of the total number of studies conducted have been collected from ISI Web Of Knowledge (http://www.digitallibrary.edu.pk/isi.html). The 941 studies were on management of e-waste, 518 studies on disposal of e-waste, 168 studies on health impacts of e-waste, and 79 studies on remediation of e-waste, while 25 and 73 studies were published on e-waste-related air and water quality, respectively, during 2000 to mid-2018. This indicates the pollution related to remediation and e-waste regarding air and water quality have been paid less attention as compared to management, disposal, and health impacts.

18.8 Conclusion

E-waste management and disposal is an important area of research especially with reference to developing countries. In Pakistan, e-waste is not considered as separate surplus as per its risks and hazards involved and combined with solid waste.

Children are more vulnerable to e-waste as they collect e-waste without using PPE. Lack of awareness can cause serious health issues, and there is terrible need to create awareness among the general public regarding risks of e-wastes. Safe disposal of e-waste is suggested along with the solid waste, and sound policies regarding EWM and their implementation is important. More work is required to protect the environment especially in the area of water and air quality; as indicated, less work has been done in both these areas.

Acknowledgment The authors highly acknowledge the resources used from the Centre for Climate Research and Development (CCRD), COMSATS University Islamabad.

References

Ahalya N, Ramachandra T, Kanamadi R (2003) Biosorption of heavy metals. Res J Chem Environ 7:71–79

Asante KA, Agusa T, Biney CA, Agyekum WA, Bello M, Otsuka M, Itai T, Takahashi S, Tanabe S (2012) Multi-trace element levels and arsenic speciation in urine of e-waste recycling workers from *Agbogbloshie*, Accra in Ghana. Sci Total Environ 424:63–73

Aucott M, McLinden M, Winka M (2003) Release of mercury from broken fluorescent bulbs. J Air Waste Manag Assoc 53:143–151

Balde CP, Wang F, Kuehr R, Huisman J (2015) The global e-waste monitor – 2014, United Nations University, IAS – SCYCLE, Bonn, Germany

Balde CP, Forti V, Gray V, Kuehr R, Stegmann P (2017) The Global E-waste Monitor – 2017, United Nations University (UNU), International Telecommunication Union (ITU) & International Solid Waste Association (ISWA), Bonn/Geneva/Vienna

Bollag JM, Bollag WB (1995) Soil contamination and feasibility of biological remediation. In: Skipper HD, Turco RF (eds) Bioremediation science and applications. Soil Science Society of America, Madison, WI, pp 1–10

Chan JKY, Xing GH, Xu Y, Liang Y, Chen LX, Wu SC, Wong CKC, Leung CKM, Wong MH (2007) Body loadings and health risk assessment of polychlorinated dibenzo-p-dioxins and dibenzofurans at an intensive electronic waste recycling site in China. Environ Sci Technol 41:7668–7674

Chung WK, King GM (2001) Isolation, characterization, and polyaromatic hydrocarbon degradation potential of aerobic bacteria from marine macrofaunal burrow sediments and description of *Lutibacterium anuloederans* gen. nov., sp. nov., and *Cycloclasticus spirillensus* sp. nov. Appl Environ Microbiol 67:5585–5592

de Garbino JP (2004) Children's health and the environment: a global perspective. A resource manual for the health sector. World Health Organization, New York

Deng XQ, Wu GQ, Huang HR (2008) Discuss on e-waste recycling system and management system. In: Proceedings of the 38th international conference on computers and industrial engineering, vols 1–3, pp 3012–3016

Esteve-Nunez A, Caballero A, Ramos JL (2001) Biological degradation of 2,4,6-trinitrotoluene. Microbiol Mol Biol Rev 65:335–352

Fang WX, Yang YC, Xu PZM (2013) PM and health risk assessment for heavy metals in a typical factory for cathode ray tube television recycling. Environ Sci Technol 47:2469–12476

Fujita N, Saito R, Watanabe K, Nagata S (2000) An essential role of the neuronal cell adhesion molecule contactin in development of the Xenopus primary sensory system. Dev Biol 221:308–320

Funcke W, Hemminghaus HJ (1997) PXDF/D in flue gas from an incinerator charging wastes containing Cl and Br and a statistical description of the resulting PXDF/D combustion profiles. Organohalogen Compd 31:93–98

Ha NN, Agusa T, Ramu K, Tu NPC, Murata S, Bulbule KA, Parthasaraty P, Takahashi S, Subramanian A, Tanabe S (2009) Contamination by trace elements at e-waste recycling sites in Bangalore, India. Chemosphere 76:9–15

Hai F (2013) Pakistan needs to regulate hazardous e-waste. https://tribune.com.pk/story/511915/pakistan-needs-to-regulate-hazardous-e-waste/. Accessed 6 July 2018

Hammes F, Verstraete W (2002) Key roles of pH and calcium metabolism in microbial carbonate precipitation. Rev Environ Sci Biotechnol 1(1):3–7. https://doi.org/10.1023/A:1015135629155

Hou B, Zhu K, Lu J, Zhao YF (2006) Research on bioremediation of petroleum contaminated soils and its perspectives. Sichuan Environ 6:131–140

Ilyas S, Anwar MA, Niazi SB, Ghauri MA (2007) Bioleaching of metals from electronic scrap by moderately thermophilic acidophilic bacteria. Hydrometallurgy 88:180–188

Iqbal M, Breivik K, Syed JH, Malik RN, Zhang G, Jones KC (2015) Emerging issue of e-waste in Pakistan: a review of status, research needs and data gaps. Environ Pollut 207:308–3018

Izatt NE, Izatt SR, Bruening RL (2012) Green procedure for the selective recovery of precious, specialty, and toxic metals from electronic wastes. In: Electronics goes green 2012+(EGG), IEEE, pp 1–6

Karigar CS, Rao SS (2011) Role of microbial enzymes in the bioremediation of pollutants: a review. Enzyme Res. https://doi.org/10.4061/2011/805187

Kasassi A, Rakimbei P, Karagiannidis A, Zabaniotou A, Tsiouvaras K, Nastis A, Tzafeiropoulou K (2008) Soil contamination by heavy metals: measurements from a closed unlined landfill. Bioresour Technol 99:8578–8584

Kathiresan K (2003) Polythene and plastic-degrading microbes in an Indian mangrove soil. Rev Biol Trop 51:629–633

Kiddee P, Naidu R, Wong MH (2013) Electronic waste management approaches: an overview. Waste Manag. https://doi.org/10.1016/j.wasman.2013.01.006

Kumari D, Qian XY, Pan X, Achal V, Li Q, Gadd GM (2015) Microbially-induced carbonate precipitation for immobilization of toxic metals. In: Laskin AI, Sariaslani S, Gadd GM (eds) Advances in applied microbiology, vol 94. Academic Press, New York, pp 79–108

Leung ZW, Cai MH, Wong P (2006) Environmental contamination from electronic waste recycling at Guiyu, southeast China. J Mater Cycles Waste Manag 8:21–33

Li H, Yu L, Sheng G, Fu J, Peng P (2007) Severe PCDD/F and PBDD/F pollution in air around an electronic waste dismantling area in China. Environ Sci Technol 41:5641–5646

Li JH, Duan HB, Shi PX (2011) Heavy metal contamination of surface soil in electronic waste dismantling area: site investigation and source-apportionment analysis. Waste Manag Res 29:727–738

Lowenstam (1981) Minerals formed by organisms. Science 211:1126–1131

Lowenstam HA, Weiner S (1989) On biomineralization. Oxford University Press, New York

Morf LS, Tremp J, Gloor R, Schuppisser F, Stengele M, Taverna R (2007) Metals, non-metals and PCB in electrical and electronic waste—actual levels in Switzerland. Waste Manag 27:1306–1316

Mustapha MU, Halimoon N (2015) Microorganisms and biosorption of heavy metals in the environment: a review paper. J Microb Biochem Technol 7:253–256. https://doi.org/10.4172/1948-5948.1000219

Muyzer G, Stams AJ (2008) The ecology and biotechnology of sulphate reducing bacteria. Nat Rev Microbiol 6:441–454

Neethu CS, Mujeeb RKM, Saramma AV, Mohamed HAA (2015) Heavy-metal resistance in Gram-negative bacteria isolated from Kongsfjord, Arctic. Can J Microbiol 61:429–435

Nnorom IC, Osibanjo O (2008) Overview of electronic waste (e-waste) management practices and legislations, and their poor applications in the developing countries. Resour Conserv Recycl 52:843–858

Okwadha GD, Li J (2011) Biocontainment of polychlorinated biphenyls (PCBs) on flat concrete surfaces by microbial carbonate precipitation. J Environ Manag 92:2860–2864
Pant D, Joshi D, Upreti MK, Kotnala RK (2012) Chemical and biological extraction of metals present in E waste: a hybrid technology. Waste Manag 32:979–990
Patel S, Kasture A (2014) E (Electronic) waste management using biological systems-overview. Int J Curr Microbiol App Sci 3:495–504
Peters-Michaud N, Katers J, Barry J (2003) Occupational risks associated with electronics demanufacturing and CRT glass processing operations and the impact of mitigation activities on employee health and safety. In: Proceedings of the electronics and the environment, IEEE international symposium, pp 323–328
Pham VA, Ting YP (2009) Gold bioleaching of electronic waste by cyanogenic bacteria and its enhancement with bio-oxidation. Adv Mater Res 71:661–664
Plepys A (2002) Implication of globalization and new product policies for the suppliers from developing countries. In: Proceedings of international symposium on electronics and the environment 2002, IEEE, San Francisco, CA, pp 53–58
Prakash A, Satyanarayana T, Johri BN (2012) Microorganisms in environmental management: microbes and environment. Springer, Dordrecht
Pramila S, Fulekar MH, Bhawana P (2012) E-waste—a challenge for tomorrow. Res J Recent Sci 1:86–93
Premalatha M, Tabassum A, Tasneem A, Abbasi SA (2014) The generation, impact, and management of E-waste: state of the art. Crit Rev Environ Sci Technol 44:1577–1678
Qasim SR, Chiang W (1994) Sanitary landfill leachate: generation, control and treatment. CRC, Boca Raton, FL
Qu W, Bi X, Sheng G, Lu S, Fu J, Yuan J, Li L (2007) Exposure to polybrominated diphenyl ethers among workers at an electronic waste dismantling region in Guangdong, China. Environ Int 33:1029–1034
Rajkumar M, Sandhya S, Prasad MNV, Freitas H (2012) Perspectives of plant associated microbes in heavy metal phytoremediation. Biotechnol Adv 30:1195–1750
Robinson HB (2009) E-waste: an assessment of global production and environmental impacts. Sci Total Environ 408:183–191
Saini R, Khanna R, Dutta RK, Cayumil R, Ikram-Ul-Haq M, Agarwala V, Ellamparuthy G, Jayasankar K, Mukherjee PS, Sahajwalla V (2017) A novel approach for reducing toxic emissions during high temperature processing of electronic waste. Waste Manag 64:182–189
Sajid M, Syed JH, Iqbal M, Abbas Z, Hussain I, Baig MA (2018) Assessing the generation, recycling and disposal practices of electronic/electrical-waste (E-Waste) from major cities in Pakistan. Waste Manag. https://doi.org/10.1016/j.wasman.2018.11.026
Scheutz C, Mosbaek H, Kjeldsen P (2004) Attenuation of methane and volatile organic compounds in landfill soil covers. J Environ Qual 33:61–71
Schinner F, Burgstaller W (1989) Extraction of zinc from industrial waste by *Penicillium* sp. Appl Environ Microbiol 55:1153–1156
Seguel A, Barea JM, Cornejo P, Borie F (2015) Role of arbuscular *mycorrhizal propagules* and glomalin related soil protein in aluminium tolerance of two barley cultivars growing in acid soils with high aluminium levels. Crop Pasture Sci 66:696–705
Singh A, Ward OP (2004) Biodegradation and bioremediation. Springer, New York
Sojinu SO, Oluwadayo O, Sonibare OE, Zeng EY (2012) Assessing anthropogenic contamination in surface sediments of Niger Delta, Nigeria with fecal sterols and n-alkanes as indicators. Sci Total Environ 441:89–96
Song X, Liu M, Wu D, Qi L, Ye C, Jiao J, Hu F (2014) Heavy metal and nutrient changes during vermicomposting animal manure spiked with mushroom residues. Waste Manag 34:1977–1983
Stewart ES, Lemieux PM (2003) Emissions from the incineration of electronics industry waste. In: Proceedings of electronics and the environment, IEEE international symposium, pp 271–275
Sthiannopkao S, Wong MH (2013) Handling e-waste in developed and developing countries: initiatives, practices, and consequences. Sci Total Environ 463–464:1147–1153

Stockholm Convention (2009a) UNEP/POPS/COP.4/SC-4/14 Listing of hexabromodiphenyl ether and heptabromodiphenyl ether
Stockholm Convention (2009b) UNEP/POPS/COP.4/SC-4/18 Listing of tetrabromodiphenyl ether and pentabromodiphenyl ether
Stockholm Convention (2010a) Technical review of the implications of recycling commercial pentabromodiphenyl ether and commercial octabromodiphenyl ether. In: 6th POP Reviewing Committee meeting (UNEP/POPS/POPRC.6/2) and Annex (UNEP/POPS/POPRC.6/INF/6)
Stockholm Convention (2010b) Annex to the technical review of the implications of recycling commercial pentabromodiphenyl ether and commercial octabromodiphenyl ether. In: 6th POP Reviewing Committee meeting (UNEP/POPS/POPRC.6/INF/6)
Surajit D (2014) Microbial biodegradation and bioremediation. Elsevier, New York
Suzuki I (2001) Microbial leaching of metals from sulfide minerals. Biotechnol Adv 19:119–132
Swedish Environmental Protection Agency (SEPA) (2011) Recycling and disposal of electronic waste Health hazards and environmental impacts. Swedish Environmental Protection Agency, Report 647
Tabak M, Clark DS, Hatchett SP, Key MH, Lasinski BF, Snavely RA, Wilks SC, Town RPJ (2005) Review of progress in fast ignition. Phys Plasmas 12:057305
Tanskanen P (2013) Management and recycling of electronic waste. Acta Mater 61:1001–1011
Townsend GT, Musson S, Jang YC, Chung IH (2004) Leaching of hazardous chemicals from discarded electronic devices. Prepared for U.S. EPA Region 4 and Region 5. Florida Center for Solid and Hazardous Waste Management, Gainesville, FL
Tripathi BD (2011) A short-term study on toxic effects of distillery sludge amendment on microbiological and enzymatic properties of agricultural soil in a tropical city. J Earth Sci Climat Change 2:106
Tsitonaki B, Petri M, Crimi H, Mosbæk RL, Siegrist PL, Bjerg (2010) In situ chemical oxidation of contaminated soil and groundwater using persulfate: a review. Crit Rev Environ Sci Technol 40:55–91
Umair S (2015) Informal electronic waste recycling in Pakistan. Licentiate Thesis, in Planning and Decision Analysis, with specialization in Environmental Strategic Analysis Printed by US-AB in Stockholm, Sweden
United Nations Environment Program (UNEP) (2007) E-waste Volume I: inventory assessment manual. United Nations Environmental Program Division of Technology, Industry and Economic, International Environmental Technology Centre, Osaka/Shiga
United Nations Environment Program (UNEP) (2009) Recycling—from e-waste to resources. United Nations Environment Program and United Nations University
Volesky B, May H, Holan ZR (1993) Cadmium biosorption by *Saccharomyces cerevisiae*. Biotechnol Bioeng 41:826–829
WAMDA (2018) The Middle East and its e-waste problems. https://www.wamda.com/2018/10/middle-east-e-waste-problem. Accessed 22 Jan 2019
Wang JP, Guo XK (2006) Impact of electronic wastes recycling on environmental quality. Biomed Environ Sci 19:137–142
Wang T, Fu J, Wang Y, Liao C, Tao Y, Jiang G (2009) Use of scalp hair as indicator of human exposure to heavy metals in an electronic waste recycling area. Environ Pollut 157:2445–2451
Wong CSC, Duzgoren-Aydin NS, Aydin A, Wong MH (2007a) Evidence of excessive releases of metals from primitive e-waste processing in Guiyu, China. Environ Pollut 148:62–72
Wong CSC, Wu SC, Duzgoren-Aydin NS, Aydin A, Wong MH (2007b) Trace metal contamination of sediments in an e-waste processing village in China. Environ Pollut 145:434–442
Xing GH, Chan JKY, Leung AOW, Wu SC, Wong MH (2009) Environmental impact and human exposure to PCBs in Guiyu, an electronic waste recycling site in China. Environ Int 35:76–82
Yang GC, Yeh CF (2011) Enhanced nano-$Fe_3O_4/S_2O_8{}^{2-}$ oxidation of trichloroethylene in a clayey soil by electro-kinetics. Separ Purif Technol 79:264–271
Zeng X, Xu X, Boezen HM, Huo X (2016) Children with health impairments by heavy metals in an e-waste recycling area. Chemosphere 148:408–415

Zhao G, Wang Z, Dong MH, Rao K, Luo J, Wang D, Zha J, Huang S, Xu Y, Ma M (2008) PBBs, PBDEs, and PCBs levels in hair of residents around e-waste disassembly sites in Zhejiang Province, China, and their potential sources. Sci Total Environ 397:46–57

Chapter 19
Biodegradation of E-waste Pollution

Aneela Iqbal, Maria Manzoor, and Muhammad Arshad

Abstract Sustainable management of electronic waste (e-waste) is gaining considerable attention throughout the world. E-waste pollutes natural resources (soil, air, surface/groundwater) directly as well as indirectly. Toxic chemicals released from e-waste are carcinogenic in nature and considered as health hazards. E-waste has very complex nature with distinct properties. In developed countries, it is either buried in the landfills or sent to less developed regions like Africa, China, India, and Pakistan. Expensive metals like gold (Au), silver and copper (Cu) are recovered and reused from e-waste. Inadequate use of personal protective gears and dangerous recovery methods risk human health as well as the ecosystem. Owing to their tendency for biomagnification along the food chain, there is a need for proper management from collection till disposal. The available practices for e-waste management include physical and chemical methods. Biological methods however offer wide range of inexpensive treatment techniques for safe removal of toxic e-waste components. The major bioremediation techniques used for e-waste management are bioleaching, biosorption, bioaccumulation, biotransformation, biomineralization, microbially enhanced chemisorption of metals, and phytoremediation. Considering the type of waste, the combination of these biological approaches has great potential to overcome problems associated with e-waste.

Keywords E-waste · Management · Bioremediation · Phytoremediation

19.1 Introduction

Growth of electronics industry is needed in today's modern world. Being innovative in application, it is becoming world's most powerful industry. Every year, electronic equipment worth billions of dollars is shipped over oceans. However, life cycle

A. Iqbal · M. Manzoor · M. Arshad (✉)
Institute of Environmental Sciences and Engineering, School of Civil and Environmental Engineering, National University of Sciences and Technology, Islamabad, Pakistan
e-mail: marshad@iese.nust.edu.pk

M. Z. Hashmi, A. Varma (eds.), *Electronic Waste Pollution*, Soil Biology 57,
https://doi.org/10.1007/978-3-030-26615-8_19

assessment of electronic products suggests these to be a hazard after their usage time. Sustainable management of electronic waste (e-waste) is gaining considerable attention throughout the world. E-waste pollutes natural resources (soil, air, surface/groundwater) directly as well as indirectly. Toxic chemicals released from e-waste are carcinogenic in nature and considered as health hazard. Owing to their tendency for biomagnification along the food chain, there is a need for proper management from collection till disposal. E-waste has very complex nature with distinct properties. Electronic waste consists of toxic chemicals mostly heavy metals and acids along with non-degradable plastics. Generally, these expired items are sent to landfills if not recycled. According to estimates, about 75% of e-waste is uncertain about its fate being either refurbished/reused for repairing or stays at homes, offices, and industries as junk material. E-waste comprised of toxic chemicals like mercury lamps, circuit boards, leaded glasses, etc. are exported for recycling to Asia (mostly China, India, and Pakistan) and Africa (NRDC 2012; Basel Action 2013). Tons of this e-waste are then dismantled (broken down into smaller pieces or items), unscrewed, shredded, or burnt. Burning generates smoke and dust heavily loaded with carcinogenic/mutagenic chemical compounds. This also leads to respiratory and skin problems. Burning circuit boards yield gold (Au), silver (Ag), cadmium (Cd), platinum (Pt), and copper (Cu). Wire coating yields polyvinyl chlorides and polychlorinated biphenyls. Printing toners produce carbon particles that may cause skin and lung cancer (Kevin et al. 2008).

Especially less developed areas in China, India, and Pakistan are prime spots for dumping e-waste. According to reports, about 70% of the e-waste is exported to China, while the remaining 30% is sent to India, Pakistan, Taiwan, and some African countries. These are considered as dumping e-waste stations of the world owing to their cheap labor, poor legislation, and lack of awareness among population. Poverty is the main driving force behind using these areas to absorb e-waste generated by developed countries like the United States, United Kingdom, and Europe (Hicks et al. 2005). In India's capital (Delhi), about 10,000–20,000 tons of e-waste is dismantled every year by locals without use of any personal protective equipment or protocols. This in turn badly affects human health and the ecosystem. Therefore, to overcome this problem, it is need of the hour to understand the risks and to review established treatment strategies (Pandve 2007; Keekeesocean 2012).

19.2 Sources, Composition, and Health Effects of E-waste

Almost every expired, non-repairable electronic item falls under the category of e-waste. It ranges from non-functional household items like cell phones, TVs, cameras, CD players, fans, air conditioners, ovens, sewing machines, washing machines, fridge, freezers, heaters, juicer blenders to office or industrial equipment like fax machines, telephones, printers, toners, cartridges, batteries, calculators, and wall clocks to name a few. Electronic equipment used in laboratories and by military is also a major source of hazardous e-waste. The scraps of personal computers are

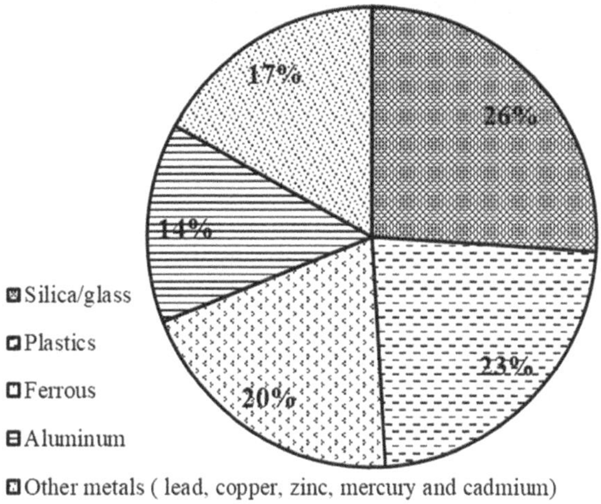

Fig. 19.1 Elemental composition of personal computer (Needhidasan et al. 2014)

Table 19.1 Metal abundance in different kinds of e-waste

	Metal abundance (%)				
E-waste	Pb	Ni	Al	Fe	Cu
Wiring board	2.6	2	7	12	16
Circuit boards	3	29	2	7	12
Computer board	1.5	1	5	7	20
Computer main board	2.2	1.1	2.8	4.5	14.3
TV board	1	0.3	10	28	10
Mobile phone	0.3	0.1	1	5	13
DVD player	0.3	0.05	2	62	5
Calculator	0.1	0.5	5	4	3
Television	0.2	0.04	1.2		3.8

Pant et al. (2014), Clark and Norris (1996), Monika (2010), Needhidasan et al. (2014)

also a major source of heavy metal pollution in the environment. The major composition of PC is given in Fig. 19.1.

Table 19.1 shows various toxic metals present in different e-waste materials. Heavy metals dominate within different elements. Hazardous metals, particularly Pb and Ni, are of major concern within selected metals.

Fe, Cu, Al, Au, and other metals constitute 60% of total generated e-waste, whereas plastics and other toxic chemicals contribute about 30 and 2.70%, respectively (Widmer et al. 2004). Exposure to these leads to various health issues like damage to the central nervous system (CNS), endocrine system function, digestive system, lungs, kidneys, and liver. A list of toxic chemicals found in e-waste and their probable health impacts has been summarized in Table 19.2.

Table 19.2 Sources and health effects of toxic chemicals present in e-waste (Puckett et al. 2018)

Hazardous elements	E-waste items	Health impacts
Beryllium (Be)	Computer motherboards	Lung cancer
		Beryllicosis
Cadmium (Cd)	Semiconductors	Kidney and liver failure
	Chip resistors	CNS damage
	CRT front panels	
Copper (Cu)	Wires	Wilsons's disease
	Circuit boards	Liver failure
	CRT front panels	Dysfunctioning of the digestive system
Chromium (Cr)	Galvanized items (steel plates)	Bronchitis
	CRT front panels	
	Decorators or hardeners (steel housing)	
Mercury (Hg)	Relay systems	CNS damage
	CRT front panels	Reported skin and lung disorders in fishes
	Circuit boards	
Lead (Pb)	Glass panels	Damage to the CNS, circulatory system, and kidneys
	Circuit boards	Retarded brain development in kids
	Gaskets of computer monitors	
	CRT front panels	
	Batteries	
Lithium (Li)	Batteries	Lung edema
		Adverse effects on nursing babies due to transfer through breast milk
Nickel (Ni)	Batteries	Dermatitis
		Asthma
Barium (Ba)	CRT front panels	Muscle fatigue
		Liver and spleen failure
		Heart attacks
Phosphorus (P)	CRT front panels	Muscle weakness
		Toxic to the heart, liver, and spleen
Plastics and PVCs	Cable coverings	Dioxins produced during burning lead to reproductive and developmental disorders
	Outer covers for computers, TVs	

19.3 Approaches for E-waste Management

In India, Environment Protection Act, 1986, makes the industry (polluter) responsible for payment of any damage to environment incurred due to processing of industry. This is widely known as "Polluter Pays Principle" or Extended Producer

Responsibility (EPR). In "International Environmental Law," it is mentioned as the "principle 16" of the "Rio Declaration on Environment and Development." According to the law, it is necessary for the government to enact legislations for release of hazardous materials to the environment and violation of legislation be treated with punishment/penalties. Although the Government of Ghana, Africa, is a signatory to every international treaty, according to estimates, about 4,000,000 tons of e-waste is treated in Ghana. Poverty being the major driving force for this transport, reusable electronic goods are resold in the markets (though without any warranty), while the unusables are dumped openly and burned. The same is observed in Guiyu, China. The city once was famous for paddy rice harvesting, but now it has been converted to a junkyard for e-waste. Due to poverty, people of the area are willing to risk their lives to dismantle e-waste. They aim to extract gold (Au), aluminum (Al), steel, and plastics, and the remaining is burnt to ashes. Smoke thus produced causes lung diseases or affects their skin. According to a study by Basel Action Network (2013) in the area, Linjaing River in Guiyu is heavily contaminated with Pb. Its concentration was estimated to range between 1.9 and 24 mg L^{-1} in comparison to 0.01 mg L^{-1} limit set by the World Health Organization (WHO). They fetch water from about 30 km away. Soil is also heavily contaminated due to open dumping of e-waste. Traces of various hazardous hydrocarbons, dioxins, and brominated chemicals were found from the area. India is not also far behind in the situation. Due to poor legislation, open scrap markets are functional in major areas (Janet et al. 2007).

Open burning/incineration, landfills, or acid baths are the available methods to dispose e-waste. Pyrolysis is also carried out in most of the cases. Pyrolysis is heating of the waste material in a closed chamber (oxygen-deficient environment) to convert it into charcoal, oil, and fumes, while in some cases, gasification is done in the presence of limited oxygen to convert the waste materials into tar, ashes, and fumes. However, incineration is the most common practice for e-waste management in these countries. Burning of plastics leads to release of various toxic chemicals like polychlorinated dibenzofurans, polyaromatic hydrocarbons, polychlorinated dibenzo-para-dioxins, to name a few (Janet et al. 2007; Leung et al. 2007). Most of these compounds are termed as carcinogenic for humans. Other gases released in the smoke include CO_2, SO_2, and NO_2 with traces of oxides of various heavy metals like Pb, arsenic (As), copper (Cu), manganese (Mn), mercury (Hg), nickel (Ni), etc. Acid bath is a method utilized to extract Cu, Pb, Au, and silver (Ag). For extraction, circuit boards are dipped in H_2SO_4 for 12 h. After complete dissolution of metals, the solution is boiled, and metal is extracted from the precipitates. The rest of the solution is mixed with scraped materials. People carry out all these activities with bare hands, without any personal protective equipment (PPE) or masks. Landfills are also in practice in the United States of America (USA) and Australia. The United States and Australia landfill about half of the e-waste and export the remaining half to Africa and Asia for disposal. E-waste landfill is not an eco-friendly option as there is a risk of leaching of heavy metals from buried waste. Acid batteries and circuits when dumped release acids like Hg, Ni, Cd, Pb, Cu, etc. These may leach down and contaminate groundwater resources (Pandve 2007).

Following are some of the safe methods which can be adopted to dispose e-waste in an eco-friendly manner (Ramachandra and Saira 2004):

1. First and foremost important is the recycling of the e-waste materials. Separate collection of e-waste should be encouraged by industries. Strict action should be taken against those who illegally collect e-waste and dismantle it.
2. There should be proper storage sites for collected e-waste until it is ready for extraction or reuse.
3. Strict rules and regulations should be in practice with mandatory PPEs, gloves, and glasses.
4. Basel Convention and laws relevant to the e-waste management should be implemented without any political pressure or interest.
5. There should be a ban on easy but dangerous methods for e-waste management like open dumping, incineration, acid baths, and landfill to avoid problems associated with all these disposal methods.
6. Researchers should be encouraged to find alternative for dangerous chemicals for use in the electronic products and also find alternate methods for extraction and safe waste disposal.
7. There should be proper monitoring about transportation of hazardous e-waste within the states or across the countries.
8. Use of refurbished item should be encouraged. Industries should be supported to produce by recycling or refurbishing and provided with e-Stewards certificates.
9. Awareness about e-waste and its hazardous impacts should be raised especially among poor communities.
10. Governments should allocate special funds for e-waste management to abide by the international laws and legislations within their jurisdiction.

19.4 Role of Biotechnology in E-waste Management

E-waste contains lager amount of organic and inorganic waste particularly heavy metals are major component of e-waste. For instance, circuit boards contain 28–30% of metal components. Of which 10–20% is copper, 1–5% is lead, and 0.3–0.4% are different but precious metals such as gold, silver, and platinum (Karwowska et al. 2014). To remediate these contaminants and their constituents, various conventional techniques that were in practice previously include physical and chemical techniques. Physical techniques include physical removal and excavation of contaminated soil and dumping in landfills. However, chemical methods included hydrolysis, precipitation chemical stabilization, and mining. These methods are not suitable for practical application since they involve high energy and cost and liberate secondary pollution and physical disturbance of remediation site. In this scenario, biotechnology offers a wide range of remediation techniques. All these techniques are environmentally friendly and required less energy inputs. Major biotechnology tools are bacteria and fungi with oxidoreductase activity.

19.5 Biodegradation of E-waste

Biodegradation is the chemical dissolution of materials by bacteria or other biological means and is one of the most important environmental processes affecting the breakdown of organic compounds. Organic material can be degraded aerobically with oxygen, or anaerobically, without oxygen. Biosurfactant, an extracellular surfactant secreted by microorganisms, enhances the biodegradation process. Biodegradable matter is generally organic material such as plant and animal matter and other substances originating from living organisms (Borthakur and Singh 2016).

Some microorganisms have a naturally occurring, microbial catabolic diversity to degrade, transform, or accumulate a huge range of compounds including hydrocarbons (e.g., oil), polychlorinated biphenyls (PCBs), polyaromatic hydrocarbons (PAHs), pharmaceutical substances, radionuclides, pesticides, and metals. Decomposition of biodegradable substances may include both biological and abiotic steps. Heavy metals are not biodegradable. They can only be transformed from one chemical state to another which could be less injurious to environment. Even then biological agents can effectively remediate e-waste through a range of techniques including bioleaching, bioaccumulation, biosorption, and bioprecipitation (Patel and Kasture 2014).

19.6 E-waste Bioremediation Techniques

19.6.1 Bioleaching

Bioleaching is the process wherein microorganisms solubilize the insoluble states of metals to soluble ones. In this process, the availability and solubility are increased by microbial reaction. This involves different functional activities and secretions that trigger metal solubility. Bioleaching is of two types: direct bioleaching and indirect bioleaching (Tichy et al. 1998).

19.6.1.1 Direct Bioleaching

In case of direct bioleaching, extracellular compounds are produced by microorganisms such as organic acids that in turn oxidize metals' insoluble forms to oxidized ions that are more soluble. Once the metals are soluble, these can be easily extracted and further purified for reuse purposes. Although it seems a promising method, finding appropriate kind of bacteria is a challenging task. These bacteria must not only be producing desired compounds but also toxicities encountered due to exposure to e-waste (Tichy et al. 1998).

19.6.1.2 Indirect Bioleaching

On other hand, in indirect bioleaching, bacteria with metal-oxidizing bacteria are used. The oxidation process liberated free metal cations in solution as a result of microbial-induced metal oxidation (Tichy et al. 1998). This process could be effective if there is availability of right combination of bacteria. It could be a bit complicated due to usage of multiple types of bacteria, as they must be tolerant to toxicities and then specific to the target elements/metals (Tichy et al. 1998).

19.6.2 Biosorption

It is a remediation process by which microorganisms remove metals from medium by metal sorption. Metals are concentrated by microorganisms on their out surface. This is attributed to structural characteristics of binding microorganisms. Biosorption is not an active metabolism. The adsorbed metals can be released later on in the presence of other anions with higher tendency to the attached adsorbed metals (Volesky and Holan 1995).

19.6.3 Bioaccumulation

It is the ability of microorganisms to absorb metal contaminants and concentrate it inside the cell structure. This involves detoxification mechanism and pile up of metals inside cell vacuole. It is an active metabolism where cell energy is required to absorb and detoxify metal inside the cell. This way, metals are removed from the environment and become part of living cells (Das and Adholeya 2012). The organic contaminants are either broken down outside of cell and used as carbon source or taken up and converted to secondary metabolites inside the cytoplasm. Inorganic metals are not biodegradable and, therefore, are converted to less toxic forms or immobilized inside the cell biomass. Accessibility to the target elements from e-waste could be an issue for the bioaccumulating microorganisms. For that purpose, pre-treatments could be needed in certain cases (Hou et al. 2006).

19.6.4 Biotransformation

It refers to the microbially mediated chemical state transformation of metals that are less toxic to the environment. The oxidation state of metals modified as a result of addition or removal of electrons by oxidizing or reducing bacteria is termed as biotransformation. In this process, the contaminant is not broken down, but the chemical state of substance is altered in the presence of biochemical agents. As a result, the chemical characteristics of the targeted metals are modified (Das and Adholeya 2012). Biotransformation is of two types: direct biotransformation and indirect biotransformation.

19.6.4.1 Direct Biotransformation

It is also known as enzymatic reduction or enzymatic biotransformation. Wherein the sole agents of biochemical conversion of oxidation state of metals are enzymes produced by microorganisms. In this process, multivalent toxic metals are reduced. The extracellular enzymes donate electrons that bring reduction of these metal contaminants outside of cell structure (Tabak et al. 2005).

19.6.4.2 Indirect Biotransformation

It is a type of bioremediation process in which metal-reducing and sulfate-reducing bacteria are used to stabilize multivalent toxic heavy metals. Metals are immobilized in sedimentary and subsurface environment by action of reducing bacteria (Tabak et al. 2005).

19.6.5 Biomineralization

Biomineralization is the most widely used technique for e-waste management that involved removal of toxic heavy metals, plastics, and other organic components either by precipitation or degradation into principal components. By precipitation, the toxic heavy metals are converted to relatively stable forms, and the organic compounds are broken down to less toxic and more stable forms. Biomineralization of heavy metals (nickel, copper, lead, cobalt, zinc, and cadmium) by formation of stable complexes with removal efficiency of 88–99% was exhibited by urease-producing bacteria (Li et al. 2013). The attachment of metal ions to inorganic anions or ligands released from microorganisms to form precipitation is the basic mechanism involved in biomineralization (Patel and Kasture 2014).

19.7 Bioremediation Agents and Role in E-waste Management

The major agents that are responsible for bioremediation of e-waste and toxic components are versatile group of bacteria, fungi, and plants.

19.7.1 Role of Bacteria

Environmental bacteria play an important role in biomanagement and remediation of e-waste besides regulating important ecological processes, extending from nutrient mineralization to waste recycling in natural contaminated systems. Different studies have been conducted to demonstrate the significance of bacteria in decontamination of e-waste and with success. Use of soil bacteria for biomining and bioextraction of precious metals from natural reservoirs to contaminated e-waste soil has been in practice. For example, bioleaching of gold from e-waste has efficiently been observed using *Chromobacterium violaceum*, *Pseudomonas aeruginosa*, *Pseudomonas fluorescens*, and *Escherichia coli* (Li et al. 2015). The highest gold extraction and recovery have been reported in *C. violaceum* (Natarajan and Ting 2015). *Chromobacterium* species have also been reported to have high gold tolerance and have been used as an effective inoculant for bioleaching of gold (Madrigal-Arias et al. 2015). Bioleaching of different heavy metals from printed circuit boards has also been found in *Acidithiobacillus ferrooxidans* and *Acidithiobacillus thiooxidans* and was applied as a bioremediation technique for decontamination of e-waste (Nie et al. 2015). Addition of metal chelation compound EDTA along with bacteria *S. plymuthica* has been found to be an efficient hybrid technique for e-waste metal extraction (Pb, Ba, Ca, Si, and Cd) (Pant et al. 2014). Bacteria using different techniques for bioremediation of e-waste are summarized in Table 19.3.

Besides metals, other major components of e-waste include organic compounds particularly persistent organic pollutants (POPs) and phenolic compounds. Bacteria also play important role in degradation of these toxic POPs. In a study conducted, polybrominated diphenyl ethers (PBDE) and tetrabromobisphenol were efficiently degraded by *Rhodococcus* sp. and by *Ochrobactrum* sp., respectively (Borthakur and Singh 2016).

19.7.2 Role of Fungi

Fungi are considered as major biological agents responsible for biodegradation of organic wastes and are widely used for degradation of hazardous environmental pollutants and, therefore, have huge potential to contribute significantly toward the pollution-abatement aspects of e-waste. Among fungi, saprophytes are of particular importance since these groups of fungi are associated in decomposition and mineralization process. As far as e-waste degradation is concerned, many studies revealed the beneficial role of fungi in e-waste management. *Aspergillus* species have been found to have greater tolerance against gold and used extensively to extract gold through fungus-induced bioleaching (Madrigal-Arias et al. 2015). *Aspergillus niger*,

Table 19.3 Bioremediation of e-waste by bacteria

Bioremediation techniques	Bacterial strains	Metals	References
Bioleaching	*Acidithiobacillus thiooxidans*	Arsenic, lead	Stephen and Macnaughtont (1999)
	Micrococcus roseus	Cadmium	
	Thiobacillus ferrooxidans	Arsenic, lead	
Biosorption	*Bacillus sphaericus*	Chromium	Hu et al. (1996) and Atkinson et al. (1998)
	Myxococcus xanthus	Uranium	
	Pseudomonas aeruginosa	Cadmium, uranium	
	Streptoverticillium cinnamoneum	Lead	
Bioaccumulation	*Bacillus circulans*	Chromium	Srinath et al. (2002)
	Bacillus megaterium	Chromium	
	Deinococcus radiodurans	Uranium	
	Micrococcus luteus	Uranium	
	Anaeromyxobacter sp.	Uranium	Lovley and Coates (1997) and Francis (1998)
	Clostridium sphenoides	Uranium	
	Halomonas sp.	Uranium	
	Serratia sp.	Chromium	
Biomineralization	*Bacillus fusiformis*	Lead	Tabak et al. (2005)
	Cupriavidus metallidurans	Cadmium	
	Desulfotomaculum	Arsenic	Achal et al. (2012)
	Auripigmentum		
	Sporosarcina ginsengisoli	Arsenic	Benzerara et al. (2011)

Penicillium simplicissimum, and their mixtures were also found efficient in bioremediation of heavy metals from printed circuit boards through process of bioleaching (Nie et al. 2015). Different fungi associated with bioremediation of e-waste are given in Table 19.4.

Aspergillus niger and *Penicillium simplicissimum* were found efficient in mobilizing high concentration of Cu, Cd, Sn, Al, Ni, Pb, and Zn from e-waste showing their potential utility for treating e-waste contaminated soil. These fungi appeared as a promising biological technique in metallurgical processes nowadays and have significant contribution in management of e-waste polluted sites (Madrigal-Arias et al. 2015; Nie et al. 2015).

Table 19.4 Bioremediation of e-waste by fungi

Bioremediation techniques	Fungal strains	Metals	References
Bioleaching	*Aspergillus fumigatus*	Arsenic	Ren et al. (2009)
	Aspergillus niger	Cadmium, lead	
Biosorption	*Rhizopus arrhizus*	Uranium	Ahalya et al. (2003)
	Saccharomyces cerevisiae	Cadmium	
Bioaccumulation	*Aspergillus niger*	Chromium, lead	Malik (2004)
	Monodictys pelagica	Chromium, lead	
Biotransformation	*Rhizopus oryzae*	Chromium	Malik (2004)
	Fusarium oxysporum	Cadmium	
Biomineralization	*Aspergillus flavus*	Lead	Govarthanan et al. (2012)

19.7.3 Role of Plants

Plants have evolved great capacity to tolerate and decontaminate the soil by sequestering metals inside plant biomass (Arshad et al. 2016; Gul et al. 2018; Manzoor et al. 2018). Plants in combination with endophytes (microbes residing in plants) are being considered promising for effective phytoremediation of organic pollutants (Iqbal et al. 2018). Soil contaminated with industrial pollution and e-waste has elevated concentrations of toxic heavy metals. Plants can naturally grow on such contaminated soils and remediate the soil. There are different strategies in phytoremediation based upon mode of action of plants. The categories relevant to e-waste include phytoextraction and phytostabilization. These methods are established procedures for heavy metal removal from soil. Some examples are reported in Table 19.5. Since soil is a major sink for heavy metals from e-waste, application of plants is very relevant for bioremediation of e-waste (Sinha et al. 2010). There is need to explore further potential of plants in biomanagement of e-waste in the near future.

19.8 Environmental Significance of Biodegradation of E-waste

Microorganisms nowadays are widely used for solubilization of metals and their consequent recovery from electronic waste. Environmentally friendly and cost-effective nature of bioleaching have highlighted the potential application of these bacteria and fungus in bioleaching of metals from e-waste polluted soil as well as

Table 19.5 Phytoremediation techniques for e-waste management

Phytoremediation techniques	Plants	Metals	References
Phytoextraction	*Pelargonium* spp.	Lead	Arshad et al. (2008, 2016), Manzoor et al. (2018) Sinha et al. (2010), Han et al. (2016)
	Vetiveria zizanioides	Lead	
	Agrostis tenuis	Lead	
	Pteris vittata	Arsenic	
	Thlaspi caerulescens	Cadmium, nickel, zinc	
	Alyssum spp.	Nickel	
	Lecythis ollaria	Selenium	
	Astragalus racemosus	Selenium	
	Macadamia neurophylla	Manganese	
	Maytenus bureaviana	Manganese	
	Ipomea alpine	Copper	
	Crotalaria cobalticola	Cobalt	
Phytostabilization	*Populus canadensis*	Cadmium, lead	David et al. (1997) and Wilde et al. (2005)
	Atriplex lentiformis	Uranium	
	Chrysopogon zizanioides	Arsenic, lead	
	Mediterranean L. albus	Arsenic, cadmium	

mining. The technology thus appeared as a promising technique for metallurgical processes (Nie et al. 2015). In general, very limited work has been reported in literature related to application of biological methods for e-waste management and remediation. The potential is far high and there is a need to develop effective methodologies.

19.9 Conclusions

Biotechnological processes have huge potential to address environmental issues imposed by e-waste. These technologies are efficient and adequate and can be modified based on purpose and type of management practices. The most significant aspect is the environment-friendly characteristics of these biological agents. The technologies are cost-effective and generate no secondary pollution and cause no physical disturbance to the environment.

References

Achal V, Pan X, Fu Q, Zhang D (2012) Biomineralization based remediation of As (III) contaminated soil by *Sporosarcina ginsengisoli*. J Hazard Mater 201:178–184

Ahalya N, Ramachandra TV, Kanamadi RD (2003) Biosorption of heavy metals. Res J Chem Environ 7(4):71–79

Arshad M, Silvestre J, Pinelli E, Kallerhoff J, Kaemmerer M, Tarigo A, Shahid M, Guiresse M, Pradère P, Dumat C (2008) A field study of lead phytoextraction by various scented *Pelargonium* cultivars. Chemosphere 71(11):2187–2192

Arshad M, Merlina G, Uzu G, Sobanska S, Sarret G, Dumat C, Silvestre J, Pinelli E, Kallerhoff J (2016) Phytoavailability of lead altered by two *Pelargonium* cultivars grown on contrasting lead-spiked soils. J Soils Sediments 16(2):581–591

Atkinson BW, Bux F, Kasan HC (1998) Considerations for application of biosorption technology to remediate metal-contaminated industrial effluents. Water SA 24(2):129–135

Basel Action (2013) Exporting harm: the high-tech trashing of Asia [video file]. http://www.youtube.com/watch?v=yDSWGV3jGek. Accessed July 2018

Benzerara K, Miot J, Morin G, Ona-Nguema G, Skouri-Panet F, Férard C (2011) Significance, mechanisms and environmental implications of microbial biomineralization. Comptes Rendus Geosci 343(2):160–167

Borthakur A, Singh P (2016) Microbial biotechnology for sustainable electronic waste management. In: Conference Proceedings – Asia-Pacific Conference on Biotechnology for Waste Conversion 2016, Hong Kong SAR, People's Republic of China, 5–8 Dec 2016, pp 314–317

Clark DA, Norris PR (1996) *Acidimicrobium ferrooxidans* gen. nov., sp. nov. mixed-culture ferrous iron oxidation with *sulfobacillus* species. Microbiology 142:785–790

Das M, Adholeya A (2012) Role of microorganisms in remediation of contaminated soil. In: Prakash A, Satyanarayana T, Johri BN (eds) Microorganisms in environmental management. Springer, New York, pp 81–112

David ES, Ilya R, Robert DS (1997) Phytoremediation of metals: using plants to remove pollutants from the environment. Curr Opin Biotechnol 8:221–226

Francis AJ (1998) Biotransformation of uranium and other actinides in radioactive wastes. J Alloys Compd 271:78–84

Govarthanan M, Lee KJ, Cho M, Kim JS, Kamala-Kannan S, Oh BT (2012) Significance of autochthonous *Bacillus* sp. KK1 on biomineralization of lead in mine tailings. Chemosphere 90(8):2267–2272

Gul I, Manzoor M, Silvestre J, Rizwan M, Hina K, Kallerhoff J, Arshad M (2018) EDTA-assisted phytoextraction of lead and cadmium by Pelargonium cultivars grown on spiked soil. Int J Phytoremediation. https://doi.org/10.1080/15226514.2018.1474441

Han YH, Fu JW, Chen Y, Rathinasabapathi B, Ma LQ (2016) Arsenic uptake, arsenite efflux and plant growth in hyperaccumulator *Pteris vittata*: role of arsenic-resistant bacteria. Chemosphere 144:1937–1942

Hicks C, Dietmar R, Eugster M (2005) The recycling and disposal of electronic waste in China – legislative and market response. Environ Impact Assess Rev 25:459–471

Hou B, Zhu K, Lu J, Zhao YF (2006) Research on bioremediation of petroleum contaminated soils and its perspectives. Sichuan Environ 6(24):131–140

Hu MZC, Norman JM, Faison BD, Reeves ME (1996) Biosorption of uranium by *Pseudomonas aeruginosa strain* CSU: characterization and comparison studies. Biotechnol Bioeng 51 (2):237–247

Iqbal A, Arshad M, Hashmi I, Karthikeyan R, Gentry TJ, Schwab AP (2018) Biodegradation of phenol and benzene by endophytic bacteria from refinery wastewater fed *Cannabis sativa*. Environ Technol 39(13):1705–1714

Janet KY, Xing CG, Xu Y, Liang Y, Chen LX, Wu SC et al (2007) Body loadings and health risk assessments of polychlorinated dobenzo-p-dioxines and dibezofurans at an intensive electronic waste recycling site in China. Environ Sci Technol 41:7668–7674

Karwowska E, Andrzejewska-Morzuch D, Lebkowska M, Tabernacka A, Wojtkowska M, Telepko A et al (2014) Bioleaching of metals from printed circuit boards supported with surfactant producing bacteria. J Hazard Mater 264:203–210

Keekeesocean (2012) E waste in India short documentary [video file]. http://www.youtube.com/watch?v=sFfaYc_pIx8

Kevin B, Labunska I, Santillo D, Johnston P (2008, August 10) Chemical contamination at e-waste recycling and disposal sites in Accra and Korforidua, Ghana. Greenpeace Toxic Technical Note

Leung AO, Lutisemburg WJ, Wong AS, Wong MH (2007) Spatial distribution of polybrominated diphenyl ethers and polychlorinated dibenzo-p-dioxins and bibenzofurans in soil and combined residue at Guiyu: an electronic-waste recycling site in southeast China. Environ Sci Technol 41:2730–2737

Li M, Cheng X, Guo H (2013) Heavy metal removal by biomineralization of urease producing bacteria isolated from soil. Int Biodeter Biodegrad 76:81–85

Li JY, Liang CJ, Ma CJ (2015) Bioleaching of gold from waste printed circuit boards by *Chromobacterium violaceum*. J Mater Cycles Waste 17:529–539

Lovley DR, Coates JD (1997) Bioremediation of metal contamination. Curr Opin Biotechnol 8 (3):285–289

Madrigal-Arias JE, Argumedo-Delira R, Alarcon A, Mendoza-Lopez MR, Garcia-Barradas O, Cruz-Sanchez JS et al (2015) Bioleaching of gold, copper and nickel from waste cellular phone PCBs and computer goldfinger motherboards by two *Aspergillus niger* strains. Braz J Microbiol 46:707–713

Malik A (2004) Metal bioremediation through growing cells. Environ Int 30(2):261–278

Manzoor M, Gul I, Silvestre J, Kallerhoff J, Arshad M (2018) Screening of indigenous ornamental species from different plant families for Pb accumulation potential exposed to metal gradient in spiked soils. Soil Sediment Contam 27(5):439–453

Monika JK (2010) E-waste management: as a challenge to public health in India. Indian J Commun Med 35(3):382–385

Natarajan G, Ting YP (2015) Gold biorecovery from e-waste: an improved strategy through spent medium leaching with pH modification. Chemosphere 136:232–238

Needhidasan S, Samuel M, Chidambaram R (2014) Electronic waste—an emerging threat to the environment of urban India. J Environ Health Sci Eng 12(1):12–36

Nie HY, Yang C, Zhu NW, Wu PX, Zhang T, Zhang YQ et al (2015) Isolation of *Acidithiobacillus ferrooxidans* strain Z1 and its mechanism of bioleaching copper from waste printed circuit boards. J Chem Technol Biot 90:714–721

NRDC (2012) Natural Resources Defense Council, E-waste: saving developing communities from our electronic junk. http://www.nrdc.org/living/stuff/e-wastesaving- developing-communities-electronicjunk.asp

Pandve HT (2007) E-waste management in India: an emerging environmental and health issue. Indian J Occup Environ Med 11(3):116

Pant D, Singh P, Upreti MK (2014) Metal leaching from cathode ray tube waste using combination of *Serratia plymuthica* and EDTA. Hydrometallurgy 146:89–95

Patel S, Kasture A (2014) E (electronic) waste management using biological systems-overview. Int J Curr Microbiol Appl Sci 3(7):495–504

Puckett J, Byster L, Westervelt S, Gutierrez R, Davis S, Hussain A (2018) Exporting harm: the high-tech trashing of Asia. Basal Action Network, Seattle. http://www.ban.org. Accessed Aug 2018

Ramachandra TV, Saira VK (2004) Environmentally sound option for e-waste management. Envis J Human Settlements. http://ecsenvironment.com/images/Library/Journals/Environmentally%20Sound%20Options%20For%20E-Wastes%20Management%20Ramachandra%20TV%20%20Saira%20Varghese%20K.pdf
Ren W, Li P, Geng Y, Li X (2009) Biological leaching of heavy metals from a contaminated soil by *Aspergillus niger*. J Hazard Mater 167(1):164–169
Sinha RK, Valani D, Sinha S, Singh S, Herat S (2010) Bioremediation of contaminated sites: a low-cost nature's biotechnology for environmental cleanup by versatile microbes, plants & earthworms. In: Solid waste management environmental remediation. Nova Science, Hauppauge, NY, pp 1–72
Srinath T, Verma T, Ramteke PW, Garg SK (2002) Chromium (VI) biosorption and bioaccumulation by chromate resistant bacteria. Chemosphere 48(4):427–435
Stephen JR, Macnaughtont SJ (1999) Developments in terrestrial bacterial remediation of metals. Curr Opin Biotechnol 10(3):230–233
Tabak HH, Lens P, van Hullebusch ED, Dejonghe W (2005) Developments in bioremediation of soils and sediments polluted with metals and radionuclides–1. Microbial processes and mechanisms affecting bioremediation of metal contamination and influencing metal toxicity and transport. Rev Environ Sci Biotechnol 4(3):115–156
Tichy R, Rulkens WH, Grotenhuis JTC, Nydl V, Cuypers C, Fajtl J (1998) Bioleaching of metals from soils or sediments. Water Sci Technol 37(8):119–127
Volesky B, Holan ZR (1995) Biosorption of heavy metals. Biotechnol Prog 11(3):235–250
Widmer R, Oswald HK, Sinha DK, Schnellmann M, Heinz B (2004) Global perspectives on e-waste. Environ Imp Assess Rev 25:436–458
Wilde EW, Brigmon RL, Dunn DL, Heitkamp MA, Dagnan DC (2005) Phytoextraction of lead from firing range soil by *Vetiver* grass. Chemosphere 61(10):1451–1457

Chapter 20
Enzymatic Role in the Degradation of E-waste Pollution

Ankita Srivastava, Sukriti Jaiswal, Niharika Chandra, and Sunil Kumar

Abstract E-waste pollution due to heavy metals has got worldwide attention. The undesirable persistent property of heavy metals poses a serious threat to the environment as well as plants and animals. It may also lead to several diseases in humans. It has a wide range of scope to stress on cost-effectiveness, suitability, and sustainability of the techniques, which mitigate the effect of change in environment. Contamination of food products also influences anthropogenic change on the environment and exploration of the above prospects. Bioremediation could be the cleaner, safer, cost-effective, and environment-friendly tool for decontamination of a wide range of pollutants. Various biological agents like bacteria, yeast, fungi, algae, and higher plants are used in bioremediation as main tools in treating oil spills and heavy metals pollution. In order to regulate increasing pollution and environmental problems, an endless search for new biological agents is still required. Microorganisms have wide capacity of regulating pollution, but the exact mechanisms are still unknown. Therefore, it is indeed required to review available options to control environmental pollution. The role of enzymes in this way is one of the most recent technology for the management of e-waste.

Keywords Electronic waste · Phytoremediation · Toxic metals · Recycling · Microbial remediation

---

A. Srivastava · S. Kumar (✉)
Faculty of Bio-Sciences, Institute of Bio-Sciences and Technology, Shri Ramswaroop Memorial University, Barabanki, Uttar Pradesh, India
e-mail: sunil.bio@srmu.ac.in

S. Jaiswal · N. Chandra
Faculty of Biotechnology, Institute of Bio-Sciences and Technology, Shri Ramswaroop Memorial University, Barabanki, Uttar Pradesh, India

M. Z. Hashmi, A. Varma (eds.), *Electronic Waste Pollution*, Soil Biology 57,
https://doi.org/10.1007/978-3-030-26615-8_20

20.1 Introduction

Waste can be defined as material that is rejected or no longer in use. At present a huge quantity of waste is produced, which requires serious attention. The undesired growth in the waste has led to the poor quality of our lives. On the other hand, this has led to multiple problems, like huge amount of hazardous and other wastes generated from electric products "electronic waste (e-waste)." The presence of several poisonous chemicals and noxious substances in the disposed e-waste, i.e., electronic gadgets, is causing environmental and health problems (Needhidasan et al. 2014).

Previously it has been reported that Europeans produce 20 kg whereas Americans produce 7 kg of e-waste/person in a year. However, Indians produce 1,46,000 tons of e-waste in a year. Therefore, it is very much needed to take sufficient actions to regulate e-waste rise and to avoid the increased rate of toxic waste generated in the environment and for health hazards (Monika and Kishore 2010).

The metals present in the environment can be degraded and detoxified by microbial organisms. It is well known that adsorption, methylation, demethylation, oxidation, and reduction reactions work as controller of metal toxicants (Jan et al. 2015). The microorganisms require these metal ions for their physiological and metabolic process (Ayangbenro and Babalola 2017). Bacteria precipitate these metals into metal ions within the cells (Kotrba et al. 1999). Bacteria after degradation of metals produce chemical substances like phenols and catechol from siderophore for ion uptake process (Howard 1999). Bacteria can degrade metals like chromium, mercury, and selenium by oxidation or reduction reactions (Tchounwou et al. 2012). This process is categorized into two reactions: assimilatory and dissimilatory reactions. Bacterial assimilation is involved in the transformation of metals (Merchant and Helmann 2012). Metals that can act as a terminal electron acceptor can increase the physiological and metabolic activity of microbes. The metals do not play any role in dissimilatory reaction (Kracke et al. 2015). Enzymes produced by bacteria take part in oxidation or reduction of metals (Finnegan and Chen 2012). The serine hydrolase enzyme has been isolated on a large scale from plastic-eating fungi, i.e., *Pestalotiopsis microspora*, which breaks down plastics in landfills. Hence, this chapter is entirely focused on illustrating the roles of various microbial enzymes, which can be used to degrade/control the hazardous e-waste.

20.2 E-waste

The term e-waste stands for electronic waste which includes all the varieties of electronic devices that have been used or replaced by their advanced versions (Pinto 2008).

The rapid advancing technology has led to the creation or say explosion of e-waste since more and more numbers of devices are being discarded due to their

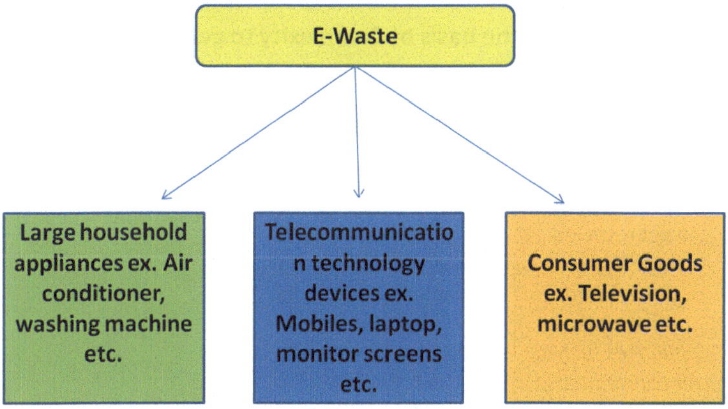

Fig. 20.1 The three major divisions of e-waste on the basis of detectable components

damaged parts, undesirable requirement, and availability of better models. Other than the electronic device itself, its manufacturing components and the accessories associated with their enhanced use are also considered as e-waste (Needhidasan et al. 2014).

Some of the common daily life examples of e-waste include televisions, from older box model to flat screen of newer versions, DVD and CD players, VCRs, radio, old mobile phones, damaged charger, earphones, batteries, disposed appliances such as iron, mixer grinder, hair straightener, bulbs, tube lights, remote, refrigerator or parts, filters of purifier, etc. Most of the used electrical appliances that are supposed to be reused, recycled, or further sold secondhand, etc. are also counted under the label of e-waste. The list of e-waste is endless and so is the severity of facing its management (Osibanjo and Nnorom 2007). Since much of the e-waste consists of hazardous substances such as heavy metals which could cause serious consequences to the people involved in its management and treatment as well as to the common mass (Monika and Kishore 2010).

The vast range of e-waste has been organized into three major divisions on the basis of some common elements that are usually used in their manufacturing process (Annamalai 2015) (Fig. 20.1). These elements are easily distinguishable which simplifies their detection and separation processes. These elements include plastics, rubber, glass, metals, circuits, chips, wire and cables, heating and cooling elements, concrete, textile, batteries, etc. (Wong et al. 2007).

Another major classification of e-waste can be done on the basis of their complexity to get recycled due to various parameters as shown in Fig. 20.2.

E-waste is made up of thousands of various substances which can be classified as hazardous and nonhazardous substances (Table 20.1) (Needhidasan et al. 2014). These include glass, plastics, metals such as Fe, Cu, and Al and noble metals (Ag, Au, Pt, Pd, etc.), wood, circuits, rubber, etc. The very existence of heavy metals in e-waste such as Hg, Pb, As, Cd, Cr, etc. above their threshold value categorizes them as hazardous substances (Wong et al. 2007).

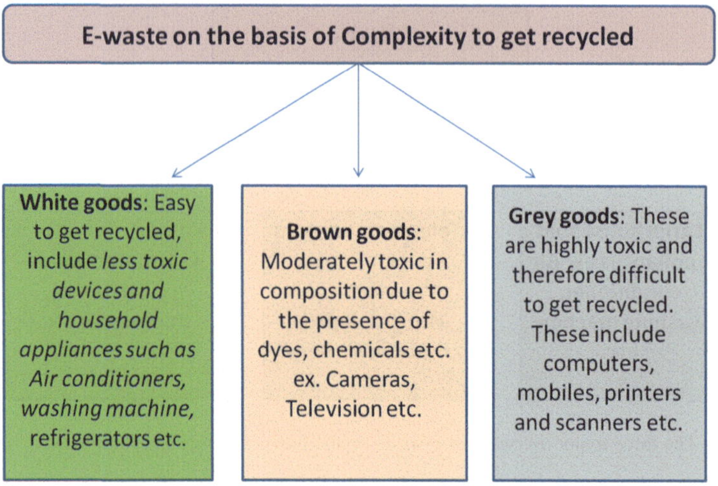

Fig. 20.2 Classification of e-waste can on the basis of their complexity to get recycled

Table 20.1 E-waste composition (Pinto 2008)

Constituents of e-waste	Percentage (%) of the constituents present
Iron and steel	50
Plastics	21
Nonmetals	13
Others	16

20.3 Problems Associated with E-waste

E-waste being nonbiodegradable does not rot or decompose by itself. Getting rid of e-waste is one of the major problems being faced these days across the globe. These wastes when dumped in landfills release harmful leachate that leads to pollution of groundwater (Needhidasan et al. 2014). Moreover, the sludge obtained from melting of e-waste, particularly the circuits and computer chips, when comes in contact with the soil results in its acidification. This has caused a huge problem in an advanced area in Hong Kong called Guiyu, which is facing severe water scarcity due to its contamination by the leachate and sludge disposal of illegal e-waste treatment carried out here.

Various environmental hazards can occur due to inappropriate disposal of e-waste in landfills. For instance, many toxic elements may leach out polluting the groundwater such as mercury from broken circuits, polychlorinated biphenyls (PCBs) released from condensers, polybrominated diphenyl ethers (PBDE), as well as

Table 20.2 Harmful consequences of e-waste on human health (Monika and Kishore 2010)

Types of e-waste	Components	Serious health consequences
Soldering used in printed circuit boards (PCB), glass panels, and computer gaskets	Lead	• Nervous system, blood system, and kidney damage • Brain development in children is affected
Resistors in chip and semiconductors	Cadmium	• Irreversible toxic effects • Gets accumulated in liver and kidney • Neural damage
Relays and switches, PCB	Mercury	• Long-term brain damage • Bioaccumulation in fishes results in respiratory and skin disorders
Untreated and galvanized steel plates from corrosion, a decorator and steel hardener	Hexavalent chromium	• Asthmatic bronchitis • Damage to DNA
Cabling and computer coverings	Plastics including PVC	Dioxin produced by burning causes • Problems in reproduction and development • Damage to immune system • Interference with regulatory hormones
Plastic casing of electronic equipment	Brominated flame retardants	• Endocrine system functions gets damaged
Front panel of Cathode Ray Tubes	Barium	Short-term revelation causes: • Weakness in muscle • Heart, liver, and spleen get damaged
Motherboard	Beryllium	• Lung cancer • Fumes and dust inhalation causes berylliosis • Skin diseases, e.g., warts

cadmium from plastics and high amount of lead from glass and cathode ray tubes (Kaw and Kannan 2017; Wu et al. 2008). Often uncontrolled fire starts at landfills which expose metals, chemicals like harmful dioxins, furans, and PCBs, and plastics to be burned releasing their toxic fumes in the open air causing serious health problems. Disposal of e-waste by incineration releases poisonous smoke and fumes which leads to air pollution and respiratory disorders. Furthermore, if the e-waste is disposed together with the household waste, it will harm both the health and the ecosystem. Table 20.2 shows some of the harmful effects of e-waste components on health (Monika and Kishore 2010).

20.4 E-waste Management and Control

Out of entire e-waste being generated, about 75% are stored in houses and offices and eventually end up mixing with household waste due to lack of proper waste management techniques. At industry level, several waste reduction techniques and sustainable design of product can help in e-waste management from the very beginning of its generation (Kumar et al. 2017). Waste reduction techniques in industry involve inventory maintenance, modification of production process, reduction in volume, and recovery and reuse of waste (Needhidasan et al. 2014).

20.4.1 Inventory Maintenance

Appropriate monitoring of the components and materials that are used in the production and manufacturing process is an essential way for the reduction of waste generation. By decreasing the quantity of toxic materials required in the process as well as the amount of surplus raw materials available, e-waste generation can be reduced. Two ways to achieve this task are by establishing review for material purchase and control procedures and devising an inventory tracking system (Findlay et al. 2015).

20.4.2 Production-Process Modification

By changing the process of production such as upgrading the material to form the product or by using efficient input materials or both, waste generation can be reduced. Potential waste reduction techniques can be categorized into the process of improvement of operating and maintenance procedures, material change, and also process-equipment modification (Datta et al. 2018).

20.4.3 Volume Reduction

These techniques aim at e-waste volume reduction and hence decrease the cost of its disposal. The techniques to reduce the volume of the waste stream are categorized into two different groups: segregation of source and concentration of waste. Segregation of wastes is a simple and economical method for reducing waste. Concentration of waste increases the probability of that material to be recycled or reused. Methods involved in the above process include gravity and filtration in vacuum, reverse osmosis, and freeze vaporization (Yoada et al. 2014).

20.4.4 Recovery and Reuse

One could avoid costs of waste disposal, reduce costs of raw material, and can create income from commercial e-waste using this method. Recovery of waste can be done on-site, off-site, or *via* exchange programs among industries. A large number of physical and chemical methods are available to retrieve waste material such as reverse osmosis, electrolysis, condensation, electrolytic recovery, filtration, and centrifugation (Yoada et al. 2014).

20.4.5 Sustainable Product Design

Following factors can help reduce the waste hazard at the stage of product designing itself (Yoada et al. 2014):

- Rethink the product design: Product having fewer amounts of toxic materials. For example, materials are reduced in newer computer designs which are flatter, light weight, and more compact.
- Use of renewable substances and energy: Bio-plastics are made with plant-based chemicals or polymers instead of petrochemicals and are biodegradable. Biological toners, adhesives, inks, etc. are being used. Solar computers are also available but are highly expensive (Babu et al. 2013).
- Use of nonrenewable safer substances: Product made from such materials is reused and repaired or modified for their upgradation.

20.5 E-waste Management Problem in India

In order to tackle the harmful effects of e-waste, an effective approach for the management and control of e-waste was required. The late 1980s witnessed a strict regulation of environmental policies in developed countries which led to a high price hike for disposal of waste following which the traders began exporting the e-waste to developing countries as an inexpensive way to dump their wastes. These incautious activities led to the formation and endorsement of the strategies and mandates at the Basel Convention which gives support and directs on legal as well as technical matter, collects statistical data, and instructs on appropriate management of toxic waste (Monika and Kishore 2010).

Despite attaining great heights in terms of technological advancements, developing countries like India and other Asian nations are very close to dangers associated with the disposal of e-waste from developed countries like the USA. Recent studies reveal that most of the recyclable e-waste in the USA is exported to Asia for their disposal or to be recycled without any concern regarding the health of workers or environment due to the fact that Asian nations have low-cost labor, poor implementation of environmental policies, and poor occupational standards (Monika and Kishore 2010; Ohajinwa et al. 2017). The severity of these issues is yet to be recognized. However, Toxic Links India and other such organizations are collecting data regarding the matter of concern in order to stop and control this toxic trade (Needhidasan et al. 2014).

20.6 Management Options

Since the problem associated with e-waste management has reached an alarming level, it can be tackled collectively by the support of individuals and government. Some steps which can be taken at government and public levels are discussed below.

20.6.1 Management Options for the Government

Regulatory agencies should be set up in every district to stringently monitor the e-waste disposal. Furthermore, laws, regulations, and administrative policies for toxic and hazardous waste management should be reviewed and strictly reinforced (Pandve 2010). Research in the field of development and raising standards for waste management, its disposal, and environmental monitoring should be encouraged and funded by the governing bodies (Datta et al. 2018). NGOs and other organizations working on solving the waste management issue must be encouraged and supported. Government should adopt the polluter pays principle and extended producer responsibility methods for effective e-waste management. Also, strict laws and regulations must be enforced against the dumping of the toxic waste in the nation by the outsiders (Pandve 2010).

20.6.2 Responsibilities of the Citizen

Prevention of waste is always the best way to waste management options including recycling of the waste. Therefore, working electronics should be donated for their reuse and kept away from waste management for longer duration. These wastes must never be mixed and discarded along with household wastes. One should prefer buying only those electronics that are recyclable, energy efficient, easily upgradable, and composed of less toxic components, require minimal packaging, and have been certified by regulatory authorities for quality assurance (Needhidasan et al. 2014).

20.7 Bioremediation of E-waste Pollution

All the methods discussed above that are being currently used for the control and management of e-waste have some of their drawbacks associated with them whether it is the cost of waste treatment, workers' health and safety concerns, requirement of landfills, ground percolations, or air or water pollution (Monika and Kishore 2010; Pandve 2010). A newer and better approach to solve the problem of e-waste management is bioremediation of e-waste. Bioremediation is a process that involves the use of microorganisms (either occurring naturally or due to stimulated growth) to treat, break down, and detoxify the environmental pollutants, thereby cleaning the contaminated site (Ojuederie and Babalola 2017). It is a low-cost and environment-friendly biotechnology approach, which involves the use of microbes and microbial enzymes to degrade the pollutants into less toxic or nontoxic materials (Azubuike et al. 2016). Several recent studies have reported that a very high number of enzymes obtained from different microbes such as bacteria, fungi, and enzymes from plants play a significant role in the biodegradation of hazardous organic pollutants. Extensive research work has been done to know the mechanisms of enzymes related to bioremediation activity such as those of oxidoreductases and hydrolases (Karigar and Rao 2011).

However, bioremediation process is largely dependent on the enzymatic activity of microbes, which react with pollutants converting them into safe products, but this is only effective in the environmental conditions that facilitate the growth as well as activity of the microbes. Therefore, this process often requires the alternation of the conditions of environment to promote the microbial growth and its activity (Karigar and Rao 2011).

The use of plants for treating pollutants through the process of bioremediation is referred to as phytoremediation. Phytoremediation is a green technology process that helps in the detoxification and treatment of toxic chemicals in soil, surface and groundwater, air, etc. (Lone et al. 2008). For instance, *Arabidopsis thaliana*, a genetically modified plant expressing two bacterial genes, one gene converts arsenate into arsenite while the second gene binds and stores arsenite into vacuoles, has also been used in the removal of arsenic from the contaminated site (Azubuike et al. 2016).

Bioremediation besides being an effective way to tackle waste management faces some problems such as it's a really slow process. Also not all bacteria and fungi have the ability to decompose pollutants. Most of the known variants of microbe capable of effective bioremediation still require laboratory conditions for their optimal activity due to the hindrance in growth caused by changing pH, temperature, soil structure, humidity, nutrient availability, nonuniform availability of contaminants, and presence of harmful substances in the natural environment (Dua et al. 2002). Mostly bioremediation process is an aerobic process but growth under anaerobic condition also provides with the decomposition of the recalcitrant molecules. The remediation of organo-pollutants, recalcitrant compounds, and lignin depends on the role of various intracellular and extracellular enzymes present in bacteria and fungi (Dashtban et al. 2010).

20.8 Enzymatic Bioremediation of E-waste

20.8.1 Multicopper Oxidase (MCO)

Multicopper oxidase (MCO) is an enzyme belonging to laccase family. This enzyme is used for oxidation detection which functionally works as a biosensor and can be used in bioremediation (Piscitelli et al. 2010; Sirim et al. 2011). MCOs are capable of oxidizing its substrate on accepting electrons at a mononuclear copper center and then transferring them to a tri-nuclear copper center (Bento et al. 2010). This dioxygen binds to the tri-nuclear center and following the transfer of four electrons is reduced to two molecules of water. MCO has great potential in several aspects like gold (Au) adsorption that is almost tenfold higher than other proteins. By zeta-potential analysis it is revealed that recombinant MCO is anionic and Au adsorption on the protein is by electrostatic interactions (Fig. 20.3). This is an attempt in order to explore the activity of MCO to adsorb Au which may become a novel application for bio-recovery of gold in the industry (Wang et al. 2009).

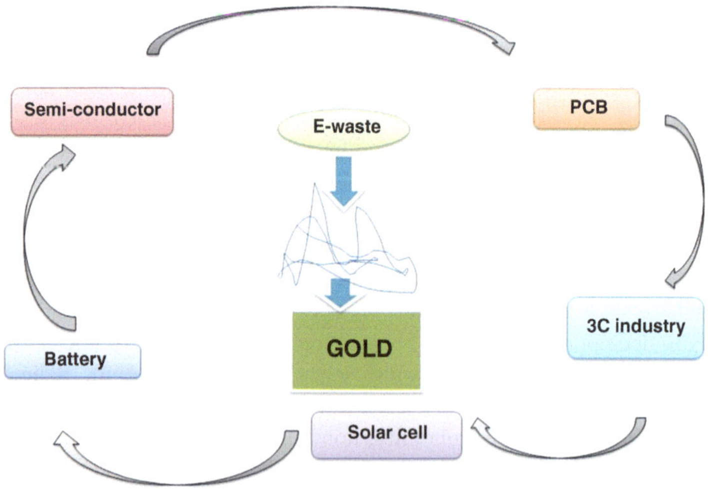

Fig. 20.3 Electrostatic interaction

20.8.2 Phosphatase

In different studies, the level of heavy metals and PCBs has influence on the activity of phosphatase enzyme. This can also be obtained from microbial community of paddy soils from the immediate vicinity of e-waste (Tang et al. 2014). Urease has the highest enzyme activity compared to that of phosphatase and catalase. It shows the inconsistent trend in activity of soil enzyme as evident in different sampling sites. The microbial analysis confirmed that there was no correlation between the pollutants and microbial community. The phosphatase activity was found to be decreased in soils that were contaminated with both PAHs and heavy metals (Tang et al. 2014).

20.8.3 Cyanidase

Cyanide dihydratase or cyanidase is found to be in a range of microorganisms which can metabolize cyanide. Additionally, there are other enzymes which convert cyanide to nontoxic products. The nitrilases are cyanide-degrading enzymes and can have different properties to hydrolyze cyanide without any cofactors or secondary substrates (Jaszczak et al. 2017). *P. fluorescens* has the activity of both cyanide hydratase and cyanide dihydratase. However, no enzymes yet have been purified from it. Hence, it is very much possible that this bacteria contains a single enzyme with both the activities (Kunz et al. 1992).

Furthermore, it has been reported that cyanidase detoxifies the cyanide containing extract from de-bittering apricot seeds in food industry. The diffusional-type flat membrane reactor (FMR) has been developed which is supposed to perform superior, relative to stirred tank reactor and fixed-bed reactor, configurations (Akyildiz et al. 2010). Cyanide diffuses through a semipermeable membrane so as to react with the entrapped catalyst that square measure gift behind the membrane; then the reaction product diffuses back across the membrane to the solution (Saddoud et al. 2007). From previous studies, it's calculable that three million a lot of cyanide square measure used yearly throughout the planet in several industrial processes together with the assembly of chemical intermediates, artificial fibers, rubber, and prescription drugs, likewise as in ore activity, coal process, and coat. Several plants, microbes, and insects will unharness HCN upon protein chemical reaction of sure compounds they manufacture. Since cyanide may be a metabolic substance and may be fatal to humans and alternative organisms, it's essential that it's far from effluents before discharge (Knowles 1988). Cyanidase relies on sure Gram-negative microorganism isolates from the species *Alcaligenes denitrificans* and it's ready by proprietary strategies. It's characterized by a high affinity and a high stability toward cyanide and is ready to get rid of the latter all the way down to terribly low levels (Aisenberg et al. 2004; Clermont et al. 2001)

20.8.4 Cyanide Hydratase

Cyanide hydratase is associate accelerator having living thing, inducible property with cyanide, and metallocyanides as inducers. Cyanide then undergoes reaction of reaction to formamide by cyanide hydratase that successively is reborn to formic acid and ammonia by amidase. The cyanide hydratases are rather more substrate specific as compared to nitrilases with HCN, being the foremost effective substrate for all the enzymes. It has been noted that cyanide hydratase was stable once immobilized while the accelerator from *Gloeocercospora sorghi* was rather more stable than that from *Stemphylium loti* (Basile et al. 2008). They reported that immobilized cyanide hydratase is fitted to the treatment of cyanide-containing industrial effluents (Luque-Almagro et al. 2005).

20.8.5 Azoreductase

Azoreductases square measure the enzymes that affect the dye compounds and convey regarding cleavage and ordered degradation. The initial and extreme initiative in degradation method of dyestuff is the cleavage of electrophilic azolinkage which ends up in de-colorization activity. Poisonous effluents that square measure containing radical dyes discharged from sizable amount of industries and that they adversely have an effect on water resources, soil fertility, aquatic organisms, and scheme integrity (Franciscon et al. 2012). The accelerator degradation of radical dyes might change state subtractive cleavage of radical teams below delicate conditions. Azoreductase also catalyzes the subtractive of radical linkages in benzidine-based dyes associated alternative compounds containing a radical bond to provide aromatic amines (Chen 2006). The characterization of the *P. aeruginosa* azoreductase enzymes as well as the property has been studied antecedently, including determining their thermostability, compound preference, and kinetic constants against a range of their favored substrates. Throughout growth of *P. aeruginosa* the expression levels of those enzymes might alter due to the presence of those azo-substrates; conjointly these catalyst azoreductases possibly act as NADH quinone oxidoreductases. The low sequence identities determined among NADPH quinone oxidoreductase and azoreductase enzymes suggest merging evolution (Ryan et al. 2010).

20.8.6 Nitrile Hydratase

Nitrile hydratase (NHase) may be a soluble microorganism metallo-enzyme that catalyzes the association of cyanide compounds to the corresponding amides, which can be regenerated by associate degree amidase to the corresponding acids and ammonia (Brandao et al. 2003). NHases enzymes are of great importance as biocatalysts for the economic production of amide and nicotinamide and additionally for environmental bioremediation, wherever it should be shown to be effective within the removal of nitriles from waste streams (Brandao et al. 2003). NHases are expected to possess great potential as catalysts in organic chemical process as they convert nitriles to the corresponding higher-value amides below delicate conditions. Cyanide hydratase is utilized in several producing and engineering activities; then they're common contaminants in industrial wastes additionally as effluents. Thanks to their property of high toxicity, it's necessary that nitriles are aloof from wastes before they're discharged into the atmosphere. In many studies addressing microbial detoxification of propenonitrile bearing wastes, efforts to develop microbial treatment strategies are hampered by the fact that such wastes typically contain high levels of acrylo-nitrile and alternative toxic elements that are repressive to microbic growth (Gong et al. 2012; Yamada et al. 2001).

Peroxidases are oxidoreductases that catalyze the reduction of peroxides like oxide (H_2O_2) and chemical reactions of various organic and inorganic compounds. The oxidase activity involves donating electrons that bind to different substrates like ascorbate and ferricyanides so as to break them into harmless elements (Chanwun et al. 2013). Peroxidases might embrace activity of 2 enzymes that are polymer oxidase and metallic element oxidase. Polymer oxidase (LiP) is additionally cited as di-aryl gas oxidoreductase and it's a heme-containing accelerator that catalyzes gas peroxide-dependent aerophilic degradation of polymer (Falade et al. 2017). Metallic element oxidase (MnP) is produced by *Phanerochaete chrysosporium*, and it's conjointly been ascertained to catalyze the chemical reaction of many mono-aromatic phenols and aromatic dyes; however, the completion of those reactions depends upon the presence of each powerful metallic element and bound sorts of buffers. Indeed, MnP catalyzes the chemical reaction of Mn(II) to Mn(III) within the presence of Mn(III) helpful ligands. Then the chemical reaction of organic substrates is administrated within the presence of those Mn(III) complexes (Urzua et al. 1998).

20.8.7 Polyphenol Oxidase

Polyphenol enzymes (PPO) are a cluster of proteins that exist in two forms, tyrosinase and laccase, among microorganisms, plants, and animals. These oxidoreductive enzymes stay effective during a wide selection of pH scale and temperature, significantly if they're immobilized on some carrier or matrices, and that they will degrade a large sort of mono- or di-phenolic compounds. The event of polyphenol enzyme (PPO)-based biosensors for studying phenoplast seems to be at a mature stage of technology. It has been antecedently studied that polyphenol enzyme-based biosensors are developed on the principles of amperometry (Garcia et al. 2016). These biosensors are extensively reported because of their major properties as well as high sensitivity, simplicity, fast responsiveness, and simple to miniaturize (Mehrotra 2016). The performance and stability of a biosensor depends upon the fabric used for protein immobilization method and additionally the tactic of protein immobilization (Mohamad et al. 2015). Totally different materials have been used for the method of immobilization of PPO like magnetic nanoparticles, sol–gels, self-gelatinizable graft polymer of poly (vinyl alcohol) with 4-vinylpyridine, magnetic nanoparticle-coated carbon nanotube (CNT) nano-composite, inorganic unsubstantial, filter papers, carbon paste, polymers, chitosan, alginate, clays, and self-assembled monolayers (Roth et al. 2016). Laccase and tyrosinase are phenol oxidases that catalyze the oxidization of phenoplast compounds. Tyrosinase also referred to as polyphenol enzyme, phenolase, or catecholase catalyzes 2 consecutive reactions that embrace the hydroxylation of mono-phenols with molecular element to create ortho-di-phenols and also the dehydrogenation of ortho-di-phenols with element to create o-quinones. Quinones are principally unstable and endure nonenzymatic chemical change to yield water-insoluble substances that may be simply removed by easy filtration. The immobilization of tyrosinase has the advantage of holding enzymes within the reactor and protecting them from inactivation by reaction with quinones. Immobilized tyrosinase maintained its activity up to ten cycles. Hence, it might seem that the mix of immobilized tyrosinase with chitosan is useful for the removal of deadly phenols (Abdullah et al. 2007). Tyrosinase immobilization takes place by defense methodology during a chemical compound, poly 3, 4-ethylenedioxythiophene (PEDT), for detection of phenolics. Glassy carbon conductor was additionally optimized for various parameters. Laccase is made by many fungi and appears capable of decreasing the toxicity of phenoplast compounds through a chemical action process. Also, as a result of its relative non-specificity, it will induce the cross coupling of waste material phenols with present phenols. Laccase will oxidize phenoplast compounds to their corresponding anionic free radicals that are extremely reactive (Jeon et al. 2012). Laccase-based amperometric biosensors are a necessary category of biosensors in which current is produced as a result of accelerator reaction. This alteration acts as a symbol that is measured by physical techniques and principally noble metal, gold, or carbon-based electrodes area unit made (Guo et al. 2014; Ibarra-Escutia et al. 2010).

20.9 Advantages of Enzymatic Degradation of E-waste

Enzymes offered many benefits like greater specificity, higher standardization, simple handle and store, and no dependence on microorganism growth rates (Cezairliyan and Ausubel 2017). Enzymes are biological catalysts. The use of enzymes is interesting, as it replaces the utilization of harsher chemical like solvents and enzymes additionally perform on neutral pH scale and temperature. It doesn't produce harmful chemicals, which eventually damage the surroundings, though enzymes have high value to their extraction and purification process (Ravindran and Jaiswal 2016). Extracellular enzymes are secreted by the microbes, like polymer oxidase (LiP) from white rot fungi. These are naturally made so harvested (Dashtban et al. 2010). Every accelerator has one specific role like lowering the action energy for the degradation of intramolecular bonds, and a few area unit have an effect on the big selection completely different substrate. These are ready to react with artificial and xenobiotic compounds. This reaction will amend the recalcitrant state to it that is biodegradable (Upadhyay et al. 2016). At the same time, the matter of huge quantity of venturesome waste and alternative waste generated from electronic world. This electronic waste carries toxic substances and deadly chemicals in electronic gadgets, disposal of electronic wastes changing into a giant downside from our atmosphere. Enzyme laccases are used for many applications in organic synthesis because the oxidization of practical teams still because the coupling of phenols and steroids, medical agents which have anesthetics, anti-inflammatory drugs, antibiotics, and sedatives, and the development of carbon–nitrogen bonds and in the synthesis of advanced natural product and industries of cosmetics (Hildebrand et al. 2015). Electronic waste is an electronic product that requires recycling or proper disposal. This waste consists of over one thousand completely different substances, which are harmful to nature (Pinto 2008). This metal is copper, silver, aluminum, gold, platinum, atomic number 46, etc. The disposal of e-waste is a big downside in several regions across the globe (Needhidasan et al. 2014). There are few processes that help in disposal of e-waste like microremediation. Microremediation is outlined because of the use of microorganisms to eliminate, contain, or transfer the contaminants to nonhazardous or less harmful variety of wastes with the assistance of metabolisms of microorganisms (Ayangbenro et al. 2018). The increasing applications of enzymes are creating a growing demand for biocatalysts that exhibit improved or new properties. The varied chemical science treatments like chemical precipitation, curdling, action, floatation, and membrane filtration provide many benefits like easy operation and management, flexibility to alter temperature, and area unit speedy however their advantages (Azizi et al. 2016). Polyphenol enzyme (PPO) shows a robust potential to be used as a biosensor, providing some specific benefits over alternative enzymes. These benefits embrace the flexibility of polyphenol enzyme to catalyze electron-transfer reactions while not would like of further cofactors, oxidization of phenolic resin compounds within the presence of gas, and its smart stability. Polyphenol enzyme-based biosensor systems additionally show high property, sensitivity, smart stability, and reduced assay times. Electronic wastes have globally been on the rise because of technological advancement, raised awareness on the importance of knowledge technology, and continuous improvement in style. These have necessitated the increased usage of equipment notably in developing countries wherever the importation of used electronic gadgets is on the rise in recent times without proper strategies of utilization of the wastes (Ohajinwa et al. 2017). These electronic wastes contain recalcitrant and toxicant materials that cause distinct challenges that disagree from alternative varieties of wastes to the surroundings.

20.10 Scope and Future Prospects of Enzymatic Degradation of E-waste

In terms of the scope of accelerator degradation of e-waste these days is to lower the concentration vary of organic pollutants at undetectable levels or if measurable but the bounds that are established as safe or conjointly tolerable by regulatory agencies (Alneyadi et al. 2017). Accelerator bioremediation method has been improved day-by-day victimization many molecular tools that are appropriate for fast rectification things. In human lives, the utility of plastics is ceaselessly increasing, whereas the process of degradation is changing into a serious challenge for the atmosphere nowadays (Thompson et al. 2009). For the safer disposal of plastic waste, several helpful tasks have been applied; however, they were conjointly found to be connected with environmental hazards rather than decreasing the impact of plastic waste. In the global market, these perishable plastic light-emitting diodes have emerged as serious new-generation polymers (Song et al. 2009). Conjointly the employment of those plastics in standard of living are meant to be accrued due to their property of complete likewise as economical degradation by microbial enzymes over artificial polymers. From totally different previous studies, perishable plastics degradation has accrued data concerning degrading microorganisms, their enzymes, genes, and conjointly compound structure (Tokiwa et al. 2009). The collected knowledge related to those studies is helpful for the molecular evolution and production of perishable polymers and conjointly for the union of enzymes and degrading microbes. It's been studied that perishable plastics plays a crucial role in numerous fields such as via food packaging, in cosmetics, agriculture, and in medicine devices (Babu et al. 2013). Therefore, researchers are ceaselessly working on developing latest forms of perishable plastics so as to cut back the harmful effects of plastics on human health and atmosphere, terribly rare for property disposal of perishable plastics; many new methods which are less economic and environment friendly have been developed. Therefore, the infrastructure of current waste management isn't working satisfactorily, totally different agencies as well as Environmental Protection Agency and World Food Organization acts for effective and promising development worldwide. Our accrued understanding concerning this subject can contribute toward the producing of novel perishable plastic compounds having industrial implementation and development of promising methods for polymer waste management (Rydz et al. 2014). The long-run analysis ought to be centered on many fields like accelerator removal of synthetic resin contaminants by polyphenol oxidases (PPO) involving totally different studies that deal with the basic likewise as applied aspects of various processes like isolation, purification, characterization, and immobilization of PPO from their originating sources that create this methodology as eco-friendly, cost-convenient, and extremely economical too. Many support materials and a number of other methodologies of their utilization have been explored in numerous studies. These support materials are advanced and conjointly time intense for preparation likewise as a number of these support materials are unhealthy biocompatible and that they lower the potency of immobilized catalyst. This study might facilitate to figure for development of latest technology for biosensing. Conjointly price effective and biocompatible supports may be developed. Recycling method is the key to cut back the e-waste and it has environmental advantages at each stage of the life cycle of a PC product from the material from which it is created to its final disposal. Besides reducing gas emissions which contribute to global warming, recycling decreases air pollution related to creating new merchandise from raw materials.

Acknowledgment Authors thank Shri Ramswaroop Memorial University, Barabanki, Uttar Pradesh, India, and National Institute of Technology, Raipur (CG), India, for continuous support and assistance during the course of research work and scientific writing.

Conflict of Interest None.

References

Abdullah J, Ahmad M, Heng LY, Karuppiah N, Sidek H (2007) An optical biosensor based on immobilization of laccase and MBTH in stacked films for the detection of catechol. Sensors (Basel) 7:2238–2250. https://doi.org/10.3390/s7102238

Aisenberg G, Rolston KV, Safdar A (2004) Bacteremia caused by Achromobacter and Alcaligenes species in 46 patients with cancer (1989–2003). Cancer 101:2134–2140. https://doi.org/10.1002/cncr.20604

Akyildiz BN, Kurtoglu S, Kondolot M, Tunc A (2010) Cyanide poisoning caused by ingestion of apricot seeds. Ann Trop Paediatr 30:39–43. https://doi.org/10.1179/146532810X12637745451951

Alneyadi AH, Shah I, AbuQamar SF, Ashraf SS (2017) Differential degradation and detoxification of an aromatic pollutant by two different peroxidases. Biomol Ther 7:E31. https://doi.org/10.3390/biom7010031

Annamalai J (2015) Occupational health hazards related to informal recycling of E-waste in India: an overview. Indian J Occup Environ Med 19:61–65. https://doi.org/10.4103/0019-5278.157013

Ayangbenro AS, Babalola OO (2017) A New strategy for heavy metal polluted environments: a review of microbial biosorbents. Int J Environ Res Public Health 14:E94. https://doi.org/10.3390/ijerph14010094

Ayangbenro AS, Olanrewaju OS, Babalola OO (2018) Sulfate-reducing bacteria as an effective tool for sustainable acid mine bioremediation. Front Microbiol 9:1986. https://doi.org/10.3389/fmicb.2018.01986

Azizi S, Kamika I, Tekere M (2016) Evaluation of heavy metal removal from wastewater in a modified packed bed biofilm reactor. PLoS One 11:e0155462. https://doi.org/10.1371/journal.pone.0155462

Azubuike CC, Chikere CB, Okpokwasili GC (2016) Bioremediation techniques-classification based on site of application: principles, advantages, limitations and prospects. World J Microbiol Biotechnol 32:180. https://doi.org/10.1007/s11274-016-2137-x

Babu RP, O'Connor K, Seeram R (2013) Current progress on bio-based polymers and their future trends. Prog Biomater 2:8. https://doi.org/10.1186/2194-0517-2-8

Basile LJ, Willson RC, Sewell BT, Benedik MJ (2008) Genome mining of cyanide-degrading nitrilases from filamentous fungi. Appl Microbiol Biotechnol 80:427–435. https://doi.org/10.1007/s00253-008-1559-2

Bento I, Silva CS, Chen Z, Martins LO, Lindley PF, Soares CM (2010) Mechanisms underlying dioxygen reduction in laccases. Structural and modelling studies focusing on proton transfer. BMC Struct Biol 10:28. https://doi.org/10.1186/1472-6807-10-28

Brandao PF, Clapp JP, Bull AT (2003) Diversity of nitrile hydratase and amidase enzyme genes in *Rhodococcus erythropolis* recovered from geographically distinct habitats. Appl Environ Microbiol 69:5754–5766

Cezairliyan B, Ausubel FM (2017) Investment in secreted enzymes during nutrient-limited growth is utility dependent. Proc Natl Acad Sci USA 114:E7796–E7802. https://doi.org/10.1073/pnas.1708580114

Chanwun T, Muhamad N, Chirapongsatonkul N, Churngchow N (2013) *Hevea brasiliensis* cell suspension peroxidase: purification, characterization and application for dye decolorization. AMB Express 3:14. https://doi.org/10.1186/2191-0855-3-14

Chen H (2006) Recent advances in azo dye degrading enzyme research. Curr Protein Pept Sci 7:101–111

Clermont D, Harmant C, Bizet C (2001) Identification of strains of Alcaligenes and Agrobacterium by a polyphasic approach. J Clin Microbiol 39:3104–3109

Dashtban M, Schraft H, Syed TA, Qin W (2010) Fungal biodegradation and enzymatic modification of lignin. Int J Biochem Mol Biol 1:36–50

Datta P, Mohi GK, Chander J (2018) Biomedical waste management in India: Critical appraisal. J Lab Physicians 10:6–14. https://doi.org/10.4103/JLP.JLP_89_17

Dua M, Singh A, Sethunathan N, Johri AK (2002) Biotechnology and bioremediation: successes and limitations. Appl Microbiol Biotechnol 59:143–152. https://doi.org/10.1007/s00253-002-1024-6

Falade AO, Nwodo UU, Iweriebor BC, Green E, Mabinya LV, Okoh AI (2017) Lignin peroxidase functionalities and prospective applications. MicrobiologyOpen 6:e00394. https://doi.org/10.1002/mbo3.394

Findlay R, Webb A, Lund J (2015) Implementation of advanced inventory management functionality in automated dispensing cabinets. Hosp Pharm 50:603–608. https://doi.org/10.1310/hpj5007-603

Finnegan PM, Chen W (2012) Arsenic toxicity: the effects on plant metabolism. Front Physiol 3:182. https://doi.org/10.3389/fphys.2012.00182

Franciscon E, Grossman MJ, Paschoal JA, Reyes FG, Durrant LR (2012) Decolorization and biodegradation of reactive sulfonated azo dyes by a newly isolated *Brevibacterium* sp. strain VN-15. Springerplus 1:37. https://doi.org/10.1186/2193-1801-1-37

Garcia LF, Benjamin SR, Antunes RS, Lopes FM, Somerset VS, Gil ES (2016) *Solanum melongena* polyphenol oxidase biosensor for the electrochemical analysis of paracetamol. Prep Biochem Biotechnol 46:850–855. https://doi.org/10.1080/10826068.2016.1155060

Gong JS, Lu ZM, Li H, Shi JS, Zhou ZM, Xu ZH (2012) Nitrilases in nitrile biocatalysis: recent progress and forthcoming research. Microb Cell Factories 11:142. https://doi.org/10.1186/1475-2859-11-142

Guo M, Wang H, Huang D, Han Z, Li Q, Wang X, Chen J (2014) Amperometric catechol biosensor based on laccase immobilized on nitrogen-doped ordered mesoporous carbon (N-OMC)/PVA matrix. Sci Technol Adv Mater 15:035005. https://doi.org/10.1088/1468-6996/15/3/035005

Hildebrand A, Kasuga T, Fan Z (2015) Production of cellobionate from cellulose using an engineered *Neurospora crassa* strain with laccase and redox mediator addition. PLoS One 10: e0123006. https://doi.org/10.1371/journal.pone.0123006

Howard DH (1999) Acquisition, transport, and storage of iron by pathogenic fungi. Clin Microbiol Rev 12:394–404

Ibarra-Escutia P, Gomez JJ, Calas-Blanchard C, Marty JL, Ramirez-Silva MT (2010) Amperometric biosensor based on a high resolution photopolymer deposited onto a screen-printed electrode for phenolic compounds monitoring in tea infusions. Talanta 81:1636–1642. https://doi.org/10.1016/j.talanta.2010.03.017

Jan AT, Azam M, Siddiqui K, Ali A, Choi I, Haq QM (2015) Heavy metals and human health: mechanistic insight into toxicity and counter defense system of antioxidants. Int J Mol Sci 16:29592–29630. https://doi.org/10.3390/ijms161226183

Jaszczak E, Polkowska Z, Narkowicz S, Namiesnik J (2017) Cyanides in the environment-analysis-problems and challenges. Environ Sci Pollut Res Int 24:15929–15948. https://doi.org/10.1007/s11356-017-9081-7

Jeon JR, Baldrian P, Murugesan K, Chang YS (2012) Laccase-catalysed oxidations of naturally occurring phenols: from in vivo biosynthetic pathways to green synthetic applications. Microb Biotechnol 5:318–332. https://doi.org/10.1111/j.1751-7915.2011.00273.x

Karigar CS, Rao SS (2011) Role of microbial enzymes in the bioremediation of pollutants: a review. Enzym Res 2011:805187. https://doi.org/10.4061/2011/805187

Kaw HY, Kannan N (2017) A review on polychlorinated biphenyls (PCBs) and polybrominated diphenyl ethers (PBDEs) in South Asia with a focus on Malaysia. Rev Environ Contam Toxicol 242:153–181. https://doi.org/10.1007/398_2016_14

Knowles CJ (1988) Cyanide utilization and degradation by microorganisms. Ciba Found Symp 140:3–15

Kotrba P, Doleckova L, de Lorenzo V, Ruml T (1999) Enhanced bioaccumulation of heavy metal ions by bacterial cells due to surface display of short metal binding peptides. Appl Environ Microbiol 65:1092–1098

Kracke F, Vassilev I, Kromer JO (2015) Microbial electron transport and energy conservation – the foundation for optimizing bioelectrochemical systems. Front Microbiol 6:575. https://doi.org/10.3389/fmicb.2015.00575

Kumar S et al (2017) Challenges and opportunities associated with waste management in India. R Soc Open Sci 4:160764. https://doi.org/10.1098/rsos.160764

Kunz DA, Nagappan O, Silva-Avalos J, Delong GT (1992) Utilization of cyanide as nitrogenous substrate by *Pseudomonas fluorescens* NCIMB 11764: evidence for multiple pathways of metabolic conversion. Appl Environ Microbiol 58:2022–2029

Lone MI, He ZL, Stoffella PJ, Yang XE (2008) Phytoremediation of heavy metal polluted soils and water: progresses and perspectives. J Zhejiang Univ Sci B 9:210–220. https://doi.org/10.1631/jzus.B0710633

Luque-Almagro VM et al (2005) Bacterial degradation of cyanide and its metal complexes under alkaline conditions. Appl Environ Microbiol 71:940–947. https://doi.org/10.1128/AEM.71.2.940-947.2005

Mehrotra P (2016) Biosensors and their applications – a review. J Oral Biol Craniofac Res 6:153–159. https://doi.org/10.1016/j.jobcr.2015.12.002

Merchant SS, Helmann JD (2012) Elemental economy: microbial strategies for optimizing growth in the face of nutrient limitation. Adv Microb Physiol 60:91–210. https://doi.org/10.1016/B978-0-12-398264-3.00002-4

Mohamad NR, Marzuki NH, Buang NA, Huyop F, Wahab RA (2015) An overview of technologies for immobilization of enzymes and surface analysis techniques for immobilized enzymes. Biotechnol Biotechnol Equip 29:205–220. https://doi.org/10.1080/13102818.2015.1008192

Monika, Kishore J (2010) E-waste management: as a challenge to public health in India. Indian J Community Med 35:382–385. https://doi.org/10.4103/0970-0218.69251

Needhidasan S, Samuel M, Chidambaram R (2014) Electronic waste – an emerging threat to the environment of urban India. J Environ Health Sci Eng 12:36. https://doi.org/10.1186/2052-336X-12-36

Ohajinwa CM, Van Bodegom PM, Vijver MG, Peijnenburg W (2017) Health risks awareness of electronic waste workers in the informal sector in Nigeria. Int J Environ Res Public Health 14:911. https://doi.org/10.3390/ijerph14080911

Ojuederie OB, Babalola OO (2017) Microbial and plant-assisted bioremediation of heavy metal polluted environments: a review. Int J Environ Res Public Health 14:E1504. https://doi.org/10.3390/ijerph14121504

Osibanjo O, Nnorom IC (2007) The challenge of electronic waste (e-waste) management in developing countries. Waste Manage Res 25:489–501. https://doi.org/10.1177/0734242X07082028

Pandve HT (2010) Some initiative in e-waste disposal, management and recycling. Indian J Occup Environ Med 14:20–21. https://doi.org/10.4103/0019-5278.64611

Pinto VN (2008) E-waste hazard: the impending challenge. Indian J Occup Environ Med 12:65–70. https://doi.org/10.4103/0019-5278.43263

Piscitelli A, Pezzella C, Giardina P, Faraco V, Giovanni S (2010) Heterologous laccase production and its role in industrial applications. Bioeng Bugs 1:252–262. https://doi.org/10.4161/bbug.1.4.11438

Ravindran R, Jaiswal AK (2016) Microbial enzyme production using lignocellulosic food industry wastes as feedstock: a review. Bioengineering (Basel) 3:E30. https://doi.org/10.3390/bioengineering3040030

Roth HC, Schwaminger SP, Peng F, Berensmeier S (2016) Immobilization of cellulase on magnetic nanocarriers. ChemistryOpen 5:183–187. https://doi.org/10.1002/open.201600028

Ryan A, Wang CJ, Laurieri N, Westwood I, Sim E (2010) Reaction mechanism of azoreductases suggests convergent evolution with quinone oxidoreductases. Protein Cell 1:780–790. https://doi.org/10.1007/s13238-010-0090-2

Rydz J, Sikorska W, Kyulavska M, Christova D (2014) Polyester-based (bio)degradable polymers as environmentally friendly materials for sustainable development. Int J Mol Sci 16:564–596. https://doi.org/10.3390/ijms16010564

Saddoud A, Hassairi I, Sayadi S (2007) Anaerobic membrane reactor with phase separation for the treatment of cheese whey. Bioresour Technol 98:2102–2108. https://doi.org/10.1016/j.biortech.2006.08.013

Sirim D, Wagner F, Wang L, Schmid RD, Pleiss J (2011) The laccase engineering database: a classification and analysis system for laccases and related multicopper oxidases. Database 2011: bar006. https://doi.org/10.1093/database/bar006

Song JH, Murphy RJ, Narayan R, Davies GB (2009) Biodegradable and compostable alternatives to conventional plastics. Philos Trans R Soc Lond Ser B Biol Sci 364:2127–2139. https://doi.org/10.1098/rstb.2008.0289

Tang X, Hashmi MZ, Long D, Chen L, Khan MI, Shen C (2014) Influence of heavy metals and PCBs pollution on the enzyme activity and microbial community of paddy soils around an e-waste recycling workshop. Int J Environ Res Public Health 11:3118–3131. https://doi.org/10.3390/ijerph110303118

Tchounwou PB, Yedjou CG, Patlolla AK, Sutton DJ (2012) Heavy metal toxicity and the environment. EXS 101:133–164. https://doi.org/10.1007/978-3-7643-8340-4_6

Thompson RC, Moore CJ, vom Saal FS, Swan SH (2009) Plastics, the environment and human health: current consensus and future trends. Philos Trans R Soc Lond Ser B Biol Sci 364:2153–2166. https://doi.org/10.1098/rstb.2009.0053

Tokiwa Y, Calabia BP, Ugwu CU, Aiba S (2009) Biodegradability of plastics. Int J Mol Sci 10:3722–3742. https://doi.org/10.3390/ijms10093722

Upadhyay P, Shrivastava R, Agrawal PK (2016) Bioprospecting and biotechnological applications of fungal laccase. 3 Biotech 6:15. https://doi.org/10.1007/s13205-015-0316-3

Urzua U, Kersten PJ, Vicuna R (1998) Manganese peroxidase-dependent oxidation of glyoxylic and oxalic acids synthesized by *Ceriporiopsis subvermispora* produces extracellular hydrogen peroxide. Appl Environ Microbiol 64:68–73

Wang YQ, Liang WS, Geng CY (2009) Coalescence behavior of gold nanoparticles. Nanoscale Res Lett 4:684–688. https://doi.org/10.1007/s11671-009-9298-6

Wong MH et al (2007) Export of toxic chemicals – a review of the case of uncontrolled electronic-waste recycling. Environ Pollut 149:131–140. https://doi.org/10.1016/j.envpol.2007.01.044

Wu JP, Luo XJ, Zhang Y, Luo Y, Chen SJ, Mai BX, Yang ZY (2008) Bioaccumulation of polybrominated diphenyl ethers (PBDEs) and polychlorinated biphenyls (PCBs) in wild aquatic species from an electronic waste (e-waste) recycling site in South China. Environ Int 34:1109–1113. https://doi.org/10.1016/j.envint.2008.04.001

Yamada H, Shimizu S, Kobayashi M (2001) Hydratases involved in nitrile conversion: screening, characterization and application. Chem Rec 1:152–161

Yoada RM, Chirawurah D, Adongo PB (2014) Domestic waste disposal practice and perceptions of private sector waste management in urban Accra. BMC Public Health 14:697. https://doi.org/10.1186/1471-2458-14-697

Chapter 21
Managing Electronic Waste Pollution: Policy Options and Challenges

Muhammad Abid, Farhad Zulfiqar, and Mohsin Raza

Abstract This chapter focuses on the current status and management of e-waste in Pakistan, with an emphasis on describing the key sources, impacts, and challenges in the management of e-waste in the country. Further, this study also discussed in detail different policy options and mechanisms that could be used to sustainably manage e-waste in Pakistan. The findings of the study highlight that the domestic production and imports of e-waste in Pakistan are rapidly growing due to increasing local consumption and low restrictions on the import of used electronic equipment. It has been observed that the recycling of e-waste is being done without any precautionary measures and is becoming a leading source of many environmental and health problems in most of the urban hubs. Lack of awareness about e-waste among the general public, importers, as well as scrap dealers and lack of regulation and implementation constraints are the major challenges in the management of e-waste in the country. Further, it is found that there is no inventory of e-waste in Pakistan available that also limits authorities in taking serious actions against unsustainable recycling of e-waste. In order to sustainably manage e-waste in Pakistan, there is a serious need for effective legislation and policy framework for refurbished and secondhand imports. Further, Pakistan also needs to take measures to fulfill its international commitments on Basel convention and e-waste filtration prevention. Further, stakeholder training and awareness on the hazardous impacts of e-waste on local environmental and health need to be conducted at different scales.

Keywords Electronic-waste · E-waste management · E-waste challenges · Sustainable measures · Pakistan

M. Abid (✉) · M. Raza
Centre for Climate Research and Development, COMSATS University Islamabad, Islamabad, Pakistan
e-mail: muhammad.abid@comsats.edu.pk

F. Zulfiqar
Department of Economics, COMSATS University Islamabad, Islamabad, Pakistan

M. Z. Hashmi, A. Varma (eds.), *Electronic Waste Pollution*, Soil Biology 57,
https://doi.org/10.1007/978-3-030-26615-8_21

21.1 Introduction

The electronics industry is one of the largest and fastest growing industry, contributing to the socioeconomic and technological development of the world economies (Radha 2002). However, this consumer-oriented technological growth coupled with innovation in the technological world has resulted in rapid product obsolescence and a new environmental challenge called electronic waste or e-waste. The e-waste that consists of obsolete electronic devices has become an emerging problem particularly in developed economies as well as a business opportunity of increasing significance particularly for developing economies (Widmer et al. 2005). The e-waste consists of various hazardous materials which during recycling not only harm the environment but may also have negative impacts on human health.

Therefore, measures need to be taken to minimize the emission of hazardous compounds potentially causing environmental and health impacts in Pakistan. Pakistan is highly vulnerable to e-waste pollution because recycling is often carried out informally and particularly e-waste is imported from developed countries (Iqbal et al. 2017). Pakistan has been identified as one of the major importers of e-waste with an estimated generation of an estimated 317 kt of domestic e-waste (Iqbal et al. 2015). Karachi the most urbanized and industrialized coastal city hosts most of the industrial and trade activities through its two seaports. Tons of imported, old, and obsolete EEE come to Karachi from all around the world, which, in turn, is traded from vendors to scrap dealers and dismantlers who recycle this e-waste to extract valuable material from the waste without realizing the negative impacts of this recycling process on local environment and human health (Siddique et al. 2009). Unsustainable traditional ways of recycling including open-air burning, acid baths, dismantling, or shredding and melting are also causing more negative impacts on humans and the environment. Further, the remaining leftover waste is often dumped in landfills or surrounding water bodies again causing indirect negative impacts on humans and the environment (Imran et al. 2017).

Pakistan, which is already struggling hard to manage its solid waste, is likely to face a serious challenge regarding the management of e-waste and emissions caused by the recycling of e-waste, potentially causing adverse health and environmental impacts. E-waste particularly computer waste from developed countries find an easy way into developing countries like Pakistan due to its weak policies. In order to reduce the emissions from the recycling of e-waste, effective policies need to be designed keeping in view the imports of e-waste and its impacts on human health and the environment. This chapter particularly highlights the associated issues, challenges, and strategies to address the emerging problem of e-waste pollution in the context of Pakistan.

21.2 E-waste in Pakistan

The problems associated with e-waste in Pakistan started evolving in the early 1960s with the increase in average GDP and economic development (Hussain 2012). Further, increasing population and purchasing capacity have led to an increase in demand for electronics goods including home appliances, telecommunication, IT, and computers over the years. Majority of the components of electronic items are imported or smuggled and then assembled locally (Bashar 2000). Rapidly growing population and increasing rate of urbanization suggest an increase in the generation of e-waste in Pakistan. Solid waste management, which is already a challenging task in Pakistan, has become even more challenging by the invasion of e-waste generated locally as well as imported from abroad (Khan et al. 2012).

Pakistan is the sixth most populated country in the world but is considered as least developed country as it is still growing and has limited infrastructure development. Most of the population still live under poverty and cannot afford to purchase new and advanced electrical items, so they mostly go for used and secondhand products which are imported from developed countries. However, due to the lack of effective policies, there is no inventory of domestic e-waste nor of the illegal import of e-waste. Below we have discussed the different kinds of e-waste and flow of e-waste in Pakistan.

21.2.1 Domestic E-waste

Over the years, the retail and purchasing capacity of the population in Pakistan has been increased which resulted in the increased use of electronics items (International 2015). The consumer electronics market in Pakistan grows at the rate of 13% to 3.3 billion USD by 2016. Particularly, the use of smartphones and other electronic gadgets has tremendously increased. For instance, in October 2018, the total mobile phone subscriptions in Pakistan were recorded as high as 152 million, which shows high consumption of mobile phones which eventually increases the amount of e-waste generated (PTA 2018). Similarly, the purchase of TVs, LEDs, and monitors is growing at the rate of 12% due to the availability of replacements and innovations in technology. Computer sales increased to 430 million USD in 2015 from 409 million USD in 2014 (Iqbal et al. 2015). Rapid urbanization and increased exposure to digital sources of information about new electronic products are some of the major drivers of increased consumption of electronic products. The increased consumption will lead to a higher amount of domestic e-waste generated. According to an estimate, the total domestic e-waste generated in 2014 was around 316 kt with 1.4 kg e-waste per capita which is 50% higher than the e-waste generated in 2005 (210 kt) (Balde et al. 2015; Breivik et al. 2014). Although the data reflect only domestic generation and do not include illegal imports, they show an increasing trend for domestic e-waste generation.

21.2.2 Imports from Aboard

Despite the existence of Basel Convention on the control of transboundary movements and disposal of hazardous waste, the transfer of e-waste from developed countries like the USA, Canada, Australia, the EU, Japan, and Korea to Asian countries such as China, India, and Pakistan remains relatively high (Terazono et al. 2006; Cobbing 2008). According to a rough estimation, the average annual import of e-waste to Pakistan is around 95.4 kt (mostly computers and related products). According to the World Bank, IT accounts for 4% of the total goods imported in 2013, whereas about 50% of PCs produced are smuggled or brought through illegal means (Ahmed 2014). Therefore, no accurate official data or record of EEE imports is available for Pakistan. The available data on imported e-waste from legal means show that about 70–80% of e-waste is imported from the USA, 10–15% from the UK, and the rest from Canada, the UAE, and Singapore. Particularly, Dubai and Singapore also serve as predistribution hubs of e-waste coming from the EU and USA to South Asian countries like Pakistan and India (Cobbing 2008; Iqbal et al. 2015).

21.2.3 E-waste Flows

Figure 21.1 shows a flowchart of key sources and generation of e-waste in Pakistan. Generally, e-waste in Pakistan is generated from three main sources: domestic manufacturing, domestic consumption, and imports. After disposal, e-waste is collected by scrap dealer and vendors followed by dismantling and extracting of valuable and precious materials using various illegal methods. In Pakistan, e-waste is generated from three key sources: domestic manufacturing, domestic consumption, and import. After disposal, e-waste is collected by scrap dealers and vendors, who sometimes dismantle the waste in several parts which are, in turn, sold to extractors and dismantlers. Extractors and dismantlers by using illegal means then treat the waste to extract precious and valuable materials and discarded waste is then often disposed of either in landfills or water bodies (Fig. 21.1). Sometimes scrapers and dismantlers reassemble different parts of old equipment for resale. Laborers may not be fully aware of the potentially harmful consequences of recycling or dismantling electronics, thus exposing themselves to high toxicity (Saeed 2013). Crude recycling techniques are adopted like physical dismantling, open burning, acid bath, and the use of blowtorches to extract valuable metals, as all these procedures are very cost-efficient (Fig. 21.1). The scale of informal recycling is growing day by day in Pakistan and has expanded all over the country as large- and small-scale businesses (Umair et al. 2013). Up until now, there is no formal e-waste recycling facility available in Pakistan; all the e-waste is recycled through illegal and/or informal means.

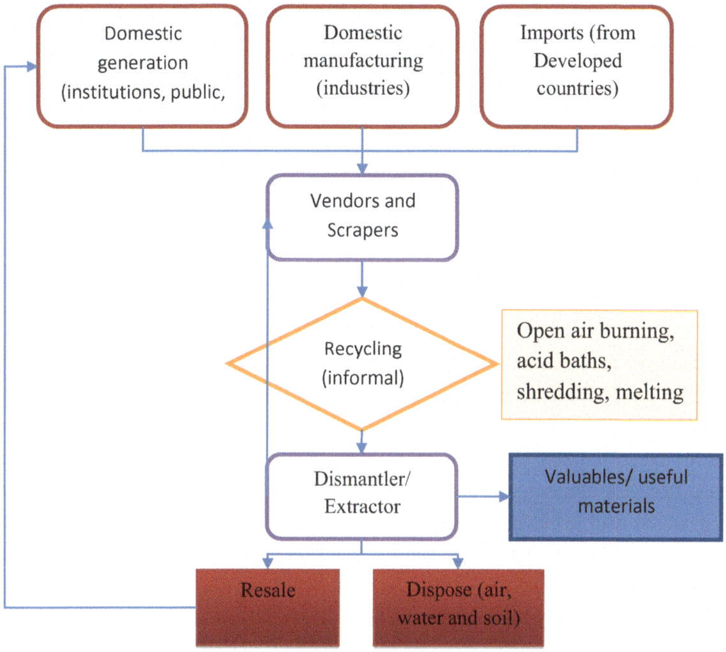

Fig. 21.1 The flow of e-waste in Pakistan from generation to disposal (adapted from Iqbal et al. 2015)

As discussed earlier, the e-waste is imported from abroad with the label of "secondhand equipment" (Syed 2013). Karachi is one of the major recycling waste sites in Pakistan followed by Lahore, Faisalabad, Peshawar, Gujranwala, and Rawalpindi, but the recycling capacity of Karachi sites is far more than other cities. The main reason for Karachi being the major recycling hub in Pakistan is its closeness to the sea where the containers full of e-waste are imported from all around the world. In Karachi, Lyari, Shershah, and Jacob lines are the largest dumpsites in Karachi for e-waste (Syed 2013). The recycling and dumping of e-waste at major sites in Karachi are found to have severe negative impacts on Indus Delta and mangroves. Many studies have shown that mangroves of Indus Delta are highly polluted with metals (Siddique et al. 2009). Similar issues have been found in the case of Lahore, the second largest and metropolitan city of Pakistan after Karachi. Lahore is constantly struggling to cope with rapidly growing urbanization and extreme problems of air, water, and soil pollution in the city (Nawaz et al. 2015). E-waste and the recycling of imported e-waste have been a important concern for the authorities in Lahore. Hafeez Center, Hall Road, Beadon Road, Misri Shah, Mayo Hospital, and Pakistan Mint are the main e-waste markets in Lahore where e-waste is recycled, and valuables are extracted without any precautionary environmental and health safety measures. Few years ago, a dozen of factories have been sealed in Lahore due to the burning of batteries in order to extract lead and cadmium (Raza 2010). Other major cities also have e-waste recycling industries but on a very small scale. For instance, in Rawalpindi College Road is famous for its scrap and secondhand electronics and other recycling activities. The use of traditional recycling methods without any precautionary measures has been reported in the market causing asthma in workers at warehouses (Saeed 2013). Child labor in the e-waste recycling business is another important issue to be considered (Saeed 2013).

21.3 Impacts of E-waste in Pakistan

Electronic waste or e-waste enters Pakistan in the guise of secondhand ICT equipment. Due to low penetration of ICT in Pakistan, there is a huge vacuum that needs to be filled. Given the low socioeconomic condition of Pakistan, secondhand ICT equipment makes its way into Pakistan. This benefits society in interconnectedness which is a positive aspect of secondhand ICT devices. The discarded ICT devices become e-waste and go to recycling. The recycling of e-waste is both advantageous and problematic. Recycling extracts precious metals from e-waste that are valuable in the market. However, the extraction process involves the release of toxic chemicals which has environmental and health repercussions. The recycling industry in Pakistan is not structured, thus existing informally. The informal recycling of e-waste is a valuable activity that earns profits to its stakeholders. Due to the nonregulation status of this industry, it negatively impacts natural environment and health of workers involved in this sector.

21.3.1 Environmental Impact

In the absence of stringent laws and prevailing primitive techniques, recycling process in Pakistan leads to adverse environmental impacts on land, water, and air. Traditional recycling methods are used in Pakistan that includes dismantling, crushing, acid dipping, and burning. Toxic elements are released during recycling activities such as lead, cadmium, barium, mercury, and chromium. The working environment gets polluted due to these recovering activities. The recycling business is usually carried out in small places (shops, homes) without any ventilation. Local air gets polluted with poisonous gases that are produced from burning of monitors, cables, and wires for copper recovery; toxic fumes suffocate the environment causing breathing problems to workers and surrounding communities (Imran et al. 2017). Increased level of PM2.5 is also reported in e-waste recycling zones in different cities of Pakistan (Iqbal et al. 2017). Leftover material after recycling or dismantling is dumped underground which contaminates the soil with toxic metals. Toxic elements in the contaminated soil could enter food production. The acid is discarded into open stream after an acid-dipping process which pollutes the water resources. One such practice is carried out in Karachi, as the biggest hub of e-waste recycling in Pakistan, where used acid is thrown in Lyari River. Lyari River flows through mangroves before falling into the Arabian Sea. Studies show the presence of high metal content in mangroves located along Lyari River (Siddique et al. 2009). Such working condition poses a serious threat to the natural environment.

21.3.2 Health Impact

The crude methods used in the recycling industry of e-waste have negative impacts on the health of workers involved in the process, and people living in the neighborhood also get affected. Often the recycling business is run by families in which people of all age participate. Women and children are highly vulnerable to the working conditions. People have no knowledge on health and safety measures of dealing with hazardous substances. Hazardous substances present in e-waste include lead, flame retardants (FRs), polyclinic aromatic hydrocarbons (PAHs), polychlorinated biphenyls (PCBs), and organochlorines (OCs). Workers take no safety measures and work without wearing gloves or protective clothing. Working people use bare hands for dismantling, burning, crushing, acid dipping, and other processes. The toxic particles stuck in hand during dismantling increase the chances of ingestion and could cause skin irritation. The cables and wires are protected with a lead casing; for copper recovery lead covering is burnt that produces toxic fumes which cause health problems upon breathing. Inhalation of lead causes blood pressure to rise and could affect memory in young children. Workers are reported to have suffered from coughing, headaches, and skin diseases (Imran et al. 2017).

21.3.3 Economic Impact

An estimated 60% of the population in Pakistan lives below the poverty line. Due to limited economic resources, people find e-waste recycling business as a source of earning money.

The high-profit margin in recycling business makes it an attractive option that drove numerous people into it. Another driving force in setting up this business is that it requires limited expertise and low infrastructure. The absence of stringent environmental laws in Pakistan like other developing countries also serves in favor of this business arrangement. Recycling e-waste is a valuable activity that involves the recovery of precious metals and other usable material. The most valuable recoverable item is gold from computer chips, motherboards, and processors. Other useful materials recovered through recycling include silver, copper, aluminum, glass, ceramic, and plastic.

21.4 Status of E-waste Management in Pakistan

21.4.1 Disposal Points and Informal Recycling Process of E-waste in Pakistan

E-waste enters the territory of Pakistan in the guise of secondhand ICT equipment through imports from the developed world. Karachi, being the seaport city of Pakistan, receives a large part of e-waste shipped from other parts of the world. Karachi is the biggest hub for e-waste processing and recycling. Other cities where recycling industry exists include Lahore, Rawalpindi, Peshawar, Multan, Sialkot, and Quetta. Some of those secondhand devices can be reused after small fixing and are sold to computer dealers for further use by customers. Devices that are beyond fixing and have completed their life become e-waste that goes to recycling business or scrap dealers who extract useful material and sell them to the respective industry. Secondhand devices usually consist of printers, personal computers, old cathode ray tube monitors, laptops, mobile phones, home appliances (TV, refrigerator, AC, washing machines, etc.), and accessories. Scrap dealers sort out e-waste into material for metal extraction and other useful items for recycling. Flowsheet of e-waste processing is provided as Fig. 21.2.

In metal extraction, gold and copper are extracted from ICT equipment. Two primitive methods are used for gold extraction. In the first method, computer chips, processors, and motherboards are added to roll mill where heavy stone balls are present. Roll mill is rotated around an axis which crushes all chips, and the sand-like mixture is formed. The mixture is then heated at high temperature and mercury is added to it which mixed with gold forms an amalgam. The amalgam, containing mercury and gold with some impurities, is taken out of the sand mixture and sent for removal of further impurities to obtain pure gold. In the second method, e-devices having gold content are dipped into concentrated acid which separates out gold and silver and settles at the bottom. The acid is then discarded, and gold is collected. For copper recovery, monitors, cables, and wires are burned to get copper wire which is sold as raw material to the industry such as fan industry in Gujranwala where copper is basic manufacturing item. Other parts of e-waste like casings, glass, keyboards, and the mouse also find their market in Pakistan.

21.4.2 Regulation

On a global scale, rules exist to control the movement of hazardous substances. A prominent one is Basel Convention (1989), to which Pakistan is also a signatory, which forbids the transport of goods containing hazardous substances from advanced countries like the EU and USA to developing countries like Pakistan. Basel Ban Amendment (1995) restricts transboundary movement of any type of hazardous material for any purpose including recycling. Also, Pakistan is party to

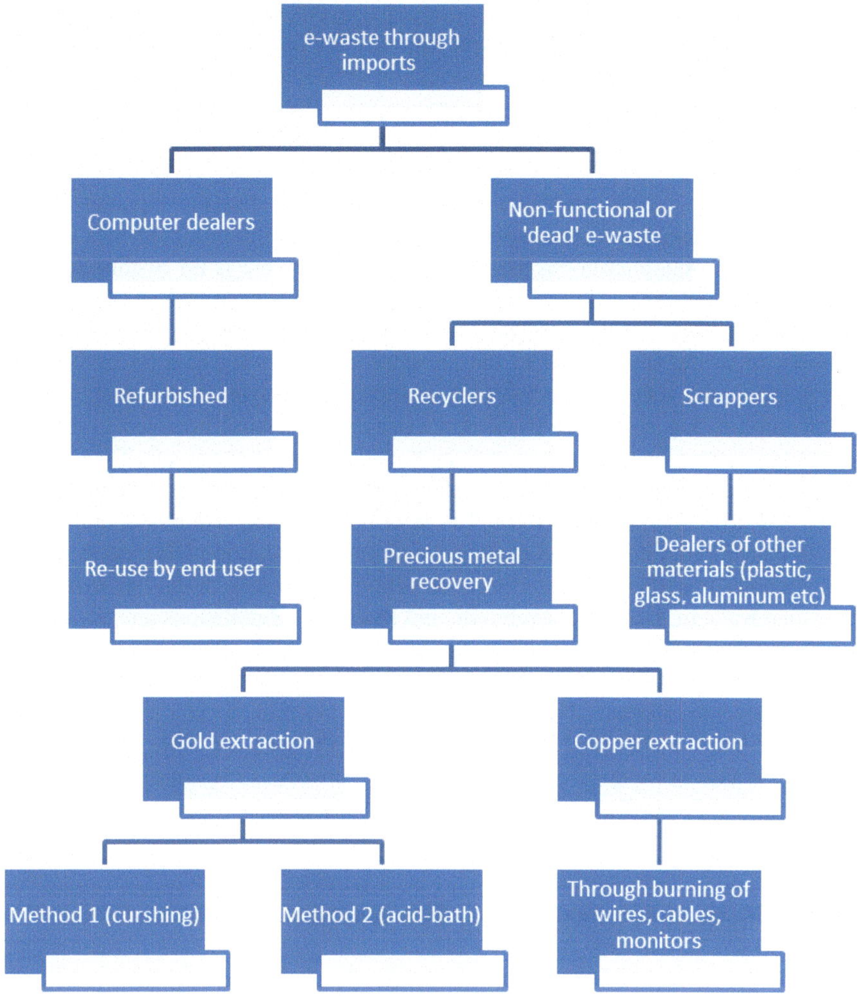

Fig. 21.2 Flowsheet of imported e-waste processing in Pakistan

Rotterdam Convention on Prior Informed Consent for Certain Hazardous Chemicals (1998), Stockholm Convention on Persistent Organic Pollutants (2001), and Vienna Convention and Montreal Protocol on Ozone Depleting Substances (ODS) that prohibit movement of hazardous substances between countries.

On a national scale, section 13 of Pakistan Environmental Protection Act 1997 (PEPA 1997) prohibits the import of any hazardous substance by any person in the country. Similarly, trade policy 2015-16 of Government of Pakistan also restricts the import of any goods prohibited under the Basel Convention (GoP 2016). It explicitly places a ban on the import of used air conditioners, refrigerators, or other home appliances. ICT equipment containing hazardous chemicals falls into the list of prohibited goods. Despite this ban, e-waste keeps on entering the Pakistani market in the guise of secondhand ICT equipment exposing the flaw in the Basel Convention and other similar acts. There is no classification of new and used ICT devices in the Basel Convention. Given the low socioeconomic condition of Pakistan and low penetration of ICT in many parts of the country, there is a huge market for secondhand electronic devices in the country. Both parties, importers and exporters, used the loophole in convention and vacuum in the local market for shipment of used ICT devices to Pakistan. Other channels are also used for importing devices such as in the form of smuggled goods via Afghanistan since there is no custom duty for land-locked countries.

Secondhand material that is nearing its end of life reaches the recycling market for recovery of useful items. Due to the presence of less stringent laws on e-waste management, the recycling industry of e-waste operates in an informal manner which implies that no proper guidelines exist on processing involved in e-waste recycling. No protocols exist about safe end-of-life disposal of e-waste. The workers involved in the informal e-waste recycling industry are unaware of occupational safety and health (OSH) guidelines. The informal recycling industry operates in a nonstructured and nonregulated manner. Weak implementation of any environmental law allows businessmen to carry out recycling activities at various spots in large cities like Karachi, Lahore, Rawalpindi, and Peshawar. Regulations are missing on the treatment of waste acid before discharging in open stream. No control is observed from local authorities or NGOs on polluting activities throughout the e-waste recycling process. These activities taking place in the city are beyond the control of local authorities. Minimal precautionary measure from businessmen observed is that burning of e-waste for copper recovery is carried out in outskirts of the city to avoid polluting city environment.

21.5 Challenges in Managing E-waste in Pakistan

The above discussion pointed out the lack of policy environment for sustainable e-waste management in Pakistan and suggested possible policy suggestions for the development of policy and legislative framework. This section deals with further challenges in sustainable management of e-waste in Pakistan.

21.5.1 Lack of Awareness About the E-waste

Various stakeholders dealing with e-waste in Pakistan are unaware of its hazardous nature (Umair 2015). Most of the hazardous e-waste includes computers and mobile phones. The importers are usually oblivious to the presence of heavy metals in the e-waste and proceed on to dismantle it in open yards. This process releases toxic materials in the air and soil which percolates in the groundwater. The labor involved in the dismantling of used e-waste rarely wears any safety equipment (Sthiannopkao and Wong 2013) and operates with bare hands and no masks. After the removal of valuable components, the remaining plastic, glass, and metal are burnt in open air, again releasing hazardous gases in the air. At most, the leftover material is buried, which eventually releases heavy metals in the underground water networks.

Therefore, this lack of awareness about the different nature of e-waste than common waste results in its mismanagement. This barrier in the sustainable management of e-waste can be overcome through capacity building of importers as well as involved labor. The awareness campaign may also involve consumers and law enforcement agencies dealing with e-waste.

21.5.2 Regulatory and Implementation Constraints

Another bottleneck problem facing sustainable e-waste management is a regulatory constraint in Pakistan. Due to the lack of legislation and policy environment dealing with e-waste, the regulatory authorities are distributed to various departments which results in conflicts and overlapping of functions. This creates a gap in coordination among various stakeholders and aligned ministries leading to inadequacy in regulations due to insufficient legal cover and eventually weak implementation. The lack of dedicated authority meant for e-waste management also results in ownership of the whole process leading to continued negligence and challenges in ensuring sustainable e-waste management. Such an authority will also be extremely useful in capacity building of all stakeholders and awareness raising.

21.5.3 Technological Constraints

The developing countries, such as Pakistan, also face technological hurdles in sustainably managing e-waste (Sthiannopkao and Wong 2013). This is another bottleneck which can only be overcome through training from the developed countries. The cooperation may be sought from the fellow member countries of Basel convention for the training of trainers. The lack of available technology exacerbates the fragile implementation mechanism in place in Pakistan due to the abovementioned poor policy framework.

21.5.4 Lack of E-waste Inventory

The research has shown that there is no system of maintaining an inventory of imported hazardous waste including e-waste (Abbas 2010). There is not even a rudimentary level of inventory management in Pakistan (Sthiannopkao and Wong

2013). As there is no inventory of what is being imported and in how much quantity, there is no way for policymakers to implement quotas based on used electronics demand in Pakistan. Therefore, it is imperative to develop and maintain an e-waste inventory for the development of policies, plans, capacity building, recycling, and imports.

Other important challenges facing sustainable e-waste management in Pakistan include limited coordination of public and private sector, funding constraints, and research-based evidence of e-waste solutions. Therefore, it can be concluded from the above discussion that coping with the challenges requires not only understanding and commitment of the e-waste problem in Pakistan but also the development of an institutional framework backed by policy and legislature. Preparation and implementation of e-waste management plan need to be prepared in consultation with the line ministries and all relevant stakeholders.

21.6 Sustainable Ways to Manage E-waste in Pakistan

The information and communication technology (ICT) has brought a technological revolution in Pakistan. However, it has also created a new hazardous stream of waste, namely electronic waste or "e-waste." It is one of the most pressing environmental problems Pakistan is likely to face in the future (Masood et al. 2014). The e-waste management involves a sophisticated technological solution which is too complicated to handle safely for a developing country such as Pakistan. The safety of workers from toxic chemicals and ensuring a safe working environment require a comprehensive policy framework. This policy framework needs to enable institutions for monitoring and evaluation of its implementation. The policy formulation process needs to be inclusive with the engagement of all the stakeholders. Therefore, we propose the following ways to manage e-waste in Pakistan sustainably.

21.6.1 Effective Legislation for E-waste Management

A policy framework is an umbrella framework under which e-waste can be effectively managed. However, a policy framework requires national priorities guided by legislation in a democratic state like Pakistan. The relevant prevailing legislation for e-waste management in Pakistan includes Pakistan environmental protection act 1997, national environment policy 2005, hospital management rules, solid waste management rules, and hazardous waste management rules. Although some of these policies are relevant to the disposal of e-waste, there is no clear guideline for managing e-waste. Therefore, an effective e-waste management policy formulation requires legislation at a federal and provincial level to provide legal cover for the development of workable strategies and projects for e-waste disposal. The lack of comprehensive e-waste management policy leads to serious short- and long-term environmental hazards through non-standardized recycling and disposal practices.

21.6.2 Policy Framework for Refurbished/Secondhand Electronics

A bulk of secondhand electronics are imported under the pretext of valuable goods. However, the majority of these electronics are obsolete and end up in recycling plants. These recycling plants do not have a legal compulsion to standardize their operations and result in releasing heavy metals used in the electronics into the soil, water, and air. The valuable components are removed, and the remaining low-value hazardous waste is either burnt up or buried without any protective recycling. This hazardous waste often includes mercury, chromium, cadmium, barium, and lead (Abbas 2010). As there is no legal obligation to separate out the heavy metals produced during e-waste recycling, the workers are inevitably exposed to toxic fumes as there is no protective equipment either due to lack of knowledge or lack of obligation on the part of their employer. The recent announcement by the government to impose a duty on the imported mobile phones may help in stemming the inflow of illegal mobile phone shipments. However, these measures need proper legislative and policy cover to be successful in effectively managing refurbished e-waste. This policy need not ban all the imported refurbished electronics; it needs a national assessment of the suitability of imports and the place of eventual use.

21.6.3 Life-cycle Assessment of E-waste

Sustainable management of e-waste is impossible without an evidence-based assessment of life cycle of electronics. Thus, government and research centers across Pakistan need to provide research-based life cycle of each electronic product. For example, the management of e-waste generated through mobile phone recycling requires assessment of the points which generate e-waste. A locally assembled mobile phone will have a different life cycle compared to an imported secondhand one. This assessment should also include the disposal scenarios after the lifetime of the product. For example, where the mobile phones will end up after their usable life—whether the usable parts will be removed, or the entire phone will be sent to recycling plants. A research-based life-cycle assessment should also provide evidence of harmful effects of e-waste on human health and environment. Therefore, a cost–benefit assessment of imported electronics will assist in making policy decisions for sustainable e-waste management in Pakistan.

21.6.4 Basel Convention and E-waste Filtration Prevention

The Basel Convention, to which Pakistan is a signatory, binds partner countries to control e-waste filtration (Katharina 2013). It requires full disclosure of the movement and management of waste including the repatriation of illegally exported hazardous waste. Therefore, sustainable management of e-waste in Pakistan requires legislation and subsequent policy framework to enforce the Basel convention. As Pakistan is a net importer of electronics, an appropriate policy framework will enable the country to impose binding restrictions on foreign exporters to streamline their product life cycle including the e-waste disposal. It provides Pakistan with an opportunity to reduce hazardous waste coming from outside and proper disposal of the e-waste generated in the country.

21.6.5 Stakeholder Training in E-waste Management

The sustainable management of e-waste requires its professional handling at each level. Currently, there is no policy in place in Pakistan dealing with educating labor handling the e-waste, and the law enforcement agencies also need capacity building regarding hazardous e-waste. Therefore, e-waste management policy should also assist in overcoming capacity constraints of allied government agencies to deal with e-waste in Pakistan. The developed countries may also assist the developing countries such as Pakistan to overcome technological constraints as well as stakeholder training. As the majority of e-waste is generated by developed countries (BAN and Greenpeace-China 2002; Gupta 2007; UNEP 2014), they have the moral obligation to assist in sustainable management in the net importers of e-waste.

21.6.6 Policy Choice of Shared Responsibility

The policy of sharing the responsibility of e-waste creation may also be experimented in Pakistan. This policy involves collecting an advance fee which will be reimbursed once the consumer deposits the used product at the formal recycling center (Yu et al. 2010). This policy is useful in curbing informal recycling places with dubious safety standards. However, further research is required in the form of pilot testing for e-waste management in Pakistan. The same model may also be applied to producers, making them responsible producers and further product labeling as green products. The government subsidy in the form of collection and disposal of e-waste is old school and burdens the exchequer. The policy of shared responsibility on the part of either consumer or producer or both is more relevant in the current economic context and will provide more opportunities for the involvement of private sectors. It will also encourage entrepreneurs and bring innovations to e-waste management.

21.7 Conclusion

This chapter gives an overview of the current situation of e-waste in Pakistan along with highlighting its environmental, health, and economic impacts. Further, this chapters also highlights the key challenges in the management of e-waste in Pakistan and provides different mechanisms and options that could be adopted in order to sustainably managing e-waste in Pakistan.

The volume of e-waste has increased over the years in Pakistan due to increasing urbanizations and imports of used electronic equipment from developed countries. Major cities in Pakistan have become the hub of e-waste recycling where the domestic and imported e-waste is dismantled, burned, or dumped that is causing severe environmental problems as well creating negative impacts on human health. The unsustainable recycling and management of e-waste in Pakistan are mainly due to lack of awareness about e-waste among the general public, importers, as well as scrap dealers. Further, lack of regulation and implementation constraints is another important factor behind the expansion of illegal and unsustainable management of e-waste in Pakistan. Further, there is no concrete inventory of e-waste in Pakistan that is necessary to develop effective policies to control the e-waste production, imports, and its recycling. In order to sustainably manage e-waste in Pakistan, there is a serious need for effective legislation and policy framework for refurbished and secondhand imports as well as life-cycle assessments need to be carried out for effective implementation of e-waste policies. Pakistan also needs to take measures to fulfill its international commitments on Basel convention and e-waste filtration prevention. Further, stakeholder training and awareness on the hazardous impacts of e-waste on local environmental and health need to be conducted at different scales.

References

Abbas Z (2010) E-waste management in Pakistan. In: Regional workshop on e-waste/WEEE management. Osaka, Japan, 6–9 July 2010

Ahmed Z (2014) Local manufacturing, benefits to industry, economy and recommendations. Pak. Today. http://www.pakistantoday.com.pk/2014/05/28/business/local-manufacturing-benefits-to-industry-economy-andrecommendations/. Accessed 13 Mar 2019

Balde CP, Wang F, Kuehr R, Huisman J (2015) The global E-waste monitor, 2014. United Nations University, IAS e SCYCLE, Bonn, Germany

BAN, Greenpeace-China (2002) Exporting harm: the high-tech trashing of Asia. Basel Action Network (BAN) and Greenpeace-China

Bashar A (2000) Electronics industry threatened by tariff discrepancies, corruption and smuggling. In: Pakistan and Gulf Economist, 19–25 June 2000
Breivik K, Armitage JM, Wania F, Jones KC (2014) Tracking the global generation and exports of e-waste. Do existing estimates add up? Environ Sci Technol 48(15):8735–8743
Cobbing M (2008) Toxic tech: not in our backyard. Uncovering the hidden flows of E-waste. Report from Greenpeace International. http://www.greenpeace.org/raw/content/belgium/fr/press/reports/toxic-tech.pdf. Accessed 25 Aug 2019
GoP (2016) Trade Policy 2015-16, Ministry of Commerce, Government of Pakistan. http://pakistanembassy.se/wordpress/wp-content/uploads/2015/08/banned-items-to-import-to-Pakistan.pdf. Accessed 25 Aug 2019
Gupta KS (2007) E-waste management: teaching how to reduce, reuse and recycle for sustainable development-need of some educational strategies. J Educ Pract 2:74–86
Hussain A (2012) A perspective on Pakistan's economic history: growth, economic structure and governance. In: Jalal A (ed) The Oxford companion to Pakistani history. Oxford University Press, Karachi
Imran M, Haydar S, Kim J, Rizwan M, Ali A (2017) E-waste flows, resource recovery and improvement of legal framework in Pakistan. Resour Conserv Recycl 125(June):131–138
International BM (2015) Pakistan consumer electronics report, 26 Feb. 2015
Iqbal M, Breivik K, Syed JH, Malik RN, Li J, Zhang G, Jones KC (2015) Emerging issue of e-waste in Pakistan: a review of status, research needs and data gaps. Environ Pollut 207:308–318
Iqbal M, Syed JH, Breivik K, Jamshed M, Chaudhry I, Li J, Zhang G (2017) E-waste driven pollution in Pakistan: the first evidence of environmental and human exposure to flame retardants (FRs) in Karachi City. Environ Sci Technol 51(23):13895–13905
Katharina KP (2013) The basel convention on the control of transboundary movements of hazardous wastes and their disposal. Proc Annu Meet (Am Soc Int Law) 107:434
Khan AA, Ahmed Z, Siddiqui MA (2012) Issues with solid waste management in South Asian countries: a situational analysis of Pakistan. J Environ Occup Sci 1(2):129–131
Masood M, Barlow CY, Wilson DC (2014) An assessment of the current municipal solid waste management system in Lahore, Pakistan. Waste Manag Res 32:834–847
Nawaz A, Shahbaz MA, Javed M (2015) Management of organic content in municipal solid waste-a case study of Lahore. Int J Environ Waste Manag 15(1):15–23
PEPA (1997) Pakistan environmental protection act, 1997, p. 29. Pakistan
PTA-Pakistan Telecommunication Authority (2018) Cellular subscriptions in Pakistan. Available https://www.pta.gov.pk/en/telecom-indicators. Accessed 01 Nov 2018
Radha G (2002) A study of the performance of the Indian IT sector' at www.nautilus.org. Accessed 21 Jun 2018
Raza A (2010) CDGL head in the sand over e-waste. News 2010. http://www.thenews.com.pk/Todays-News-5-332-CDGL-head-in-the-sand-over-ewaste
Saeed A (2013) Dangerous labour: the little known e-waste phenomenon. The Dawn, July 30, 2017. https://www.dawn.com/news/1032967. Accessed 25 Aug 2019
Siddique A, Mumtaz M, Zaigham NA, Mallick KA, Saied S, Zahir E, Khwaja HA (2009) Heavy metal toxicity levels in the coastal sediments of the Arabian Sea along the urban Karachi (Pakistan) region. Mar Pollut Bull 58(9):1406–1414
Sthiannopkao S, Wong MH (2013) Handling e-waste in developed and developing countries: initiatives, practices, and consequences. Sci Total Environ 463–464:1147–1153
Syed M (2013) The pricetag. Dawn 2013. http://www.dawn.com/news/1047748
Terazono A, Murakami S, Abe N, Inanc B, Moriguchi Y, Sakai S-i, Kojima M, Yoshida A, Li J, Yang J (2006) Current status and research on e-waste issues in Asia. J Mater Cycles Waste Manage 8(1):1–12
Umair S (2015) Informal electronic waste recycling in Pakistan. KTH Royal Institute of Technology
Umair S, Bjorklund A, Petersen EE (2013) Social life cycle inventory and impact assessment of informal recycling of electronic ICT waste in Pakistan. In: Hilty L, Aebischer E, Andersson G, Lohmann W (eds) Proceedings of the first international conference on information and communication technologies for sustainability. ETH, Zurich, pp 52–58

UNEP (2014) The basel convention on the control of transboundary movements of hazardous wastes and their disposal. Châtelaine, Switzerland
Widmer R, Oswald-Krapf H, Sinha-Khetriwal D, Schnellmann M, Heinz B¨n (2005) Global perspectives on e-waste. Environ Impact Assess Rev 25:436–458
Yu J, Williams E, Ju M, Shao C (2010) Managing e-waste in China: policies, pilot projects and alternative approaches. Resour Conserv Recycl 54:991–999

Index

M. Z. Hashmi, A. Varma (eds.), *Electronic Waste Pollution*, Soil Biology 57,
https://doi.org/10.1007/978-3-030-26615-8

The manufacturer's authorised representative in the EU is Springer Nature Customer Service Centre GmbH, Europaplatz 3, 69115 Heidelberg, Germany. If you have any concerns regarding our products, please contact ProductSafety@springernature.com

Printed and bound by CPI Group (UK) Ltd, Croydon, CR0 4YY
30/09/2026
02225657-0008